QUANTUM COSMOLOGY

Quantum Cosmology offers a guided introduction to the quantum aspects of the cosmos. Starting with an overview of early universe cosmology, the book builds up to advanced topics such as the Wheeler–DeWitt equation, gravitational path integrals, and the no-boundary proposal for the wave function of the universe. Readers will explore tunneling processes via Coleman–DeLuccia instantons, the quantum origin of primordial fluctuations, the thermodynamics of horizons, and basic notions of string cosmology. Concepts such as wormholes and semiclassical geometry are introduced with clarity and physical motivation.

The book assumes some familiarity with general relativity and quantum mechanics, but little prior knowledge of cosmology. It includes a wide range of exercises, with solutions provided. Written in a pedagogical style, it bridges the gap between undergraduate courses and the research level in this frontier area of theoretical physics.

JEAN-LUC LEHNERS studied physics and mathematics at Imperial College London and at the University of Cambridge. He obtained his Ph.D. in 2005 at Imperial College for his research in string theory. Following postdoctoral positions at the University of Cambridge, Princeton University, and the Perimeter Institute, he established the Theoretical Cosmology group at the Max Planck Institute for Gravitational Physics (Albert Einstein Institute) in Potsdam in 2010, which he ran until 2024. Dr. Lehners won two prestigious ERC Grants. His research focuses on the early universe, with a particular emphasis on quantum effects in cosmology.

QUANTUM COSMOLOGY

An Introduction

JEAN-LUC LEHNERS

Max-Planck-Institut für Gravitationsphysik
(Albert-Einstein-Institut)

Shaftesbury Road, Cambridge CB2 8EA, United Kingdom

One Liberty Plaza, 20th Floor, New York, NY 10006, USA

477 Williamstown Road, Port Melbourne, VIC 3207, Australia

314–321, 3rd Floor, Plot 3, Splendor Forum, Jasola District Centre,
New Delhi – 110025, India

103 Penang Road, #05–06/07, Visioncrest Commercial, Singapore 238467

Cambridge University Press is part of Cambridge University Press & Assessment,
a department of the University of Cambridge.

We share the University's mission to contribute to society through the pursuit of
education, learning and research at the highest international levels of excellence.

www.cambridge.org
Information on this title: www.cambridge.org/9781009516808
DOI: 10.1017/9781009516815

© Jean-Luc Lehners 2026

This publication is in copyright. Subject to statutory exception and to the provisions
of relevant collective licensing agreements, no reproduction of any part may take
place without the written permission of Cambridge University Press & Assessment.

When citing this work, please include a reference to the DOI 10.1017/9781009516815

First published 2026

Cover image: Each point in this plot represents a possible evolution of the universe, in a biaxial Bianchi
IX model with no-boundary initial conditions. The points located where lines cross provide the most
important contributions to the wave function of the universe; see Chapter 7.

A catalogue record for this publication is available from the British Library

A Cataloging-in-Publication data record for this book is available from the Library of Congress

ISBN 978-1-009-51680-8 Hardback

Cambridge University Press & Assessment has no responsibility for the persistence
or accuracy of URLs for external or third-party internet websites referred to in this
publication and does not guarantee that any content on such websites is, or will
remain, accurate or appropriate.

For EU product safety concerns, contact us at Calle de José Abascal, 56, 1°, 28003 Madrid, Spain, or
email eugpsr@cambridge.org

To Linda, Julien, Caroline & Leander

Contents

Preface *page* xi
Acknowledgments xiv
Conventions and notation xv

1 Modeling the early universe 1
  1.1 The standard cosmological model 1
    1.1.1 The cosmological principle 1
    1.1.2 Robertson–Walker universes 3
    1.1.3 The Friedmann equations 8
  1.2 A thermal history 13
    1.2.1 A relic from the big bang: the cosmic microwave background 20
  1.3 The big bang and assorted puzzles 23
    1.3.1 Singularity puzzle 23
    1.3.2 Flatness puzzle 24
    1.3.3 Horizon puzzle 25
    1.3.4 Brief list of additional puzzles 28
  Exercises 29

2 Blowing up the universe: inflation 31
  2.1 Inflation and scalar field dynamics 31
    2.1.1 Basic idea 31
    2.1.2 Gravitational dynamics of a scalar field 37
    2.1.3 Constant equation of state 38
    2.1.4 Slow-roll 40
    2.1.5 Quadratic, Higgs, and Starobinsky inflation 43
    2.1.6 End of inflation and reheating 49
  2.2 An alternative: the ekpyrotic universe 51
    2.2.1 The ekpyrotic phase 51

		2.2.2 Cyclic scenario and cosmic bounces	54
	Exercises		56
3	**Observing perturbative quantum gravity: fluctuations**		58
	3.1	Observed fluctuations in the CMB	58
		3.1.1 Historical aside: the need for seeds	67
	3.2	Cosmological perturbation theory	70
		3.2.1 Dealing with gauge transformations	72
		3.2.2 The ADM formalism	75
		3.2.3 The comoving curvature perturbation	78
		3.2.4 Quantization	81
		3.2.5 Slow-roll solutions	83
		3.2.6 Quantum to classical – a preview	86
		3.2.7 Tensor modes	88
		3.2.8 Non-Gaussian corrections	90
	3.3	Comparing theory and observations	94
	3.4	Inflationary puzzles	98
	Exercises		101
4	**Quantizing gravity: canonical approach**		102
	4.1	Hamiltonian formulation of general relativity	104
		4.1.1 Action principle, variation and boundary terms	104
		4.1.2 ADM framework and the Wheeler–DeWitt equation	105
		4.1.3 Restriction to minisuperspace	109
	4.2	How to reconstruct the universe from the wave function	112
	Exercises		117
5	**Quantizing gravity: path integral approach**		119
	5.1	Gauge fixing the sum over spacetimes	121
	5.2	Transitions in a universe with positive cosmological constant	124
		5.2.1 Dirichlet boundary conditions	126
		5.2.2 Neumann boundary conditions	141
	5.3	Relation to the Wheeler–DeWitt equation and composition law	145
	5.4	Discussion and some open questions	150
	Exercises		152
6	**Linking gravity, quantum theory, and thermodynamics**		154
	6.1	Heuristics: horizons, imaginary time, and temperature	154
	6.2	Bogolyubov transformations	160
	6.3	The Unruh effect	164

	6.4	Hawking radiation	170
	6.5	Path integrals and thermodynamics: Hawking–Page phase transition	175
	6.6	De Sitter thermodynamics and mode functions	187
		6.6.1 Mode functions on de Sitter spacetime	189
	Exercises		194
7	**Creating space and time: the no-boundary proposal**		**195**
	7.1	Motivation and examples of no-boundary instantons	195
		7.1.1 Heuristic arguments	195
		7.1.2 Basic examples and stability of perturbations	199
		7.1.3 General prescription, numerical examples, classical histories	210
	7.2	Minisuperspace implementations	220
	7.3	Further developments	236
		7.3.1 Quantum gravity and higher curvature corrections	236
		7.3.2 Allowable metrics	239
	7.4	Postdictions, predictions, and discussion	249
	Exercises		252
8	**Interpreting the wave function**		**254**
	8.1	Basic idea of decoherence	254
	8.2	Decoherence in the early universe	256
		8.2.1 Structure of the wave function	256
		8.2.2 Classicalization of the background	259
		8.2.3 Decoherence of the fluctuations	262
	8.3	Discussion	266
9	**Transitioning to different spacetimes**		**269**
	9.1	Vacuum decay via Coleman–DeLuccia instantons	269
		9.1.1 Complex time tunneling in quantum mechanics	269
		9.1.2 Vacuum decay without gravity	271
		9.1.3 Vacuum decay with gravity	276
		9.1.4 Negative modes	283
	9.2	Disappearing act: bubbles of nothing	289
	9.3	Wormholes	292
		9.3.1 Lorentzian wormholes	292
		9.3.2 Euclidean wormholes	295
	Exercises		301
10	**Going deeper yet: stringy cosmology**		**303**
	10.1	Strings require extra dimensions	304
		10.1.1 Lightning review of string theory	304

	10.1.2 Compactifications and cosmological implications	307
10.2	Braneworlds	315
	10.2.1 Dp-branes	315
	10.2.2 Brane inflation	319
	10.2.3 End-of-the-world branes and the big bang	322
10.3	Perspectives	329
	Exercises	331

Appendix A **Constants of nature and cosmological quantities**   333

Appendix B **Picard–Lefschetz theory and saddle-point approximation**   335

Appendix C **Useful functions**   342

Appendix D **Solutions to exercises**   345

Appendix E **Guide to the bibliography**   367

Bibliography	372
Index	377

Preface

The night sky has always played the role of a muse, inciting wonder and prodding people to develop theories about the universe. For a long time, the standing of such theories depended mostly on people's personal predilections. But this has changed significantly over the last 100 years. Cosmology has undergone a complete transformation from an almost purely philosophical endeavor to a ground-breaking, paradigm-shifting science.

Cosmology's progress rests on the discovery and development of both quantum theory and general relativity. Over the last century, improved observational techniques and a better theoretical understanding have allowed humans to figure out an incredible amount of knowledge about our universe, starting with the realization that the universe is truly vast; and that it is evolving, implying that it has a history that we can explore.

As physics progresses, an ever bigger set of events becomes explainable. What is noteworthy is that with each new type of explanation our perspective on the world tends to change significantly. This was the case with classical mechanics and Newton's law of gravity, which provided the first concrete hints that the heavens might be understandable. It was also the case with thermodynamics, with Maxwell's synthesis of electromagnetism, and yet again with Einstein's development of relativity – the latter implying the merging of space and time into a malleable, evolving substance.

The aim of this book is to explore what new phenomena we can understand, and how our world view is impacted, when we apply quantum theory to the universe, combining quantum principles with general relativity. Such a combination is required not only for a unified description of nature, but also because classical physics alone cannot tell us why anything happened. Classical physics gives us a myriad of solutions, but has no way of telling us which is likely and which is unlikely. This is however something that quantum theory can do, and in this manner it might provide explanations for the structure and evolution of the early universe.

A major theme will be that the two theories often thought to be most incompatible in fact appear very precisely suited for each other. Just two examples: The consistency of quantum theory (e.g., in the context of the time–energy uncertainty relation) requires the equivalence principle, as Bohr already argued in an old thought experiment. And black holes, especially their Euclidean versions, seem to "know" about quantum theory and even thermodynamics. That said, the limits of how well quantum theory and relativity can be combined without being strongly modified remains a topic of ongoing investigation. What is sure is that both theories are required to understand the structure of the universe, and only a combination of the two can tell us what happened in the early stages of cosmological evolution. Even just by contemplating the possibilities that these two theories offer, one gets a glimpse of something deep, fundamental about our world. My sincere hope is that I will be able to convey that feeling at least to some extent. Better yet, some people will hopefully be inspired to explore these issues further and extend our knowledge of quantum gravity and cosmology.

This book is intended for people who already have some knowledge of quantum mechanics (ideally also basic notions of quantum field theory) and general relativity. Very little knowledge of cosmology will be assumed, and in fact the book starts with an overview of classical early universe cosmology, leading on to a description of inflation as well as an alternative, ekpyrosis. Subsequent topics will be the semiclassical approach to gravity, the quantum origin of cosmological perturbations, the Unruh and Hawking effects, the creation of the universe according to the no-boundary proposal, quantum tunneling in the presence of gravity, wormholes, and elements of string cosmology. Much of the material we will cover can be seen as complementary to courses on quantum mechanics, general relativity, and cosmology. The book includes many topics, such as gravitational path integrals or Coleman–DeLuccia instantons, that researchers are typically expected to know something about, but that are rarely taught as they fall in between subjects, being seen as too quantum for a relativity course, and too gravitational for a course on quantum theory.

The subjects covered here have evolved very nonlinearly: Some results obtained in the 1970s or 1980s are still the state of the art, while other questions keep being developed in current research. Also, some topics (like wormholes) are undergoing a resurgence of interest. It is my hope that this book will provide the reader with the foundations needed to understand (and perhaps participate in) current research.

Writing a book feels somewhat anachronistic, but given online information overload, a book has the opportunity of providing a more streamlined approach. Since this book is intended for people who are learning the subject, I have included rather detailed explanations in many places. This means that there is probably a little more text (and more figures, and probably also more repetitions...) than usual in specialized books. This is precisely because it is not intended as a compendium of results, but rather presents what I perceive to be the most useful route to learning about these topics. At the end of chapters, there are exercises. For almost all of these, explicit solutions are collected in an Appendix (the exception being that I do not provide numerical code, since the possibilities for efficient coding are evolving fast). These exercises form an integral part of the book – they do not merely fill in some details but also illustrate additional concepts, so do not skip them!

Acknowledgments

Before embarking on this endeavor, I would first like to express my thanks.

My deep appreciation goes to my mentors Kelly Stelle, Neil Turok, Paul Steinhardt, and Burt Ovrut, who have taught me how to do physics.

Over the twenty-odd years of doing research, I have greatly benefited from discussions with my collaborators, fellow researchers, and especially with my students. In particular, I have learned from and am thankful to Andrés Anabalón, Lorenzo Battarra, Sebastian Bramberger, Wilfried Buchmüller, Alice Di Tucci, Shane Farnsworth, Job Feldbrugge, Angelika Fertig, Rhiannon Gwyn, Jonathan Halliwell, Jim Hartle, Arthur Hebecker, Michal Heller, Thomas Hertog, Anna Ijjas, Oliver Janssen, Matt Johnson, Caroline Jonas, Jussi Kalkkinen, Justin Khoury, Claus Kiefer, Axel Kleinschmidt, Michael Köhn, George Lavrelashvili, Rahim Leung, Enno Mallwitz, Paul McFadden, Vincent Meyer, Hermann Nicolai, Jim Peebles, Taotao Qiu, Jérôme Quintin, Ron Reid-Edwards, Sébastien Renaux-Petel, Laura Sberna, Marc Schneider, Paul Smyth, Stefan Theisen, Alex Vilenkin, Yannick Vreys, Dan Wesley, and Ed Wilson-Ewing.

I would like to thank Axel Kleinschmidt and the Max-Planck-Institut für Gravitationsphysik (Albert-Einstein-Institut) in Potsdam, Germany, for hospitality during the writing of this book.

And warm thanks go to my editor Vince Higgs for supporting this project right from the start.

The figures, unless otherwise noted, were produced by myself with the aid of *Inkscape* and *Mathematica*.

I dedicate this book to my wife Linda and our children Julien, Caroline, and Leander – every moment we spend together is a privilege.

Conventions and notation

General relativity conventions

$$\eta_{\mu\nu} = \text{diag}(-1,+1,+1,+1)$$
$$\Gamma^{\rho}_{\mu\nu} = \frac{1}{2} g^{\rho\sigma} \left(g_{\sigma\mu,\nu} + g_{\sigma\nu,\mu} - g_{\mu\nu,\sigma} \right)$$
$$R^{\lambda}{}_{\mu\sigma\nu} = \partial_{\sigma}\Gamma^{\lambda}_{\mu\nu} - \partial_{\nu}\Gamma^{\lambda}_{\mu\sigma} + \Gamma^{\tau}_{\mu\nu}\Gamma^{\lambda}_{\tau\sigma} - \Gamma^{\tau}_{\mu\sigma}\Gamma^{\lambda}_{\nu\tau}$$
$$R_{\mu\nu} = R^{\lambda}{}_{\mu\lambda\nu}$$

Notation

Some of the notation used in this book is:

t, physical time
τ, conformal time
σ, Euclidean time
η, Euclidean conformal time
t_q, a rescaled time coordinate
$\mathcal{N}$, number of e-folds
N, lapse
$\tilde{N}$, rescaled lapse
g_{ij}, spatial part of 4-metric
h_{ij}, spatial 3-metric
γ_{ij}, tensor perturbation
a, scale factor
$q = a^2$, scale factor squared
H, Hubble rate in physical time
$\mathsf{H} = \sqrt{\Lambda/3}$, (constant) de Sitter expansion rate in the flat slicing

$\mathcal{H}$, context dependent: Hubble rate in conformal time, or Hamiltonian
z, context dependent: spatial coordinate, redshift, or integer
ϕ, context dependent: various scalar fields, or angular coordinate
$\mathcal{T}$, temperature
σ^α with $\alpha, \ldots = (0,1)$, string worldsheet coordinates
$g_{\alpha\beta}$, string worldsheet metric
$\gamma_{\alpha\beta}$, pull-back of spacetime metric onto worldsheet
$G_{\mu\nu}$, target space (spacetime) metric
$T = \frac{1}{2\pi\alpha'}$, string tension

1
Modeling the early universe

Cosmology is the science of the universe treated as a whole. It is the science that tries to understand the origin (if there was one), the shape, and the evolution of the universe, including predictions for its future. We will start with a condensed overview of how the universe can be modeled on the largest scales accessible to observations, using general relativity. Given the scope of this book, we will put an emphasis on features that are relevant for understanding the early universe.

1.1 The standard cosmological model

1.1.1 The cosmological principle

The universe cannot be reproduced in the laboratory, and moreover we are severely limited in our ability to travel across the universe. In fact, all our knowledge about the universe must be inferred from the information that arrives here on Earth now. This, however, is not as limiting as it may sound, because of the finiteness of the speed of light. Light originating from distant places, but which we are observing now, shows us these distant places not as they are now, but as they were when the light was emitted. In this way, when looking further into the universe, we see further into the past, and this allows us to reconstruct the history of our universe (up to a point). Of course, we do not directly observe *our* past, but rather the past of increasingly distant regions. This strategy then allows us to reconstruct our history to the extent that the universe is evolving in the same way in different places, i.e., to the extent that the universe is spatially *homogeneous*.

Observations of the distribution of galaxies in the universe show that these are distributed along filaments, with huge voids of over 100 million light-years in diameter in between; see Fig. 1.1. These observations also show that on even bigger scales, namely bigger than about 300 million light-years, the universe becomes very similar in all directions. In other words, on these scales the universe becomes *isotropic*.

Figure 1.1 The distribution of galaxies in a slice of the universe 6 billion light-years away. The slice is 6 billion light-years wide, 4.5 billion light-years high, and half a billion light-years thick. Each dot represents the position of one galaxy, with brighter dots indicating a closer distance to us. Small gray patches indicate regions without data. One of the goals of cosmology is to understand the statistical properties of the distribution of galaxies in the universe. *Credit: Daniel Eisenstein and the SDSS-III collaboration.*

Thus, motivated by observations, we are led to postulate the *cosmological principle*: the universe is spatially *homogeneous* and *isotropic* on the largest observable scales. Some comments:

- If the cosmological principle were not at least approximately true, we would know very little about the distant universe. Only the realization that the universe appears to obey the same laws and have the same rough properties over large distances has allowed us to gain knowledge about the behavior of the distant universe.
- The cosmological principle certainly breaks down on small scales, but it may very well also break down on scales much larger than the currently observable part of the universe.
- The cosmological principle treats space and time on different footings. This is because the universe is only homogeneous and isotropic in a frame

that is *comoving* with typical galaxies. In other words, in the universe there is a preferred frame, and the universe only appears isotropic to the extent that one is at rest with respect to that frame. A further subtlety is that the existence of a preferred frame does not imply that far-away galaxies must be approximately at rest with respect to each other – on the contrary, there is overwhelming evidence that the universe is expanding. The preferred frame can then be imagined to follow (and largely to be defined by) this expansion.

- Despite its name, the cosmological principle is simply an observed feature of our universe and does not stem from fundamental physical principles. It is one of the main aims of early universe cosmology to find an *explanation* for why the universe is so homogeneous and isotropic.

1.1.2 Robertson–Walker universes

We can use general relativity to describe the evolution of the universe, as general relativity is the (classical) dynamical theory of space, time, and matter. We will later incorporate quantum effects and see that they too likely played an important role in shaping the universe.

At first, we will assume exact spatial homogeneity and isotropy. This is of course an idealization, but the cosmological principle implies that this should be a good approximation in describing the universe on large scales. This assumption is very restrictive, and in fact one can prove that it allows for only three distinct spatial geometries, namely flat 3-dimensional space, a 3-sphere (with constant positive curvature), or a hyperbolic 3-space (with constant negative curvature). We denote the 3-dimensional line element by dl^2. Then, using cartesian coordinates x, y, z, the line element of flat space is simply given by

$$dl_{\text{flat}}^2 = dx^2 + dy^2 + dz^2. \tag{1.1}$$

In polar coordinates, defined via $x = r\sin\theta\cos\phi, y = r\sin\theta\sin\phi, z = r\cos\theta$, with $r \geq 0, 0 \leq \theta \leq \pi, 0 \leq \phi < 2\pi$, the line element becomes

$$dl_{\text{flat}}^2 = dr^2 + r^2(d\theta^2 + \sin^2\theta\, d\phi^2). \tag{1.2}$$

We can obtain the metric for the 3-sphere by using the fact that it can be embedded in 4-dimensional flat space. If we denote the fourth spatial coordinate by w, then the line element for 4-dimensional Euclidean space is $dl_4^2 = dx^2 + dy^2 + dz^2 + dw^2$, and within this space the unit 3-sphere is defined via the equation

$$x^2 + y^2 + z^2 + w^2 = 1. \tag{1.3}$$

Differentiating, we learn that $x\,dx + y\,dy + z\,dz + w\,dw = 0$, or

$$dw = \pm \frac{x\,dx + y\,dy + z\,dz}{\sqrt{1 - x^2 - y^2 - z^2}}. \tag{1.4}$$

We can now obtain the metric for the 3-sphere by plugging this expression for dw into the 4-dimensional line element, thus restricting the line element to the 3-sphere. The result is

$$dl^2_{\text{3-sphere}} = dx^2 + dy^2 + dz^2 + \frac{(x\,dx + y\,dy + z\,dz)^2}{1 - x^2 - y^2 - z^2}. \tag{1.5}$$

In a similar way one can obtain the metric of 3-dimensional hyperbolic space by embedding the hyperboloid $x^2 + y^2 + z^2 - w^2 = -1$ into 4-dimensional Minkowski space $dl^2_4 = dx^2 + dy^2 + dz^2 - dw^2$. All three metrics can be conveniently summarized by (where Latin indices run over the spatial coordinates)

$$dl^2_3 = dx^2 + dy^2 + dz^2 + \frac{K(x\,dx + y\,dy + z\,dz)^2}{1 - K(x^2 + y^2 + z^2)} \equiv g_{(3)ij} dx^i dx^j, \tag{1.6}$$

or, in polar coordinates,

$$dl^2_3 = \frac{1}{1 - Kr^2} dr^2 + r^2(d\theta^2 + \sin^2\theta\,d\phi^2), \tag{1.7}$$

where $K = 0, +1, -1$ for the flat, spherical, and hyperbolic spaces, respectively.[1]

Above, we derived the metric on the 3-sphere only for the case where it has unit radius. We can allow for an arbitrary, and in general time-dependent, size by multiplying the line element by a factor $a^2(t)$, and likewise for the hyperbolic space. In the case of flat space, the multiplicative factor $a^2(t)$ has no absolute meaning, and only relative values at different times are meaningful. The full 4-dimensional *Robertson–Walker* (sometimes *Friedmann–Lemaître–Robertson–Walker*) metric can then be obtained by adding a time direction, yielding

$$ds^2 = -dt^2 + a^2(t)\left[\frac{1}{1 - Kr^2} dr^2 + r^2(d\theta^2 + \sin^2\theta\,d\phi^2)\right] \quad \text{(FLRW)}. \tag{1.8}$$

The function $a(t)$ is known as the *scale factor* of the universe. Its meaning can be clarified by using the metric above to calculate the spatial distance between an observer at the origin $r = 0$ and one at $r = R$ (at fixed time t,

---

[1] In the 2-dimensional analogs of these spaces, for a triangle we have that $\sum$ angles $> \pi$ for $K = +1$, whereas $\sum$ angles $= \pi$ for $K = 0$, and $\sum$ angles $< \pi$ for $K = -1$.

and with the angular coordinates also being fixed), which is given by

$$\int_{r=0}^{r=R} ds = \int_{r=0}^{r=R} a(t) \frac{1}{\sqrt{1-r^2}} dr = a(t) \begin{cases} \operatorname{arcsinh} R & (K=-1), \\ R & (K=0), \\ \arcsin R & (K=+1). \end{cases} \quad (1.9)$$

Thus observers at fixed r, θ, ϕ only change their relative distance via the global, space-independent scale factor. In other words, the scale factor describes the overall expansion or contraction of the universe. The coordinates r, θ, ϕ are so-called *comoving* coordinates, and the result above implies that the proper distance between two comoving observers evolves as $a(t)$. Moreover, for a comoving observer, proper time is measured according to $\int \sqrt{-ds^2} = \int dt = \Delta t$, which shows that the time coordinate t is simply the *proper time* measured by comoving observers.

Sometimes, it is convenient to define a *conformal time* τ via $dt = a d\tau$, so that the metric has an overall factor of a^2:

$$ds^2 = a^2(\tau) \left[-d\tau^2 + \frac{1}{1-Kr^2} dr^2 + r^2 (d\theta^2 + \sin^2\theta \, d\phi^2) \right]. \quad (1.10)$$

The general form of the Robertson–Walker metric allows us to understand a crucial effect in cosmology, namely the shifting of spectral lines observed in the light coming from distant galaxies. When the scale factor is nonconstant, the frequency of light is in general different at the time of emission than at the time of observation. This can be seen as follows: we assume that we are observing (at time t_0) the light originating (at time t_R) from a galaxy located at $r = R$. Our own position can be chosen to be at $r = 0$. Since light rays propagate according to $ds^2 = 0$, we obtain from Eq. (1.8) that

$$dt = -a(t) \frac{dr}{\sqrt{1-Kr^2}}, \quad (1.11)$$

where the minus sign was chosen because the light is coming towards us along decreasing values of r. Hence the times of emission and observation are related to the coordinate separation R by the first equality below,

$$\int_{t_0}^{t_R} \frac{dt}{a(t)} = \int_0^R \frac{dr}{\sqrt{1-Kr^2}} = \int_{t_0+\Delta t_0}^{t_R+\Delta t_R} \frac{dt}{a(t)}. \quad (1.12)$$

Subsequent wave crests of the light wave are emitted at time intervals Δt, if the frequency of the light is $\nu = 1/\Delta t$. So now consider the emission of the next crest, at time $t_R + \Delta t_R$, and its absorption at $t_0 + \Delta t_0$. This is still equal to the same coordinate separation, hence the second equality above holds too. Subtracting the left- from the right-hand side, and working to leading order in Δt, one gets

$$\frac{\Delta t_R}{a(t_R)} = \frac{\Delta t_0}{a(t_0)}. \tag{1.13}$$

In other words, the frequency at observation is related to the frequency at emission via

$$\nu_0 = \nu_R \frac{a(t_R)}{a(t_0)}. \tag{1.14}$$

Equivalently, the wavelength at observation λ_0 is shifted from the wavelength at emission λ_R via

$$\lambda_0 = \lambda_R \frac{a(t_0)}{a(t_R)} \equiv \lambda_R(1+z), \tag{1.15}$$

where z (not to be confused with the third spatial coordinate) is conventionally called the *redshift*. The wavelength of light simply changes according to the change of the space it lives in – see also Fig. 1.2. In an expanding universe, the wavelength at observation is longer than at emission, $z > 0$, and the light is indeed *redshifted*. In a contracting universe, light decreases in wavelength, $z < 0$, and in that case light is *blueshifted*.

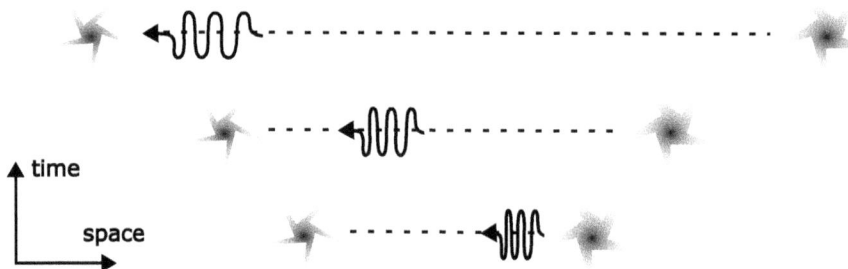

Figure 1.2 In an expanding universe, the space in between well-separated galaxies grows over time (though the space in the vicinity of the galaxies is not expanding). Light emitted from one galaxy is redshifted on the way because its wavelength grows in proportion to the expansion of space.

For nearby sources, we can expand $a(t) = a(t_0)[1 + H_0(t-t_0) + \cdots]$, where

$$H \equiv \frac{\dot{a}}{a}. \tag{1.16}$$

Here H is the *Hubble rate* and the subscript 0 customarily refers to a quantity evaluated at the present time. Then, keeping in mind that the speed of light is set to unity, we can identify $t - t_0$ with the proper distance D of the

1.1 The standard cosmological model

source, finding that to a first approximation the redshift increases linearly with distance,

$$z \approx H_0 \, D \qquad \text{(Hubble)}. \tag{1.17}$$

This is the famous *Hubble law*. Astronomical observations, first carried out by Vesto Slipher in the 1920s and to the required accuracy by Edwin Hubble in the 1930s, show that the light from almost all galaxies is redshifted, and that the redshift–distance relation is indeed approximately linear for galaxies that are neither too close nor too far. This provides crucial evidence that our universe is expanding. Only the light from a few nearby galaxies (such as the Andromeda galaxy) is blueshifted, which can be understood as the result of the strong gravitational interactions between these galaxies and the Milky Way, causing them to have significant *peculiar velocities* superimposed on the comoving *Hubble flow*. And at redshifts larger than about $z \approx 0.1$, nonlinear terms in the expression for the Hubble parameter become important.

The observational value of the present-day Hubble rate is a matter of ongoing research and debate. Current observational values fall in the range

$$67 \, \text{km s}^{-1} \, \text{Mpc}^{-1} \lesssim H_0 \lesssim 74 \, \text{km s}^{-1} \, \text{Mpc}^{-1}, \tag{1.18}$$

where one parsec is $1\text{pc} \approx 3.26$ light-years. Smaller values tend to arise in measurements involving the cosmic microwave background radiation (explained later), while larger values come from more local measurements, for instance using type Ia supernovae as *standard candles*, i.e., as sources of light with known luminosity (in this case explosions occurring after a critical amount of matter accretion onto a white dwarf). The best current measurements claim an uncertainty of around $1 \, \text{km s}^{-1} \, \text{Mpc}^{-1}$. The discrepancy between different observations is referred to as the Hubble tension, and is currently unsolved.

We should stress that the Robertson–Walker metric is only intended as a zeroth-order approximation to the coarse-grained structure of the universe. In thinking about the expansion/contraction of space, it is important to keep in mind that near localized sources (such as stars, or even entire galaxies) the metric is locally not of Robertson–Walker form, but more closely approximates the Schwarzschild form

$$\mathrm{d}s^2 = \left(1 - \frac{2M}{r}\right) \mathrm{d}t^2 + \frac{\mathrm{d}r^2}{\left(1 - \frac{2M}{r}\right)} + r^2 (\mathrm{d}\theta^2 + \sin^2\theta \, \mathrm{d}\phi^2), \tag{1.19}$$

where M is the mass of the source and r is a radial coordinate originating at the center of mass. In the Schwarzschild metric, when $r > 2M$ space is not expanding and, as a consequence, in our solar system space is neither expanding nor contracting. As an improved approximation, one could think of the entire universe as a kind of "Swiss cheese," with the cheese being well described by the Robertson–Walker metric, and the (small) holes by the Schwarzschild metric. This avoids some standard misconceptions about the expansion of space.

1.1.3 The Friedmann equations

General relativity specifies how spacetime is affected by the matter and energy content of the universe. More concretely, it relates the spacetime curvature, in the form of the Einstein tensor $G_{\mu\nu}$, to the matter content of the universe via its energy–momentum tensor $T_{\mu\nu}$, according to the Einstein equations

$$G_{\mu\nu} = \frac{8\pi G}{c^4} T_{\mu\nu} \quad \text{(Einstein)}, \tag{1.20}$$

where Greek indices run over all spacetime coordinates. We will mostly use natural units where $8\pi G = 1$, $c = 1$. In addition, the Einstein tensor obeys the Bianchi identity, from which it follows by consistency that the energy–momentum tensor is covariantly conserved,

$$\nabla^\mu G_{\mu\nu} = 0 \quad \text{(Bianchi)}, \qquad \nabla^\mu T_{\mu\nu} = 0 \quad \text{(conservation)}. \tag{1.21}$$

We will need to derive appropriate forms for both $G_{\mu\nu}$ and $T_{\mu\nu}$ in the cosmological context.

Let us start with the spacetime. Here the Einstein tensor is defined in terms of the Ricci curvature tensor $R_{\mu\nu}$ according to

$$G_{\mu\nu} \equiv R_{\mu\nu} - \frac{1}{2} g_{\mu\nu} R. \tag{1.22}$$

The cosmological principle led to the Robertson–Walker form of the metric, Eq. (1.8). In Euclidean coordinates this metric can also be written as

$$ds^2 = -dt^2 + a^2(t) g_{(3)ij} dx^i dx^j, \tag{1.23}$$

where the spatial 3-metric $g_{(3)}$ was defined in Eq. (1.6). It is a straightforward exercise to derive the associated connections and curvature tensors,

$$\Gamma^0_{ij} = \frac{\dot{a}}{a} g_{ij} = a\dot{a} g_{(3)ij}, \tag{1.24}$$

$$\Gamma^k_{0j} = \frac{\dot{a}}{a} \delta^k_j, \tag{1.25}$$

$$\Gamma^k_{ij} = \Gamma^k_{(3)ij} = K g_{(3)ij} x^k, \tag{1.26}$$

$$\Gamma^0_{00} = \Gamma^0_{0j} = \Gamma^k_{00} = 0, \tag{1.27}$$

1.1 The standard cosmological model

$$R_{00} = -3\tfrac{\ddot{a}}{a}, \qquad (1.28)$$

$$R_{ij} = (\tfrac{\ddot{a}}{a} + 2\tfrac{\dot{a}^2}{a^2} + 2\tfrac{K}{a^2})g_{ij} = (a\ddot{a} + 2\dot{a}^2 + 2K)g_{(3)ij}, \qquad (1.29)$$

$$R_{0j} = 0, \qquad (1.30)$$

$$R = 6(\tfrac{\ddot{a}}{a} + \tfrac{\dot{a}^2}{a^2} + \tfrac{K}{a^2}). \qquad (1.31)$$

The assumptions of spatial isotropy and homogeneity also lead to strong restrictions on the form of the energy–momentum tensor on large scales. Isotropy, i.e., invariance under spatial rotations, implies that (at a comoving point) the spatial part T_{ij} of the energy–momentum tensor is proportional to the metric g_{ij}, and that the mixed, vectorial component T_{0i} vanishes (otherwise a particular direction would be singled out). Homogeneity means that we must impose these conditions everywhere in space, implying that the function of proportionality in T_{ij} can depend on time alone (otherwise, different spatial locations would not be equivalent). Likewise, T_{00} can depend on time alone. The conventional definitions are

$$T_{00} \equiv \rho(t), \qquad T_{0i} = 0, \qquad T_{ij} \equiv p(t)g_{ij}, \qquad (1.32)$$

where $\rho(t)$ is the proper *energy density* and $p(t)$ the *pressure*. This particular form of the energy–momentum tensor, imposed on us by the assumptions of isotropy and homogeneity, corresponds to that of an ideal fluid. The energy density and the pressure are related to each other via the *equation of state*

$$p = w\rho. \qquad (1.33)$$

In many cases of interest, w is constant, and we will assume this to be the case unless otherwise noted.

What are the consequences of the covariant conservation $T^{\mu\nu}{}_{;\nu} \equiv \nabla_\nu T^{\mu\nu} = 0$ of the energy–momentum tensor? The $\mu = i$ component of this equation corresponds to momentum conservation, and it is automatically satisfied in a Robertson–Walker background. The $\mu = 0$ component leads to

$$\dot{\rho} + 3H(\rho + p) = 0. \qquad (1.34)$$

This equation is known as the *equation of continuity*. Note that it shows that in general, in a dynamical universe, the energy density is *not* conserved. For a constant equation of state, the equation of continuity can be integrated to yield

$$\rho \propto \frac{1}{a^{3(1+w)}}. \qquad (1.35)$$

Thus, if we know the equation of state of a certain matter component, then we know how its energy density (and thus its relative importance) scales as the universe contracts or expands. The most commonly considered ideal fluids are

- Pressure-free matter/dust, $w = 0$. Ordinary baryonic matter as well as dark matter fall into this category. Their energy density simply scales inversely to the volume of a given region of space, $\rho \propto a^{-3}$, as expected.
- Relativistic particles/radiation, $w = \frac{1}{3}$. The energy density of radiation or a gas of relativistic particles scales as a^{-4}. This means that in an expanding universe, the energy density of radiation falls off faster than the volume of space increases. Heuristically one can understand the exponent -4 as follows: The number of photons in a given co-moving volume scales with the volume, as a^{-3}. However, the wavelength of a photon also scales with the scale factor (this is the redshift), and thus its frequency/energy scales as a^{-1}. This additional factor of a^{-1} then leads to the overall scaling as a^{-4}.
- Cosmological constant, $w = -1$. The energy density of a cosmological constant is, as the name suggests, constant over time and unaffected by cosmic expansion/contraction. For this reason it is also often described as vacuum energy. Note that $p = -\rho$ implies that the energy–momentum tensor is proportional to the metric $T_{\mu\nu} \propto g_{\mu\nu}$, which is consistent with the Lorentz invariance of the vacuum. Another name that is used to describe a cosmological constant, or in fact any type of matter with an equation of state that is close to $w = -1$, is *dark energy*. While observations suggest that dark energy currently provides the dominant contribution to the overall energy density of the universe, this type of matter remains by far the least understood.

There is one important omission in this list: scalar fields, which are often used in modeling the dynamics of the universe. We will treat these in detail in Chapter 2.

We are finally ready to derive the *Friedmann equations*, which arise by inserting the homogeneous and isotropic expressions for curvature and energy–momentum into the Einstein equations. Using the expressions just derived, one gets

$$H^2 + \frac{K}{a^2} = \tfrac{1}{3}\rho \qquad \text{(Friedmann)}, \qquad (1.36)$$

$$\frac{\ddot{a}}{a} = -\tfrac{1}{6}(\rho + 3p) = -\tfrac{1}{6}\rho(1 + 3w) \quad \text{(acceleration)}. \qquad (1.37)$$

The first of these equations is sometimes referred to as simply the *Friedmann equation*, while the second one is often called the *acceleration equation*. These equations determine the evolution of the universe on the largest observable scales.

One should think of ρ and p above as denoting the total energy density and pressure – these then consist of a sum over all the individual matter types that are present in the universe. Before considering a mixture of ingredients, though, it is useful to look at a few solutions to these equations when only a single matter component is present. We will consider the examples of matter types enumerated above, for the case of a spatially flat universe ($K = 0$):

- Pressure-free matter/dust, $w = 0$. The Friedmann equation then implies that the scale factor evolves as $a(t) \propto t^{\frac{2}{3}}$, while the acceleration equation shows that the universe is decelerating. (Including integration constants, the solution can be written as $a(t) = \frac{(t-t_\star)^{2/3}}{(t_0-t_\star)^{2/3}}$, where t_0 is the time today, and $t_\star$ the time of the big bang, with $a_0 = 1$.)
- Relativistic particles/radiation, $w = \frac{1}{3}$. In this case the scale factor evolves as $a(t) \propto t^{\frac{1}{2}}$, and this also corresponds to decelerated expansion.
- Cosmological constant, $w = -1$. The Friedmann Eq. (1.36) implies that $\rho > 0$ since $K = 0$. In this case the scale factor evolves exponentially, $a(t) \propto e^{Ht}$, and the Hubble parameter is constant. This solution is known as the *de Sitter* universe. According to Eq. (1.37), in the de Sitter solution the expansion is accelerating. Above, we mentioned that matter with equation of state close to -1 is also sometimes called dark energy. Here, we can refine that definition. The acceleration equation implies that, for a positive energy density, the universe is accelerating whenever $w < -\frac{1}{3}$. We have presently entered such an era of accelerated expansion.

The solutions for $K = 0$ are those that are used most often in cosmology, as current observations indicate that our universe is very close to spatially flat. Indeed, whether the universe has positive or negative spatial curvature can be deduced by comparing its density to the *critical density*, cf. Eq. (1.36),

$$\rho_{\rm crit} = 3H^2 \,. \tag{1.38}$$

And observations show that our universe's energy density is very close to the critical density, which is given by

$$\rho_{\rm crit,0} = 3H_0^2 \approx \frac{3}{8\pi G}(70\,{\rm km\,s^{-1}\,Mpc^{-1}})^2 \approx 10^{-26}\,{\rm kg\,m^{-3}}\,, \tag{1.39}$$

which amounts to only about one hydrogen atom per cubic meter of volume. As a side remark, let us stress that even spatially flat universes ($K = 0$) are curved in a 4-dimensional sense – their curvature arises from the time evolution of the scale factor.

But even in a universe that appears spatially flat on large scales, the fact that the matter in the universe is distributed inhomogeneously on small scales indicates that on those scales the average curvature must necessarily be positive or negative. Hence it is certainly also important to look at solutions with $K = \pm 1$. The corresponding solutions are more easily found using conformal time τ. In terms of conformal time, the Friedmann equations read

$$a'^2 + Ka^2 = \frac{1}{3}\rho a^4, \tag{1.40}$$

$$a'' + Ka = \frac{1}{6}(\rho - 3p)a^3, \tag{1.41}$$

where a prime denotes a derivative with respect to conformal time.

For $K = +1$, the spatial sections of the universe are spheres, and this is known as the *closed* model of the universe. The Friedmann Eq. (1.40) shows that the expansion can come to a halt, even when the energy density is positive. After such a halt, the universe re-collapses, as can be seen in the solutions obtained in the presence of dark matter or radiation

$$a(\tau) = \begin{cases} 1 - \cos\tau & w = 0 \quad 0 < \tau < 2\pi, \\ \sin\tau & w = \frac{1}{3} \quad 0 < \tau < \pi. \end{cases} \tag{1.42}$$

These solutions are used in describing gravitational collapse of overdense regions.

For $K = -1$, the universe is called *open*, and the corresponding solutions are

$$a(\tau) = \begin{cases} \cosh\tau - 1 & w = 0 \quad 0 < \tau < \infty, \\ \sinh\tau & w = \frac{1}{3} \quad 0 < \tau < \infty. \end{cases} \tag{1.43}$$

For those solutions, the universe expands forever. Open-universe solutions can serve as useful approximations to the void regions of the universe.

Before proceeding with a more quantitative analysis of the Friedmann equations, let us highlight their main philosophical consequence. They imply that the universe is not static. Space and time are evolving quantities, even on the largest observable scales. This marks a major shift in our thinking about nature: up until about 100 years ago, people were mostly convinced (on philosophical or religious grounds) that the universe on the whole had to be static, immutable, and infinite. The discovery of the expansion of the universe, going hand in hand with the elaboration and vindication of general relativity, has brought with it a major conceptual paradigm shift, as the universe is now seen as an evolving object in itself. The universe has a history, which cosmologists are trying to unravel.

1.2 A thermal history

To obtain a realistic large-scale model of the universe, we must include all of the relevant matter/energy constituents. For the standard, so-called ΛCDM model, these are radiation (mostly photons, but can also include massive particles moving at relativistic speeds), pressure-free matter (cold dark matter and ordinary baryons), and a cosmological constant Λ. We should note that there are good theoretical reasons for why dark energy might not be a simple cosmological constant, but current observations are still compatible with a cosmological constant and hence, for simplicity, we will stick with Λ for now. The Friedmann equation (1.36) then reads

$$3H^2 = \frac{\rho_{r,0}}{a^4} + \frac{\rho_{m,0}}{a^3} - \frac{3K}{a^2} + \Lambda. \qquad (1.44)$$

If we choose units in which the current scale factor is $a_0 = 1$, then $\rho_{r,0}$ and $\rho_{m,0}$ denote the current energy densities in radiation and pressure-free matter.

As the universe expands, it is clear that matter types that scale with a less negative power of the scale factor eventually dominate over those that scale with more negative powers of a. This leads to a series of epochs, during each of which a different energy component dominates and drives the dynamics. A useful way to rewrite Eq. (1.44) is obtained by dividing by the current critical density $3H_0^2$, resulting in the two equivalent expressions, either in terms of scale factor a or redshift z,

$$\left(\frac{H}{H_0}\right)^2 = \frac{\Omega_r}{a^4} + \frac{\Omega_m}{a^3} + \frac{\Omega_K}{a^2} + \Omega_\Lambda \qquad (1.45)$$

$$= \Omega_r(1+z)^4 + \Omega_m(1+z)^3 + \Omega_K(1+z)^2 + \Omega_\Lambda, \qquad (1.46)$$

where the current fractional energy densities are conventionally defined as

$$\Omega_r = \frac{\rho_{r,0}}{3H_0^2}, \quad \Omega_m = \frac{\rho_{m,0}}{3H_0^2}, \quad \Omega_K = \frac{-K}{H_0^2}, \quad \Omega_\Lambda = \frac{\Lambda}{3H_0^2}. \qquad (1.47)$$

By definition these sum to unity,

$$\Omega_r + \Omega_m + \Omega_K + \Omega_\Lambda = 1. \qquad (1.48)$$

It is useful to highlight how theory and observations are checked against each other in practice. Unlike for most laboratory experiments, it is not the case here that the theory makes a number of predictions, then an experiment is set up in order to check them. Rather, in a work of constant refinement, theory and observations are matched to each other. It is in this way that, on the one hand, the current and past abundances of various matter types are determined, and on the other hand the Einstein/Friedmann equations are verified. Consider for instance Fig. 1.3, which illustrates this process. The figure contains measurements of the apparent magnitude (brightness) of type Ia supernovae, whose intrinsic brightness is well-enough understood to act as reference light sources, at many redshifts. The apparent magnitude however depends on the expansion history $H(z)$ of the universe. And this history can then be fitted from those arising by assuming general relativity and various matter contributions, as exemplified in the figure.

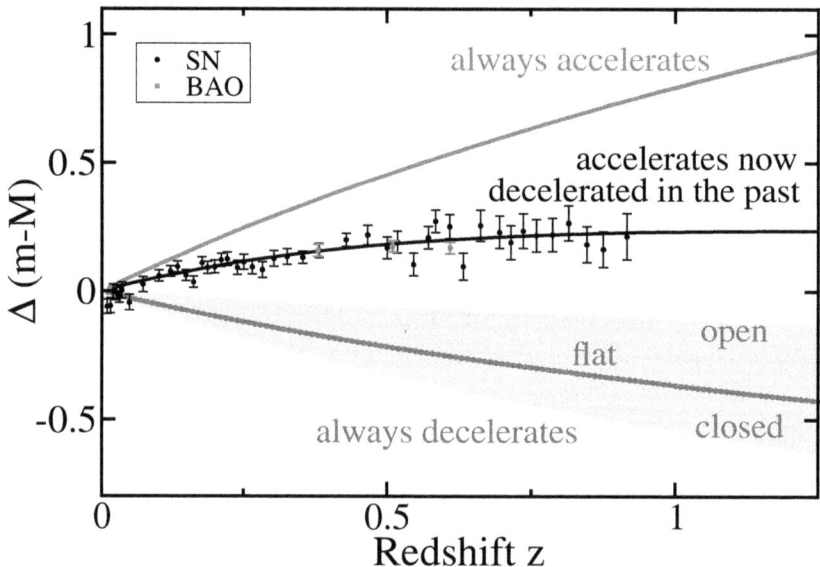

Figure 1.3 A plot of the change in apparent magnitude of cosmological sources (type Ia supernovae from Scolnic et al. (2015) in black, and baryon acoustic oscillations from BOSS Collaboration (2017) in gray) versus redshift. This figure does not contain the most up-to-date data, but illustrates well how the underlying cosmological model is fitted to the data. The reference model is an open empty universe with $\Omega_K = 1$. Models without dark energy ($\Omega_\Lambda = 0$) and various amounts of matter ($0.3 \leq \Omega_m \leq 1.5$) are shown in the shaded area, and are seen to be ruled out. A pure vacuum energy dominated universe with $\Omega_\Lambda = 1$ is shown by the top curve and is also ruled out. Favored is a model with $\Omega_m = 0.3$ and $\Omega_\Lambda = 0.7$, depicted here by the black middle line. Figure reproduced from Huterer and Shafer (2018). © IOP Publishing. Reproduced with permission. All rights reserved.

We should stress that this particular measurement is just one of many different probes of both the expansion history and the matter content of the universe. These measurements are remarkably consistent with each other (they tend to agree on the level of a few percent), and they lead to the following current composition (inferred here from 2018 Planck satellite measurements combined with baryon acoustic oscillation (BAO) measurements, with 1σ errors – see Planck Collaboration; Aghanim et al. (2020a)):

$$\Omega_\Lambda = 0.689 \pm 0.006\,, \quad H_0 \equiv 100 \cdot h = (67.7 \pm 0.4)\,\mathrm{km\,s^{-1}\,Mpc^{-1}}\,,$$
$$\Omega_\mathrm{m} = 0.311 \pm 0.006\,, \quad \Omega_\mathrm{r} = (9.18 \pm 0.17) \times 10^{-5}\,. \tag{1.49}$$

In this fit, it was assumed that the universe is spatially flat, $\Omega_\mathrm{K} = 0$. When the spatial curvature is allowed to vary, one finds what amounts to an upper bound,

$$\Omega_\mathrm{K} = 0.000\,7 \pm 0.001\,9\,. \tag{1.50}$$

Let us take stock then. The main ingredient in today's universe is dark energy, causing the universe to accelerate its expansion. In fact, from Eq. (1.37) we can easily determine the redshift at which acceleration began, when $\ddot{a} = 0$, giving

$$z_\mathrm{acc} \approx 0.64\,. \tag{1.51}$$

This was when the universe had about 61% of its current (linear) size, and was about 7.6 billion years old (we will see below how to calculate ages).

Before the dark energy phase, there was a phase of matter domination, during which the expansion decelerated. For completeness we should point out that out of the current pressure-free matter only about a fifth is thought to be made of ordinary baryons, making up gas and dust clouds, stars, planets, and us. The rest is dark matter, whose effects are seen via gravity (e.g., due to gravitational lensing of light emitted by distant galaxies, and bent by intervening dark matter), but which has not been detected directly. Understanding dark energy and dark matter better are major goals of cosmology, but they have affected the very early universe less strongly, since the matter phase was preceded by a radiation-dominated phase.

The switch between matter and radiation domination occurred at the so-called time of equality, when the size of the universe was

$$a_\mathrm{eq} = \frac{1}{1 + z_\mathrm{eq}} = \frac{\Omega_\mathrm{r}}{\Omega_\mathrm{m}} \approx \frac{1}{3\,388}\,, \tag{1.52}$$

compared to its current size, and its age was about $51\,000$ years. The radiation density is composed mostly of the cosmic microwave background (discussed below), but also contains contributions from neutrinos, which are very light particles and hence behave relativistically. The main physical characteristic during the radiation phase is the temperature. Recall that photons have energy $E = h\nu = hc/\lambda = k_B \mathcal{T}$, where ν is the frequency, λ the wavelength, k_B Boltzmann's constant, and $\mathcal{T}$ the absolute temperature, where we have momentarily restored units. Thus the wavelength can be expressed as $\lambda = hc/(k_B \mathcal{T})$. As the universe expands or contracts, the wavelength changes with the scale factor $\lambda \propto a(t)$, and thus the temperature evolves as

$$\mathcal{T}(t) = \mathcal{T}_0 \frac{a_0}{a(t)}. \tag{1.53}$$

This relation implies that the universe was *hotter* in the past, hence the name *hot big bang model*. In the very early universe, photons and other matter particles were tightly coupled, implying that in fact all matter constituents were hot. For this epoch, there exists a useful relation between temperature and time; the Friedmann equation implies that

$$H^2 = \frac{1}{4t^2} = \frac{1}{3}\rho \approx \mathcal{T}^4, \tag{1.54}$$

where we have neglected a factor of order unity relating the energy density in radiation to the fourth power of temperature. The relation above, namely $\mathcal{T} \approx 1/\sqrt{t}$, holds in natural units, and if we restore units, we get the formula

$$\mathcal{T}_{\text{MeV}} \approx \frac{1}{\sqrt{t_{\text{sec}}}}, \tag{1.55}$$

where the temperature is expressed in MeV and the time in seconds (counting from a putative big bang). For completeness we should mention that, going to earlier times and hence hotter temperatures, more and more massive particles start behaving relativistically and hence the number of species contributing to the radiation density goes up as we approach the big bang.

In fact we are now in a position to calculate the age of the universe or, more precisely, the time that has elapsed since the big bang. The Friedmann Eq. (1.45) can be recast as the differential

$$dt = \frac{da}{H_0 \sqrt{\Omega_r a^{-2} + \Omega_m a^{-1} + \Omega_K + \Omega_\Lambda a^2}}. \tag{1.56}$$

Integrating from $a = 0$ to $a = a_0 = 1$ gives us the time since the big bang, namely

$$t_{\text{BB}} \approx 0.954 \frac{1}{H_0} \approx 13.8 \times 10^9 \text{ years} \qquad \text{(age of universe)}. \tag{1.57}$$

This may be compared with the age of the solar system, which formed about 4.65 billion years ago. Integrating (1.56) up to the scale factors a_{eq} and a_{acc} gives the ages mentioned above. Note that, for quick order-of-magnitude estimates, it is useful to memorize that the time since the big bang and the

1.2 A thermal history

size of the observable universe (on the order of 10 billion years and 10 billion light-years) are approximately given by 10^{60} Planck times and 10^{60} Planck lengths, respectively.

The high temperatures in the early universe had a marked effect on the behavior of matter. Switching from our backwards-looking perspective to now counting the time forward from the (putative) origin at $t = 0, a = 0$, we may list some of the major events in the history of the universe:

$t < 10^{-14}$ s, $\mathcal{T} > $ TeV

We still know very little with certainty about this epoch, as the corresponding particle physics currently cannot be reproduced in accelerator experiments. From our current laws of physics, we can assume that the electroweak symmetry was unbroken during this early phase, and in fact the gauge group describing all matter interactions could have been much bigger (such as $SO(10)$, allowing a grand unified description of particle physics). Also, supersymmetry may have played a role at those early times, but it remains too early to tell. Much of the rest of this book will be about this epoch.

10^{-14} s $< t < 10^{-10}$ s, $100 \, \text{GeV} < \mathcal{T} < 10 \, \text{TeV}$

During this time, the electroweak symmetry $SU(2) \times U(1)$ was broken, and the W, Z bosons became massive. The particle physics that took place at these energies is currently being probed by the Large Hadron Collider at CERN, and is, to a large extent, well understood.

$t \sim 10^{-5}$ s, $\mathcal{T} \sim 200 \, \text{MeV}$

As the universe cooled, free quarks and gluons bound into baryons (particles with three quarks, such as protons and neutrons) and mesons (particles with a quark and an antiquark).

$t \sim 0.2$ s, $\mathcal{T} \sim 1 - 2 \, \text{MeV}$

The cross sections of the weak interactions became very small as the temperature dropped to these scales, and thus the neutrinos decoupled from the rest of the matter in the universe. This neutrino background evolved independently from that time on. In the long-term future, one may hope that it will be possible to build a neutrino telescope (perhaps on the far side of the Moon) that can measure these primordial neutrinos directly. This would give us direct information about the physical conditions at these times close to the big bang.

As a byproduct, the ratio of neutrons to protons also *froze out* around this time. Up until this time, protons could easily convert into neutrons, and vice versa, via the exchange of neutrinos, electrons, and positrons,

$$p + e^- \leftrightarrow n + \nu, \qquad p + \bar{\nu} \leftrightarrow n + e^+. \tag{1.58}$$

The neutron is heavier than the proton by an amount $Q = m_n - m_p \approx 1.3\,\text{MeV}$, where $m_{n,p}$ denote the masses of a neutron and a proton, respectively. Standard thermodynamics then implies that, as long as the two species are still in equilibrium, i.e., as long as the reactions above are efficient, the number densities of neutrons (n_n) and protons (n_p) are related by

$$\frac{n_n}{n_p} \approx e^{-Q/\mathcal{T}}. \tag{1.59}$$

The reactions above require an energy $m_n - m_p - m_e \approx 1.3 - 0.5 \approx 0.8\,\text{MeV}$ to be efficient, where m_e denotes the mass of an electron. Below this temperature, the ratio of neutrons to protons gets frozen in, with $n_n/n_p \approx e^{-1.3/0.8} \approx 1/5$. This has important implications for big bang *nucleosynthesis*, as we will see shortly.

$t \sim 1\,\text{s}$, $\mathcal{T} \sim 0.5\,\text{MeV}$

When the temperature drops to the value of the rest mass of electrons and positrons, these start annihilating each other into photons, while the converse process, spontaneous creation of electron–positron pairs out of photons, becomes rare. The photons that are produced in this way are in thermal equilibrium, and their temperature is slightly higher than that of the neutrinos, which already decoupled earlier. The end result of electron–positron annihilation is that about one electron was left over for each billion photons. Thus there must have been a very slight excess of matter over antimatter in the early universe. This asymmetry remains largely unexplained.

$t < 5\,\text{min}$, $\mathcal{T} \sim 0.05\,\text{MeV}$

Above, we saw that the ratio of neutrons to protons froze out at a value of about 1/5. However, neutrons are only stable when they are locked inside an atomic nucleus. Free neutrons decay into a proton, an electron, and a neutrino with a half life of about 15 min. Over the next few minutes, these decays reduced the neutron-to-proton number density fraction to

$$\frac{n_n}{n_p} \approx \frac{1}{7}. \tag{1.60}$$

Meanwhile, neutrons can combine with protons to form deuterons

$$p + n \to d + \gamma, \tag{1.61}$$

while, in turn, deuterons can combine via the following chain to end up as helium nuclei

$$d + d \to {}^3He + n, \qquad {}^3He + d \to {}^4He + p. \qquad (1.62)$$

Assuming that all neutrons end up in helium atoms (which is in fact a good approximation), we would deduce that the total mass fraction in helium should be

$$Y = \frac{n_{He}m_{He}}{n_n m_n + n_p m_p} \approx \frac{(\frac{1}{2}n_n)(4m_n)}{(n_n + n_p)m_n} = \frac{2n_n/n_p}{1 + n_n/n_p} \approx \frac{1}{4}. \qquad (1.63)$$

Helium is also produced via nuclear fusion in stars. However, looking at stars that are further and further away (and thus looking further into the past) observations show that helium levels decrease and level off around a value of 1/4. Measurements of other light elements, such as deuterium and lithium, are also in reasonably good agreement with theoretical estimates of their production via primordial nucleosynthesis. This is striking evidence that the universe was once hot and dense, and provides strong support for the hot big bang model.

$t \approx 10^5$ years, $\mathcal{T} \sim$ eV

After a time of about 51 000 years (determined by integrating (1.56) up to $a_{eq} = 1/(1 + z_{eq})$), matter came to dominate over the radiation energy density, and the universe started expanding according to $a \propto t^{2/3}$. By the time the universe grew by a further factor of 3 in each direction, the helium nuclei and protons combined with the electrons to form helium and hydrogen atoms. This event is known as *recombination* and will be discussed in more detail below. The name recombination is somewhat misleading, as this is the first time after the big bang that atoms formed.

$t \approx 10^9$ years, $\mathcal{T} \sim$ meV

As we will discuss, small fluctuations in the distribution of matter and radiation were present already at early times in the history of the universe. Under the influence of gravity, small initial overdensities grew and eventually collapsed into stars and galaxies.

About 8 billion years after the big bang, the dark energy came to dominate the energy density in the universe, and the expansion started accelerating (at $z = z_{acc}$). The future dynamics of the universe depends crucially on the nature of dark energy. We will speculate about possible futures at various stages in this book.

1.2.1 A relic from the big bang: the cosmic microwave background

At the time of radiation–matter equality, the universe can still be described as a mixture of mostly photons, electrons, protons, and helium ions. Via Thomson scattering, the photons constantly bounce off the free electrons (the scattering off protons occurs much more rarely due to the significantly larger mass of the proton), which means that they cannot travel far in a straight line and hence the universe is not transparent at that time – the term *primordial soup* is sometimes used to describe this state of affairs.

As the universe cools, at some point it becomes energetically favorable for the electrons to bind with the free protons and helium ions to form hydrogen and helium atoms, respectively. In fact, because helium has a larger ionization potential, helium atoms form first. Afterwards, there are still many free electrons left over, and these can combine with protons to form neutral hydrogen atoms. Since the binding energy of hydrogen is about 13.6 eV, a first guess is that hydrogen production ("recombination") will occur when the universe has reached that temperature. However, it actually occurs a little later, when the universe has cooled off a little more. This is because there are so many more photons than electrons (about a billion times as many), so that a small fraction of atypically energetic photons is enough to keep hydrogen ionized. In the tail of the Planck distribution of the photons, there are about $e^{-B/\mathcal{T}}$ photons with energy larger than B when the temperature is $\mathcal{T}$. Hence, for a binding energy $B \sim 13.6$ eV, there are enough energetic photons as long as the temperature is above $\mathcal{T} \approx -B/(\ln 10^{-9}) \approx 3\,000$ K. This corresponds to a redshift of $z_* \approx 1\,090$.

At that time, electrons and protons combine into neutral hydrogen atoms. Photons do not interact much with hydrogen atoms, because as a bound system hydrogen is neutral. But this means that, at this particular moment, photons can travel unhindered over large distances for the first time and the universe becomes transparent! The newly freed photons pervade the universe, flying in all directions – this is the cosmic background radiation. Since the photons, electrons, and protons were in thermal equilibrium up to this time, the newly released radiation obeys a blackbody distribution with temperature $\mathcal{T} \approx 3\,000$ K.

As the universe expands, this radiation cools, but the blackbody spectrum is preserved. In equilibrium, the energy density of the radiation is given by

$$\rho = 2 \int_0^\infty \frac{\mathrm{d}^3 p}{(2\pi)^3} f_{\mathrm{B-E}} E, \tag{1.64}$$

where p denotes the momentum of a photon, $f_{\mathrm{B-E}} = 1/(e^{E/\mathcal{T}} - 1)$ denotes the Bose–Einstein distribution appropriate for bosons (it gives the average occupation number at energy E), E is the energy of a photon and is simply $E = p$, and the overall factor of 2 arises because the photon has two helicities. This gives

$$\rho = 2 \int_0^\infty \frac{\mathrm{d}^3 p}{(2\pi)^3} \frac{p}{e^{p/\mathcal{T}} - 1} = \frac{\pi^2}{15} \mathcal{T}^4. \qquad (1.65)$$

But radiation scales as $\rho \propto a^{-4}$, and hence we find once more that the temperature of the cosmic microwave background (and of all matter types that it is in equilibrium with during the early universe) scales as

$$\mathcal{T}(t) \propto \frac{1}{a(t)}. \qquad (1.66)$$

Inside the integral above, the momentum p and the temperature $\mathcal{T}$ scale in such a way with the expansion of the universe that the functional form remains preserved, in particular $p/\mathcal{T}$ remains constant. Thus, the spectrum remains that of a blackbody, with a temperature decreasing inversely to the expansion of the universe.

When released, the cosmic background radiation was at a temperature of about $3\,000\,\mathrm{K}$ (cf. the Sun's surface temperature of $6\,000\,\mathrm{K}$). By today, the wavelength of the emitted background radiation is bigger by a factor $1 + z_* \approx 1090$ and its temperature has correspondingly shrunk to $2.73\,\mathrm{K}$. When emitted, the typical wavelength was on the order of a micrometer. Now it is on the order of a millimeter, and thus the radiation is in the microwave frequency region. This is the origin of the name *cosmic microwave background* (CMB). Using the relations above, we can also infer the number density of photons, which amounts to

$$n_\gamma = 2 \int_0^\infty \frac{\mathrm{d}^3 p}{(2\pi)^3} f_{\mathrm{B-E}} = \frac{2\zeta(3)}{\pi^2} \mathcal{T}^3 \approx 411\,\mathrm{cm}^{-3}, \qquad (1.67)$$

where the numerical evaluation corresponds to today's value. This can be contrasted with the number density of matter particles, which, as we calculated in (1.39), is a billion times lower. Thus one may think of the universe as being mostly filled with CMB radiation on average.

The cosmic background radiation indeed exists and, along with big bang nucleosynthesis, it constitutes the best evidence that we have for the hot big bang model. However, just as for nucleosynthesis, it is important to realize that it is not evidence for the big bang itself. Rather, it provides clear evidence that the universe was once very dense and more than a thousand times hotter than presently.

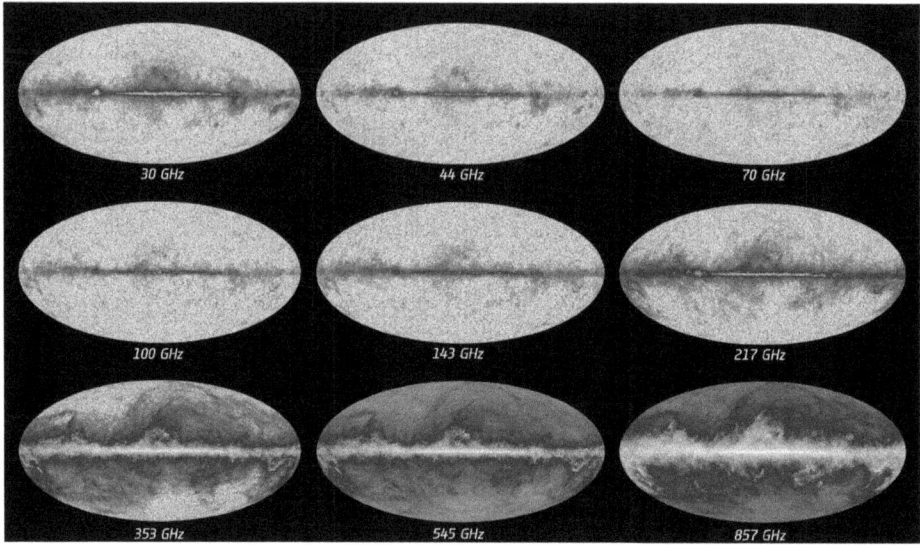

Figure 1.4 All-sky maps of the cosmic microwave background radiation, as observed by the Planck satellite in different frequency bands. The Milky Way galaxy is seen as the large horizontal structure in each picture. Foregrounds such as the Milky Way must be subtracted in order to be left purely with the primordial signals. Figure reproduced with permission from Planck Collaboration; Aghanim et al. (2020a). *Credit: ESA/Planck Collaboration.*

The CMB can be measured via radio telescopes, with the most precise measurements having been carried out by specifically dedicated satellite missions, such as the Wilkinson Microwave Anisotropy Probe (WMAP) and Planck; see Fig. 1.4. As the figure shows, foregrounds must be carefully removed before the primordial data can be analyzed. Once that is done, one finds that with high precision the temperature is the same in all directions in the sky. Thus the CMB provides the best evidence that we have for the isotropy of the universe, though only in a special frame.

The microwave background radiation itself defines a preferred frame in the universe, which can be considered to be the (comoving) "rest frame." To someone traveling at a considerable fraction of the speed of light with respect to that frame, the universe would not look isotropic at all. And in fact measurements of the CMB show a dipole with a maximal intensity of $\Delta T \approx 3.3\,\text{mK}$, which can be interpreted as indicating that we are moving with respect to the rest frame of the universe with a speed of about $370\,\text{km}\,\text{s}^{-1}$, roughly in the direction of the constellation Leo (this dipole can be thought of as a Doppler effect due to our motion relative to the background radiation). The motion of the Local Group of galaxies is even faster (about $620\,\mathrm{km\,s^{-1}}$), because of the motion of the solar system within the Milky Way and its motion, in turn, within the Local Group. When the dipole is subtracted from the data, the universe appears isotropic to one part in 10^4 in all directions. Beyond the dipole, there is a wealth of information in higher multipoles, offering us the best clues we have as to what may have preceded the hot big bang phase. This will be discussed in detail in Chapter 3.

1.3 The big bang and assorted puzzles

The hot big bang model we have just described is in very precise agreement with cosmological observations. Amongst others, it explains the observed redshifts of galaxies, the abundances of primordial elements, as well as the existence of the CMB, and it provides a consistent picture of the ages and development of astrophysical objects.

The big bang model also resolves one apparent paradox about our universe that was known for quite some time, namely the observation that the sky is dark at night (this is sometimes called *Olbers's paradox*). If stars were visible out to infinite distances, then the whole sky ought to be filled with starlight, and even when the Sun is not visible the sky should be bright. However, the big bang effectively puts a limit to how far we can see, and thus also puts a limit on the number of stars that may fill up the night sky. The big bang has left the sky filled with the cosmic background radiation discussed in Section 1.2.1, but, as we saw, it is made up of microwave radiation invisible to the naked eye. Thus, the sky indeed appears dark at night.

Nevertheless, some of the features of and assumptions inherent in the big bang model lead to the conclusion that it cannot possibly be the final word on the origin of the universe. We will go over the main open issues now.

1.3.1 Singularity puzzle

The most obvious shortcoming of the big bang model is the big bang itself. When the scale factor goes to zero, curvature invariants (whose numerical value is independent of the choice of coordinates) like the Ricci scalar (1.31), or the Riemann tensor squared, generally become infinite. For example, we have

$$R^\mu{}_{\nu\rho\sigma}R_\mu{}^{\nu\rho\sigma} = 12\frac{\ddot{a}^2}{a^2} + 12\frac{\dot{a}^4}{a^4} + 24K\frac{\dot{a}^2}{a^4} + 12\frac{K^2}{a^4} + \text{anisotropy terms}\,, \quad (1.68)$$

where we explicitly wrote out the terms involving solely the scale factor a. Clearly, during a radiation-dominated phase with $a(t) \propto a^{1/2}$, or for other power-law solutions, this quantity blows up as $a \to 0$. But this is true far more generally: The Penrose–Hawking singularity theorems (which would take us too far off course to review in detail) demonstrate that anisotropies cannot compensate for the blowup of the volume terms, but rather reinforce the singularity; the theorems imply that a curvature singularity occurs under very generic circumstances.

A singularity means trouble. It implies for instance that the matter side of the Einstein equations also blows up or becomes otherwise pathological. Infinite quantities are neither measurable nor, consequently, physical. The model simply breaks down at the big bang. In this vein, it also seems premature to assume, as has often been done in the past, that the big bang marked the origin of space and time. A more sensible conclusion is to say that, in order to explain the origin of the universe, we need something that goes beyond the classical big bang model. There is a widespread hope that quantum gravity will be able to resolve this singularity and replace it with something well defined. In later chapters, we will discuss possibilities along these lines.

It is not only the big bang itself that presents a puzzle, but there are several dynamical features of the hot big bang model that require additional explanation and justification. The two most important ones are the flatness and horizon puzzles, which we will turn to next.

1.3.2 Flatness puzzle

If we consider the Friedmann equation again, for instance in the form of Eq. (1.45), we can see that the term representing the homogeneous curvature of the universe contributes a fractional amount $|k|/(aH)^2$ to the total energy content of the universe. The dependence on the scale factor and Hubble rate implies that, as the universe expands during the radiation and matter-dominated epochs, this fractional contribution becomes larger and larger, in proportion to t and $t^{2/3}$, respectively. From Eq. (1.50) we know that present-day observations imply an upper bound

$$\frac{|K|}{(aH)^2}\Big|_{t_0} \lesssim 10^{-3}. \tag{1.69}$$

But this means then that at early times, the average curvature of the universe must have been extremely tiny.

More quantitatively, at radiation–matter equality, we must have had

$$\frac{|K|}{(aH)^2}\Big|_{t_{\text{eq}}} \lesssim 10^{-3}\frac{(aH)_0^2}{(aH)_{\text{eq}}^2} = 10^{-6}, \tag{1.70}$$

while during the radiation phase one obtains

$$\frac{|K|}{(aH)^2} \lesssim 10^{-6} \frac{a^2}{a_{\rm eq}^2} = 10^{-6} \frac{t}{t_{\rm eq}} \qquad (t < t_{\rm eq}). \tag{1.71}$$

It seems reasonable to extrapolate back at least to energies of 1 TeV, since these are the energies that can be probed by current particle physics experiments and are well understood. Then one has $|K|/(aH)^2\,|_{t_{\rm TeV}} \lesssim 10^{-31}$. This is an incredibly small quantity. How could the spatial part of the universe have been so precisely flat at very early times?

1.3.3 Horizon puzzle

Let us start by drawing a (highly simplified) conformal map of the universe. This is useful not only to appreciate the observational situation we find ourselves in, but also to understand causal structures in the universe. Recall that one may write the FLRW line element, now specializing to $K=0$ as suggested by observations, in conformal time τ as

$$\mathrm{d}s^2 = a^2(\tau)[-\mathrm{d}\tau^2 + \mathrm{d}r^2 + r^2(\mathrm{d}\theta^2 + \sin^2\theta\,\mathrm{d}\phi^2)]. \tag{1.72}$$

We can gain knowledge about causal relationships in spacetime by following the paths of lights rays (null geodesics). We can simplify the analysis by orienting the coordinates such that the light rays propagate at constant θ, ϕ values. Then the equation for null geodesics becomes remarkably simple,

$$\mathrm{d}s^2 = 0 \quad \to \quad \mathrm{d}\tau = \pm\mathrm{d}r, \tag{1.73}$$

where the sign depends on whether the light ray is outgoing or incoming. Thus if we observe an object at comoving distance r, the light was emitted at conformal time $\tau(t) = \tau(t_0) - r$.

If we now specialize to null geodesics originating at the big bang, we can evaluate the conformal time using its definition in terms of physical time t and using the Friedmann equation (1.45),

$$\tau(t) = \int_0^t \frac{\mathrm{d}t}{a}$$
$$= \int_0^a \frac{\mathrm{d}\ln(a)}{aH} \tag{1.74}$$
$$= \frac{1}{H_0} \int_0^a \frac{\mathrm{d}a}{\sqrt{\Omega_r + \Omega_m a + \Omega_\Lambda a^4}}. \tag{1.75}$$

If we evaluate the integral up to the present time, $a = a_0 = 1$, we find $\eta_{\rm BB} = 3.196/H_0$. This implies a current distance $a_0 r = r = \eta_{\rm BB}$ of about

46.2 billion light-years to the furthest region from which a particle could have reached us by today. In other words, this is the current particle horizon, see Fig. 1.5 for an illustration.

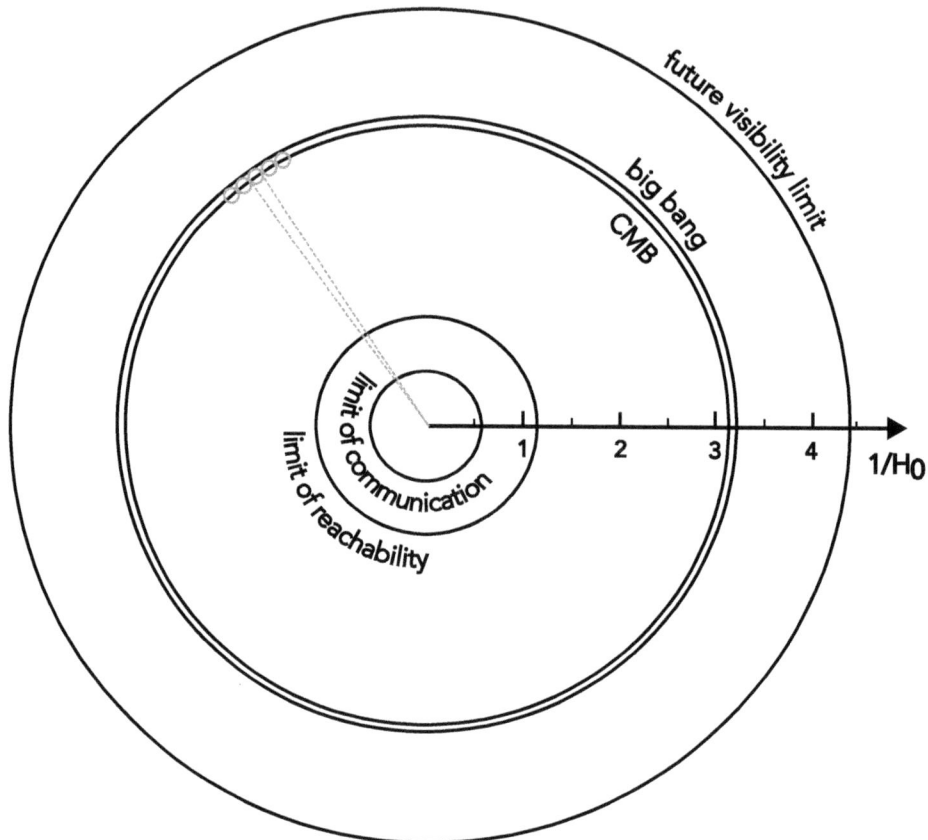

Figure 1.5 A conformal map of the observable universe. We are located in the center of the picture. Distances are given in units of the current Hubble radius $1/H_0 \approx 14.5$ billion light-years. The map shows the location of events in comoving coordinates. We can currently see out to the CMB, at comoving distance $3.13/H_0$ or redshift $z = 1089$. The small circles indicate the size of spheres of causality at last scattering. The future visibility limit (at $4.34/H_0$) is the furthest we will be able to see (in the infinite future) if the dark energy remains constant at the present value. The limit of reachability is the furthest we can send a message to, while the limit of communication is the current maximal distance to which we can send a message and hope to receive a response in the future – it is astonishingly close, at comoving distance $0.57/H_0$ or redshift $z = 0.69$.

The CMB was emitted at scale factor $a_{\text{CMB}} = 1/1090$ and consequently $r_{\text{CMB}} = 3.132/H_0$ or about 45.3 billion light-years. This is the current distance to the regions from which the CMB (the part that we see today) was emitted. When comparing with the age of the universe, one should note that this is larger than 13.8 billion light-years due to the expansion of the universe. At the time that the CMB was emitted, this region was smaller by a factor 1 090, i.e., it was a sphere with only a 41.5 million light-year radius.

The comoving distance between the big bang and the CMB was thus $\tau_{BB} - \tau_{CMB} = 0.063/H_0$. This is the distance that light rays could cover by the time of last scattering. In other words, this is the maximal region that could have been in causal contact up to that time. Corresponding regions are drawn as small gray circles in Fig. 1.5. On the current sky, they subtend an angle of $(\tau_{BB} - \tau_{CMB})/\tau_{CMB}$ radians or about 1.15 degrees. This observation should give one pause: We observe the same temperature to very good accuracy in all directions in the sky, yet the radiation is coming from regions that could not have been in causal contact with each other since the big bang. How then can they all be describing the same blackbody spectrum at the exact same temperature?

This surprising feature is known as the *horizon puzzle*. Note that it gets worse when we extrapolate further back from the CMB. If we assume that the universe was at thermal equilibrium at earlier moments then, as should be clear by inspecting Fig. 1.5, the corresponding regions of causality were even smaller, and the puzzle sharper. One may trace the origin of the problem back to the behavior of the comoving horizon $1/(aH)$, which can be seen to determine the conformal time τ via Eq. (1.74). Just as for the flatness puzzle discussed above, the problem is that during the radiation-dominated phase (and also during matter domination) the comoving horizon $1/(aH)$ steadily grows. In this way more and more regions, which up to that moment had not been in causal contact, are suddenly able to communicate. The puzzle is that, when this happens, everyone seems to know already exactly what the others have been saying all along.

Before moving on to further puzzles, let us use the opportunity to point out a few additional features of interest on the conformal map in Fig. 1.5. For instance, we can let the scale factor range from zero at the big bang all the way to infinity in the infinite future. This still results in a finite integral (1.75), yielding $\tau_\infty = 4.34/H_0$. This is the largest distance that a particle can travel from the big bang onward. Thus it is the future visibility limit, beyond which we will never be able to observe anything, if the dark energy remains constant at the present value. In the figure, this is the circle to which the big bang circle will recede in the infinite future.

If we evaluate the integral (1.75) from today ($a = 1$) up to the infinite future ($a = \infty$), we obtain a comoving distance of $1.14/H_0$. This is the furthest we can send a signal to, and it corresponds to the limit of reachability shown in the figure. If we would also like to receive a response, then that comoving distance gets halved to $0.57/H_0$, corresponding to galaxies that currently reside at a redshift of only $z = 0.69$. Anything further away than this redshift is beyond our limit of communication. This perhaps surprisingly small region constitutes the part of the universe that we can hope one day to interact with.

1.3.4 Brief list of additional puzzles

There are further aspects of the hot big bang model that seem contrived or, better said, that will require justification. We will simply list a few of these open issues.

- The CMB is isotropic to a high degree, but there are temperature fluctuations of order $\Delta\mathcal{T}/\mathcal{T} \sim \mathcal{O}(10^{-4})$, which are in fact responsible for all the gravitationally bound and collapsed structures we see in the universe today. How did these primordial temperature perturbations arise, and what determined their properties? This is the *anisotropy puzzle*.
- At high energies the gauge couplings of the standard model of particle physics approach each other (and when supersymmetry is included they appear to evolve to a common value at energies of about 10^{16} GeV). This suggests *grand unification*, namely that at high energies a larger symmetry group united the $SU(3)$, $SU(2)$, and $U(1)$ symmetries of the standard model. But if this is indeed the case, then topological defects would have formed when the symmetry was broken as the universe cooled. None such topological defects, which would manifest themselves as monopoles or cosmic strings, for example, have been observed. Why not? This the *topological defects puzzle*.
- There must have existed a slight excess of matter over antimatter in the early universe. This is inferred from the fact that our observable universe appears to be made purely out of matter, and not antimatter (otherwise we would observe the light emitted when matter and antimatter annihilate each other). To prevent a total annihilation of matter and antimatter in the early universe, there must have been a slight overabundance of matter. Various theoretical models exist that can potentially explain this so-called "baryon asymmetry." However, it remains too early to know which (if any) occurred in our universe. This is the *antimatter puzzle*.
- The second law of thermodynamics says that the entropy always grows. But then this means that in the early universe, the entropy must have

been comparatively low. In other words, the early universe must have been in a special, unlikely state. This is the *entropy puzzle*. But then what determined this state? It seems difficult for a classical theory to achieve this; rather, we may hope that a proper quantum treatment of the early universe may result in certain states being strongly favored over others.

- The hot big bang model assumes that already very early on, space and time can be treated purely classically. However, we tend to believe that nature's most fundamental laws are quantum laws. So how is it that space and time came to behave so classically? In the late universe, decoherence can be invoked to explain why macroscopic objects, and also space and time, appear classical to us. This is because the interactions of large numbers of elementary particles lead to a strong suppression of quantum interference effects. But if one thinks about the possible beginning stages of the universe, then it is not clear that decoherence can be applied, as decoherence is a process that occurs over time, and time itself may not have been classical at the creation of the universe, given that the very production of spacetime and of matter would have had to be a quantum process. This is the *classicality puzzle*.

We should point out that none of these puzzles are strict inconsistencies of the hot big bang model. Rather, they can all be seen as specifying special initial conditions for the universe to start out from. It would certainly be much more satisfying if one could explain these features by physical arguments, perhaps either being due to necessity or being caused by suitable dynamical processes occurring prior to the hot big bang. We will start exploring the most prominent suggestions for such "pre-histories" in the next chapter.

In closing, let us mention one final oddity: In the hot big bang model, dark energy plays essentially no role (and the history of the universe up to now would have been quite similar without dark energy), yet it has come to dominate the present-day energy budget of the universe. Why then is dark energy there? One feels that there must be a deeper reason.

Exercises

1.1 Estimate the density inside our Galaxy. At what redshift did the universe have the equivalent average density? For consistency of the big bang model, check that this was before the first galaxies formed.

1.2 Roughly how far away into deep space do you have to go before the expansion of space becomes significant? *Hint:* It may be useful to consider the Schwarzschild–de Sitter solution.

1.3 Calculate the current distance to galaxies at redshifts $z_g = 0.1, 1, 10$ and the cosmic time at which we are seeing them.

1.4 Give an order-of-magnitude estimate as to how many galaxies recede away from us each year such that they can no longer be communicated with.

1.5 Assume that the observable universe is the full extent of the universe, and assume radiation domination throughout its history. Then compare the age and size of the universe at temperature 1 TeV and at 1 Planck time.

2
Blowing up the universe: inflation

The puzzles of the hot big bang model strongly suggest that it does not provide us with a complete description of the early universe. Rather, the fact that the initial conditions of the hot big bang phase must have been so very special lets one hope that they might have arisen dynamically, i.e., that they are a more or less automatic outcome of a prior phase of evolution, rather than something we must put in by hand. In this chapter we will see two such proposals for prior phases of evolution: the most widely accepted one, inflation, and an alternative proposal, ekpyrosis. The reason for including an alternative is twofold – on the one hand to exemplify how completely different dynamical theories may nevertheless result in roughly the same outcome, and on the other hand to obtain a better appreciation of the strengths and weaknesses of these proposals.

2.1 Inflation and scalar field dynamics
2.1.1 Basic idea

In the previous chapter we saw that, extrapolating the hot big bang model back to early times, the spatial geometry of the universe must have been unnaturally flat, while also consisting of a myriad of causally disconnected regions that miraculously all behave in pretty much exactly the same way. Both of these oddities may be seen as being caused by the fact that the comoving Hubble horizon $1/(aH)$ grows during the radiation- and matter-dominated phases – see again Eq. (1.74) and the ensuing discussion. But then there is a fairly obvious solution to these conundrums: If there was a prior phase of evolution during which the comoving Hubble radius shrank significantly, then it might induce the required conditions.

If we assume that the universe is expanding, i.e., $H > 0$, then the condition for the comoving Hubble radius to shrink is

$$\frac{d}{dt}\left(\frac{1}{aH}\right) < 0 \quad \to \quad \frac{d}{dt}\left(\frac{1}{\dot{a}}\right) < 0 \quad \to \quad \ddot{a} > 0. \tag{2.1}$$

Thus we see that if the universe is expanding in an accelerated fashion we might have a chance of solving the flatness and horizon problems. This is precisely the idea of inflation:

Inflation is a conjectured period of accelerated expansion preceding the hot big bang evolution.

If we recall the acceleration equation (1.37), $\frac{\ddot{a}}{a} = -\frac{1}{6}(\rho + 3p)$, then we see that accelerated expansion can be achieved if we have a sufficiently *negative* pressure,

$$\ddot{a} > 0 \quad \leftrightarrow \quad p < -\frac{1}{3}\rho \quad \text{(inflation)}. \tag{2.2}$$

The strong negative pressure counteracts the energy density and effectively causes gravity to become repulsive.

The main property of inflation is that it consists of a period of accelerated expansion, typically at an extremely high rate, so that any region of the universe undergoing inflation is blown up to a huge space in a very short span of time. This blowup causes the spatial geometry to become highly smooth and flat, so that a region that is not exceptionally flat initially becomes so after a burst of inflation – see Fig. 2.1 for an illustration. During this time the comoving Hubble radius shrinks, so that one causally connected region quickly subdivides into many regions with roughly the same physical conditions. In this heuristic way we can see how the flatness and horizon puzzles may be resolved.

How much accelerated expansion do we need? To calculate this, we need a more explicit expression for the evolution of the scale factor. Recall from (1.35) that for a constant equation of state w the energy density can be solved for analytically, yielding $\rho \propto a^{-3(1+w)}$. The Friedmann equation (1.36) then implies (momentarily neglecting the K/a^2 term)

$$\dot{a}^2 \propto a^{-1-3w}, \tag{2.3}$$

$$a \propto t^{\frac{2}{3(1+w)}} \quad \text{assuming } w \neq -1, \tag{2.4}$$

$$\frac{1}{aH} \propto t^{\frac{1+3w}{3(1+w)}} \propto a^{(1+3w)/2}. \tag{2.5}$$

Thus, in an expanding universe, we see again that the comoving Hubble radius $1/(aH)$ shrinks if $w < -1/3$. This also justifies neglecting the K/a^2 term in the Friedmann equation above, since that term becomes less and

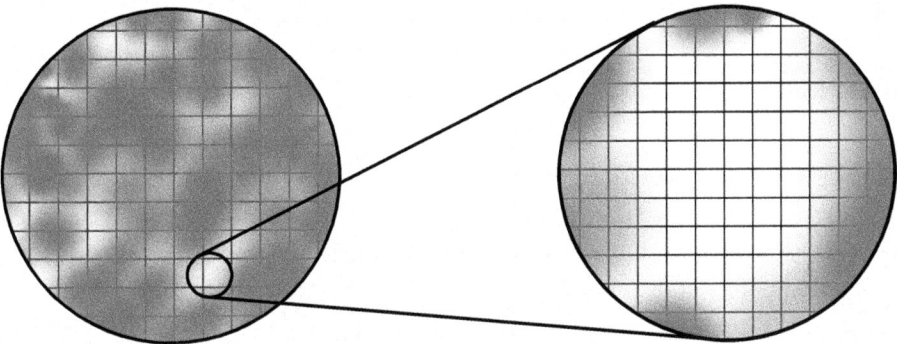

Figure 2.1 A cartoon of inflation. Even if the universe is very inhomogeneous at the beginning of inflation (left), a small region can get blown up to a very large region consisting of many, now causally disconnected, regions with similar properties (right). The grid is meant to indicate the size of a causally connected region (i.e., a Hubble radius).

less important relative to the others. We can now go back to the criterion for resolving the flatness and horizon puzzles. We should point out that the criterion depends rather sensitively on further assumptions. A popular way to estimate it is to assume that the radiation phase reached the grand unified scale, with temperatures $\mathcal{T} \sim 10^{16}$ GeV, and that the curvature term $K/(aH)^2$ in the Friedmann Eq. (1.45) was of $\mathcal{O}(1)$ at the beginning of inflation. In that case, using Eqs. (1.71) and (1.55), we require

$$\frac{(aH)_{\text{beg}}}{(aH)_{\text{end}}} < 10^{-22} \rightarrow \frac{a_{\text{end}}}{a_{\text{beg}}} > 10^{-22\frac{2}{(1+3w)}}, \qquad (2.6)$$

where beg and end refer to the beginning and end of the inflationary phase, respectively. For $w \approx -1$ this leads to the requirement

$$\frac{a_{\text{end}}}{a_{\text{beg}}} > 10^{22} \approx e^{50}. \qquad (2.7)$$

Thus it is often said that one requires at least 50 "e-folds" of inflation in order to explain the flatness and horizon puzzles dynamically. We will discuss this reasoning in more detail in Section 3.4.

Above we saw that gravity allows for accelerated expansion if the matter content effectively has an equation of state $w < -1/3$. However, not all negative values of the pressure are physically acceptable. It is widely thought that a physical theory is only stable when it satisfies the *null energy condition*,[1]

---

[1] The condition is that the null energy condition (NEC) should be satisfied on average. Quantum fluctuations can locally violate the NEC.

which for perfect fluids translates into the requirement that

$$\rho + p \geq 0 \quad \leftrightarrow \quad w \geq -1. \tag{2.8}$$

Intuitively, this makes sense: The energy density scales as $\rho \propto a^{-3(1+w)}$, and thus, for $w < -1$ we would obtain a regime in which the energy density ρ grew during the expansion of the universe. This would evidently lead to an unstable, explosive behavior. One may thus understand why large-scale violations of the null energy condition are deemed unphysical; incidentally, they also go hand in hand with violations of the second law of thermodynamics, and in quantum theories are related to the presence of ghosts.

The limiting case of $w = -1$ is however of special interest. In that case $p = -\rho$ and we have $\rho \equiv \Lambda =$ constant. This is known as a *cosmological constant*, and constitutes the simplest form of vacuum energy. The solutions to the Friedmann equation are particularly simple in that case. When the spatial sections are flat ($K = 0$), the Friedmann equation, $3H^2 = \rho = \Lambda$, immediately implies that the Hubble rate $H = \sqrt{\Lambda/3}$ is constant and that consequently, the universe expands exponentially, $a = \bar{a}\, e^{Ht}$, where $\bar{a}$ represents a reference length scale which the Friedmann equation does not fix and which can be chosen for convenience. The corresponding metric is known as the *de Sitter* metric in the flat slicing. In full it may be written as

$$ds^2 = -dt^2 + e^{2Ht} \left[dr^2 + r^2(d\theta^2 + \sin^2\theta\, d\phi^2) \right] \quad \text{with } H = \sqrt{\frac{\Lambda}{3}}$$
(dS – flat), (2.9)

where we set $\bar{a} = 1$. We may equally well solve the Friedmann equation (1.36) when the spatial sections are closed, $K = +1$, which results in the metric

$$ds^2 = -d\bar{t}^2 + \frac{3}{\Lambda}\cosh^2\left(\sqrt{\frac{\Lambda}{3}}\bar{t}\right) \left[d\chi^2 + \sin^2\chi(d\theta^2 + \sin^2\theta\, d\phi^2) \right]$$
(dS – closed). (2.10)

This is known as the de Sitter spacetime in the closed slicing, with the spatial sections being 3-spheres. Unlike the flat case, the scale of the spheres is determined by the value of the vacuum energy. Also, the Hubble rate $H = \sqrt{\frac{\Lambda}{3}}\tanh\left(\sqrt{\frac{\Lambda}{3}}\bar{t}\right)$ is not constant, though it asymptotically approaches the value $\sqrt{\Lambda/3}$.

The de Sitter metrics both describe distances on a manifold with constant curvature $R = 4\Lambda$. Since the curvature is the same everywhere, the manifold is maximally symmetric, which explains why it can be sliced with lower-dimensional symmetric spaces. For instance, one can also consider a version where the spatial sections are hyperboloids, and all these metrics are related by coordinate transformations (whose explicit expressions we will not need here). Perhaps the easiest way to see this is by considering de Sitter spacetime via its embedding in 5 dimensions. In that construction, de Sitter spacetime is expressed as the hyperboloid,

$$-z^2 + x_1^2 + x_2^2 + x_3^2 + x_4^2 = \frac{3}{\Lambda}, \tag{2.11}$$

embedded in 5-dimensional Minkowski spacetime. By direct analogy with the discussion in Section 1.1.2, it is then clear that de Sitter spacetime contains the maximally symmetric 3-dimensional subspaces discussed in connection with Robertson–Walker metrics.

However, not all slicings cover the entire manifold, see Fig. 2.2. Note that the metric with flat slicing (2.9) describes a purely expanding spacetime, and it covers only half of the maximally extended de Sitter spacetime. Meanwhile, the closed slicing (2.10) covers the entire manifold and describes a spacetime consisting of spatial 3-spheres that start out infinitely large and shrink to a minimal radius ($a(t=0) = \sqrt{3/\Lambda}$) before expanding to infinity again. When one adds perturbations, it turns out that the contracting part of the spacetime is unstable, while the expanding part (which is the part considered in inflationary models) is stable. At large times, the two metrics (2.9) and (2.10) approach each other more and more.

Before continuing, let us add an observation that will prove useful later on, when we will discuss quantum gravitational properties of de Sitter spacetime. This observation concerns the Euclidean version of de Sitter spacetime, where the time direction gets analytically continued into a spatial direction, $\bar{t} = -i(\sigma - \sqrt{\frac{3}{\Lambda}}\frac{\pi}{2})$. The de Sitter metric (2.10) then transforms into

$$ds^2 = +d\sigma^2 + \frac{3}{\Lambda} \sin^2\left(\sqrt{\frac{\Lambda}{3}}\sigma\right) [d\chi^2 + \sin^2\chi(d\theta^2 + \sin^2\theta\, d\phi^2)]$$

(dS – Euclidean), (2.12)

which can be recognized as the metric of a 4-sphere. This makes sense, as the 4-sphere is the maximally symmetric Riemannian manifold with constant positive curvature. Vice versa, one may also consider de Sitter spacetime as the pseudo-Riemannian version of a sphere. Note that the Euclidean time σ has become an angular coordinate after this analytic continuation.

Let us now go back to de Sitter spacetime in the flat slicing to illustrate

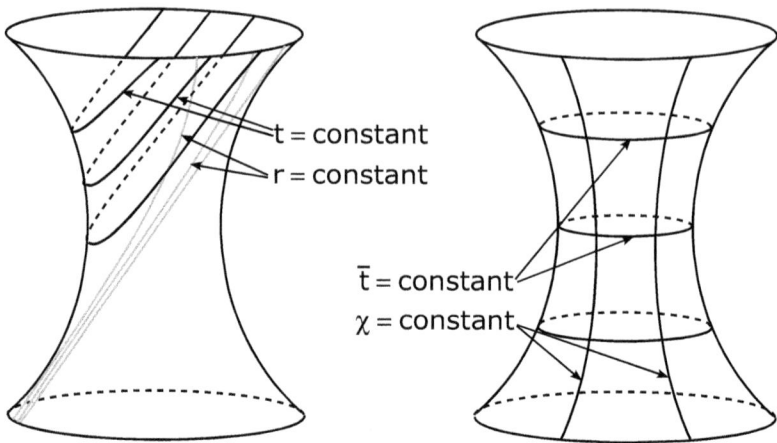

Figure 2.2 De Sitter spacetime may be visualized as a 4-dimensional hyperboloid embedded in 5-dimensional Minkowski spacetime. Two angular coordinates are suppressed here. The flat (left graph) and closed (right graph) slicings cover different parts of the manifold.

an important property, namely that it has a horizon. We can use the formula (1.74) to calculate the comoving distance that a light ray can travel from a reference scale factor a_r up to the infinite future,

$$\Delta\tau = \int_{a_r}^{\infty} \frac{\mathrm{d}a}{a^2 H} = \frac{1}{a_r H}. \qquad (2.13)$$

Thus the physical particle horizon $a_r \Delta\tau = 1/H = \sqrt{3/\Lambda}$ is finite, constant, and determined solely by the vacuum energy.[2] In this way we learn that the volume of a causally connected region in de Sitter spacetime is given by $1/H^3$, and it remains constant even though space is expanding. Incidentally, since $a \propto e^{Ht}$, the duration $1/H$ is also the time scale over which a given spatial section expands by 1 e-fold. Finally, let us highlight a slightly subtle point: Since de Sitter spacetime is maximally symmetric, each point is equivalent to any other point. Nevertheless, each point is surrounded by a horizon, whose location depends on that point. Said differently, each observer in a de Sitter universe is surrounded by their personal horizon of size $1/H$. This should be contrasted with the horizon of a black hole, whose location does not depend on the location of the observer.

The de Sitter spacetime offers a useful first approximation to inflation,

---

[2] It is sometimes said that de Sitter space expands "superluminally." This is an unfortunate choice of words, as nothing travels faster than the speed of light. It is simply the case that for objects separated by a distance larger than the horizon, there is sufficient expanding space in between so that light emitted by one object will not be able to reach the other object.

and is often used as such. However, it is clear that it can only be a first approximation, as inflation must come to an end eventually, while the de Sitter metric expands in an accelerated fashion into the infinite future. We will now see how the addition of a scalar field considerably enriches the possible dynamics.

2.1.2 Gravitational dynamics of a scalar field

So far the only scalar (spin-0) field that has been detected experimentally is the Higgs. However, extensions of general relativity, such as supergravity and string theory, typically contain many scalar fields. Such scalars could have played (and may still play) an important role in cosmology. This is because, unlike fermionic fields, they can affect spacetime everywhere. We will now present the associated formalism, and see that scalar fields are excellent candidates for driving a phase of inflation.

To this end, let us consider a scalar field $\phi(x^\mu)$, where the term "scalar" simply means that ϕ takes a numerical value at every point in spacetime. We also consider ϕ to have potential energy $V(\phi)$. With the simplest ("minimal") coupling to gravity, the action for this setup is given by

$$S = \int d^4x \sqrt{-g} \left[\frac{R}{2} - \frac{1}{2} g^{\mu\nu} \partial_\mu \phi \partial_\nu \phi - V(\phi) \right], \tag{2.14}$$

where we are still working in natural units with $8\pi G = 1$. The equations of motion are obtained by varying the action with respect to the metric $g^{\mu\nu}$ and the scalar field ϕ, yielding, respectively,

$$R_{\mu\nu} - \frac{1}{2} g_{\mu\nu} R = \phi_{,\mu} \phi_{,\nu} - \frac{1}{2} g_{\mu\nu} (\partial \phi)^2 - g_{\mu\nu} V, \tag{2.15}$$

$$\Box \phi = V_{,\phi}. \tag{2.16}$$

Here we used the customary abbreviations $(\partial \phi)^2 \equiv g^{\mu\nu} \partial_\mu \phi \partial_\nu \phi$ and $\Box \equiv \nabla^\mu \nabla_\mu$. In a flat ($K = 0$) FLRW background spacetime, these equations reduce to

$$3H^2 = \frac{1}{2} \dot{\phi}^2 + V, \tag{2.17}$$

$$\dot{H} = -\frac{1}{2} \dot{\phi}^2, \tag{2.18}$$

$$\ddot{\phi} + 3H\dot{\phi} + V_{,\phi} = 0. \tag{2.19}$$

Even these three equations are not independent of each other: One can obtain Eq. (2.18) by taking a time derivative of (2.17) and using (2.19).

The right-hand side of Eq. (2.15) is the stress–energy tensor $T_{\mu\nu}$. In the flat FLRW background it takes the form of a perfect fluid, as it must, with energy density and pressure given by

$$\rho = \frac{1}{2}\dot{\phi}^2 + V, \qquad (2.20)$$

$$p = \frac{1}{2}\dot{\phi}^2 - V. \qquad (2.21)$$

The energy density is simply the sum of the kinetic and potential energies of the scalar field, while the pressure is their difference. One may verify that these relations follow the general pattern discussed in and below Eq. (1.32). It follows that the equation of state is given by

$$w = \frac{p}{\rho} = \frac{\frac{1}{2}\dot{\phi}^2 - V}{\frac{1}{2}\dot{\phi}^2 + V}. \qquad (2.22)$$

From this definition we can see that $w \geq -1$ (with $w = -1$ arising when the kinetic energy vanishes), so that the null energy condition will automatically be satisfied. We can also define the often-used parameter ϵ via

$$\epsilon \equiv \frac{3}{2}(1+w) = \frac{1}{2}\frac{\dot{\phi}^2}{H^2}, \qquad (2.23)$$

where we made use of the Friedmann equation (2.17) to obtain the last relation. By definition $\epsilon \geq 0$, and, as the last relation shows, it tells us how much kinetic energy there is in the scalar field relative to the expansion/contraction rate of the universe.

The acceleration equation (1.37) can now be rewritten as $\ddot{a}/a = -(\rho + 3p)/6 = -(\dot{\phi}^2 - V)/3$, so in order to obtain inflation we see that we need the potential energy to dominate over (twice) the kinetic energy, $\dot{\phi}^2 < V$. An equivalent condition is that

$$\epsilon < 1, \quad \text{(condition for inflation)}. \qquad (2.24)$$

2.1.3 Constant equation of state

Let us present an example of inflationary dynamics that is useful as it allows for an exact solution with constant equation of state ϵ. It arises from considering a potential of exponential form

$$V = V_0 e^{-c\phi}, \qquad (2.25)$$

where V_0 and c are positive constants. The solution to the equations of motion is given by

$$a = a_0 t^{1/\epsilon}, \quad H = \frac{1}{\epsilon t}, \quad \phi = \frac{1}{\sqrt{2\epsilon}} \ln\left(\frac{V_0 \epsilon^2}{3-\epsilon} t^2\right), \quad V = \frac{3-\epsilon}{\epsilon^2 t^2}, \qquad (2.26)$$

where a_0 is a constant, where the origin of time has been chosen to coincide with $a = 0$, and where there are several equivalent expressions for the equation of state

$$\epsilon = \frac{c^2}{2} \quad \left(= -\frac{\dot H}{H^2} = \frac{1}{2}\frac{\dot\phi^2}{H^2} = \frac{1}{2}\frac{V_{,\phi}^2}{V^2} \right). \tag{2.27}$$

We can see that we need $c^2 < 2$ in order to obtain inflation – i.e., inflation occurs as long as the potential is flat enough. This turns out to be a general requirement, as we will see. The model above is often called *power-law inflation*, because in (2.26) the scale factor evolves as a power of the time. Note that the exponent $1/\epsilon$ is larger than 1, as required for accelerated expansion.

The solution above is a scaling solution: In the equations of motion, each term has the same time dependence ($\propto 1/t^2$), so that the relative contributions of the various terms remains unchanged over time. This self-similarity is related to the fact that a shift in ϕ can be absorbed by a redefinition of V_0. For more general potentials, ϵ will not be constant, and the equations of motion typically do not admit simple closed-form solutions. Then, if ϵ is not changing fast, one may use the above scaling solution as a useful first approximation.

The scaling solution presented above is an attractor – i.e., starting from general homogeneous and isotropic initial conditions, or from initial conditions with only small inhomogeneities and anisotropies, all solutions quickly approach the scaling solution. Here is a calculation that shows this for spatially independent initial conditions: We can solve for small perturbations around the background solution by writing

$$a = a_s(t)[1 + \delta a(t)], \qquad \phi = \phi_s(t) + \delta\phi(t), \tag{2.28}$$

where a_s, ϕ_s denote the scaling solution (2.26). Here we have defined δa as the fractional shift in a, while $\delta\phi$ is defined as the total shift in ϕ – this is because the background solution for a is a power law while that for ϕ is logarithmic, and in this manner the two perturbations will come out with comparable magnitudes. Then we plug these definitions into the equations of motion (2.17)–(2.19) and expand them to linear order in perturbations. One finds that there are two possible solutions for the perturbations,

$$\delta a, \delta\phi \propto t^{-1}, t^{1-3/\epsilon}. \tag{2.29}$$

The first solution, proportional to t^{-1}, simply corresponds to a shift in the time coordinate – i.e., it corresponds to putting the big bang or big crunch at a different time coordinate value, and thus does not tell us anything about stability. To see this, compare with the Taylor expansions of the shifted time solutions, for example $\phi(t-t_\star) = \phi(t) + \dot\phi(t)(t-t_\star) + \cdots$ around t, and note that $\dot\phi \propto 1/t$. The second solution, proportional to $t^{1-3/\epsilon}$, is the interesting one: It shows that for inflation, where t is increasing, we want the exponent to be negative, so that this perturbation decays over time. This will be the case if $\epsilon < 3$, which is certainly satisfied as the condition for inflation is $\epsilon < 1$.

These arguments justify, to some extent, the assumption we made earlier of using a flat FLRW metric. The attractor property that we just calculated shows that, even if we start from a configuration that is somewhat different from the inflationary solution, the attractor ensures that this solution is quickly reached. There is one caveat: For this to happen, we are assuming here that inflation is already underway. We will discuss the conditions needed for inflation to start in more detail later on.

In the power-law model, sufficient inflation is obtained if

$$\frac{(aH)_{\rm end}}{(aH)_{\rm beg}} > e^{50} \quad \to \quad \frac{a_{\rm end}}{a_{\rm beg}} > e^{\frac{50}{(1-\epsilon)}} \quad \text{and} \quad \phi_{\rm end} - \phi_{\rm beg} > 50\frac{\sqrt{2\epsilon}}{1-\epsilon}. \quad (2.30)$$

Thus, during inflation the universe must grow by an enormous amount, at the very least by a factor of e^{50}, while the scalar field travels a distance in field space that depends strongly on the value of ϵ, for example by about 7 Planck masses for a value of $\epsilon = 1/100$. It is not currently understood how such models can arise in a more fundamental physical theory such as string theory, i.e., how one can get an effective potential that is so flat over a sufficiently large field range. Here we get our first glimpse of the realization that the assumption of an inflationary phase before the hot big bang evolution is not a small change of the existing model, but a rather drastic extension.

2.1.4 Slow-roll

Power-law inflation is a useful model when the equation of state is constant or nearly constant. However, we know that this assumption cannot always remain true, as the power-law model leads to an eternally inflating universe, while we know that inflation, if it occurred, must have come to an end. It is however unfortunately not possible to solve the equations of motion in complete generality. That said, an important simplification arises when we are in the "slow-roll" regime, where the kinetic energy is vastly subdominant to the potential energy, $\dot\phi^2 \ll V$. Further assuming that $|\ddot\phi| \ll H|\dot\phi|$, so that the scalar field kinetic energy remains small over an extended period, implies that the equations of motion can be approximated by

$$3H^2 \approx V \qquad \text{(slow-roll)}, \qquad (2.31)$$

$$3H\dot\phi \approx -V_{,\phi}. \qquad (2.32)$$

These slow-roll equations are used extensively in inflationary cosmology. The first equation simply says that the expansion rate is determined by the potential energy of the scalar field. And the second equation says that the scalar field rolls down the potential, with the Hubble rate acting as friction slowing down the descent.

Let us be more specific about the conditions under which these slow-roll equations are applicable. In this context, because of the exponential expansion of space, it is sensible to make use of the number of e-folds of expansion $\mathcal{N}$ as a variable. As a differential, it is defined via

$$d\mathcal{N} \equiv d\ln a = H dt \qquad \text{(e-folds)}. \qquad (2.33)$$

This allows us to define the equation of state parameter ϵ, which is also known as the *first slow-roll parameter*, more generally as the rate of change of the (logarithmic) Hubble rate with respect to the number of e-folds,

$$\epsilon \equiv -\frac{d\ln H}{d\mathcal{N}} = -\frac{\dot H}{H^2} \qquad \text{(1. slow-roll parameter)}. \qquad (2.34)$$

The minus sign in the definition may also be thought of as expressing this rate of change calculating back from the end of inflation. Since the Hubble rate necessarily decreases – see Eq. (2.18) – ϵ is positive. One may then define successive slow-roll parameters by iteration; in particular, the *second slow-roll parameter* is given by

$$\eta \equiv -\frac{d\ln \epsilon}{d\mathcal{N}} = -2\frac{\ddot\phi}{H\dot\phi} - 2\epsilon \qquad \text{(2. slow-roll parameter)}. \qquad (2.35)$$

It expresses how quickly the first slow-roll parameter is changing with the expansion. We will not need higher-order slow-roll parameters here.

Why are ϵ and η called slow-roll parameters? To see this, note that our assumption that the kinetic energy is much smaller than the potential also implies, via Eq. (2.31), that it is much smaller than the square of the expansion rate. Hence $\epsilon = \dot\phi^2/(2H^2)$ is required to be small for slow-roll. Then, the second assumption made above, namely that $|\ddot\phi| \ll H|\dot\phi|$, together with the smallness of ϵ implies that η must also be small. Thus the slow-roll conditions can be neatly re-expressed as

$$\epsilon \ll 1, \quad |\eta| \ll 1 \qquad \text{(slow-roll)}. \qquad (2.36)$$

Alternatively to the expressions above, the slow-roll parameters are sometimes defined directly in terms of the shape of the potential, via

$$\epsilon_V \equiv \frac{V_{,\phi}^2}{2V^2}, \qquad \eta_V \equiv \frac{V_{,\phi\phi}}{V}. \qquad (2.37)$$

These definitions help in encapsulating the requirements of having a potential that is flat ($\epsilon_V \ll 1$) over a sufficiently large range ($|\eta_V| \ll 1$). We can relate the two definitions using the slow-roll relations (2.31) and (2.32), with the result

$$\epsilon \approx \epsilon_V, \qquad \eta \approx 2\eta_V - 4\epsilon_V. \qquad (2.38)$$

With the approximation of slow-roll, the equations of motion can also be solved by expanding the Hubble rate and the scalar field as Taylor series in time. Up to the first nontrivial term, the solution is then

$$a \approx a_0 \exp\left(\sqrt{\frac{V}{3}}t - \frac{\epsilon V}{6}t^2\right), \qquad (2.39)$$

$$\phi \approx \phi_0 - \sqrt{\frac{2\epsilon V}{3}}t, \qquad (2.40)$$

where $a_0, \phi_0 = \phi(t=0)$ are constants and where the approximation above is valid as long as $\epsilon\sqrt{V}t \ll 1$. Since we are assuming here that $\epsilon \ll 1$, the expansion is quasi-exponential in time, given that the Hubble rate is approximately constant, $H \approx \sqrt{\frac{V}{3}} - \frac{\epsilon V}{3}t$.

Finally, we can estimate the number $\mathcal{N}$ of e-folds of expansion of the scale factor as

$$\mathcal{N} = \ln\left(\frac{a_{\text{end}}}{a_{\text{beg}}}\right) = \int H\,\mathrm{d}t = \int \frac{1}{\sqrt{2\epsilon}}\mathrm{d}\phi \approx \int \frac{V}{V_{,\phi}}\mathrm{d}\phi \approx \frac{V}{V_{,\phi\phi}}, \qquad (2.41)$$

where the very last approximation assumed that $V_{,\phi\phi}$ varies little during the inflationary phase (if this is not the case, one must use the preceding expression). Thus $1/|\eta_V|$ provides a useful rough estimate for the number of e-folds, and one typically needs $|\eta_V| \lesssim \mathcal{O}(10^{-2})$ in successful models.

Complementary to this point, note that the equation above tells us that the field range traversed by the inflaton field is on the order of

$$\Delta\phi \approx \sqrt{2\epsilon}\mathcal{N}, \qquad (2.42)$$

in cases where ϵ stays roughly constant during inflation. Since $\mathcal{N}$ must be at least as large as 50 (and could be substantially larger still), we can infer that the equation of state ϵ must be extremely small if we would like to have a small field range, say $\Delta\phi \lesssim 1$. In other words, only very flat potentials allow for a small field range. This is important for model building, where it is typically the case that one only has good control over all approximations over a small field range.

2.1.5 Quadratic, Higgs, and Starobinsky inflation

There exist a huge number of inflationary models, and it would be of little use to try to list them all. That said, it is certainly useful to look at a few concrete models to gain familiarity with inflation. We will restrict ourselves to three models, quadratic inflation (because of its simplicity and historical importance), Higgs inflation (because it uses the only experimentally verified scalar particle), and Starobinsky inflation (because it provides an example of an emergent, effective scalar field). As we will discuss in Chapter 3, under certain additional assumptions both Higgs and Starobinsky inflation are in good agreement with observational data, while quadratic inflation is ruled out.

Quadratic inflation

In the historical development of inflation, one particular model played a prominent role, and that is the model of *quadratic inflation*, where the potential is provided by a quadratic function,

$$V(\phi) = \frac{1}{2}m^2\phi^2, \tag{2.43}$$

see also the left panel in Fig. 2.3. As this is simply a mass term for ϕ, this model might arise rather naturally in extensions of the Standard Model of particle physics. For this potential $\epsilon \approx V_{,\phi}^2/(2V^2) = 2/\phi^2$, and hence inflation occurs as long as $|\phi| > \sqrt{2}$ (in reduced Planck units). One might not think of a quadratic potential as being very flat, but as we can see here the relative steepness of the potential becomes small as the field value is increased. Note that η is then automatically also very small, as $\eta \approx -2\epsilon$.

The number of slow-roll e-folds is given by the integral

$$\mathcal{N} = \int \frac{V}{V_{,\phi}} d\phi = -\int \frac{\phi}{2} d\phi = \frac{1}{4}(\phi_{\text{beg}}^2 - \phi_{\text{end}}^2). \tag{2.44}$$

As we just saw, inflation ends when ϕ reaches the value $\sqrt{2}$ and thus we require $|\phi_{\text{beg}}| \gtrsim 14$ in order to have at least 50 e-folds of expansion. In this potential, and more generally in potentials given by a simple monomial, the field range must thus be substantially larger than one Planck mass. In order for classical general relativity to still be a meaningful approximation, we

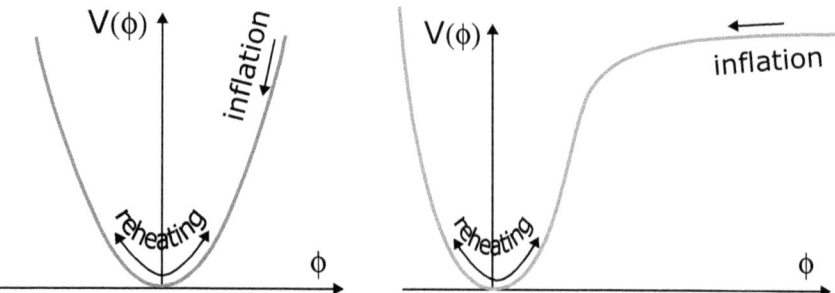

Figure 2.3 Cartoons of inflationary potentials: on the left the potential for quadratic inflation and on the right that for Starobinsky inflation in Einstein frame. For Higgs inflation, the potential has the same shape as the Starobinsky case at large field values, but near the origin it is of the standard "Mexican-hat" form familiar from Higgs physics.

must make sure that the energy scale remains below the Planck scale – i.e., we need $V_{\text{beg}} = \frac{1}{2}m^2\phi_{\text{beg}}^2 \ll 1$, implying $m \ll 1/10$. Thus, the inflaton field must be light compared to the Planck scale.

One great advantage of quadratic inflation is that inflation naturally comes to an end, once the field has rolled down sufficiently far (that is, below about $\phi \approx |\sqrt{2}|$). At that point, the scalar field starts oscillating around the minimum of the potential. The oscillations gradually become smaller in amplitude, since the expansion of the universe acts like a friction force. Meanwhile the inflaton field can decay into other types of particles, thereby filling the universe with radiation and matter. We will discuss this process of *reheating* separately, below in Section 2.1.6.

Higgs inflation

So far, only one scalar particle has been experimentally detected, and that is the Higgs boson. The Higgs field $h(x^\mu)$ (of which the particle is an excitation) is responsible for breaking the electroweak symmetry in the Standard Model of particle physics. Its Lagrangian (in unitary gauge) is given by

$$\mathcal{L}_h = -\frac{1}{2}(\partial h)^2 - \frac{\lambda}{4}(h^2 - v^2)^2, \qquad (2.45)$$

where the functional form of the potential is dictated by renormalizability. Experiments have revealed the values of the coupling constant to be $\lambda \approx 0.13$ and the vacuum expectation value to be $v \approx 246\,\text{GeV}$. The fact that the potential minimum resides at v and not at zero is responsible for the breaking of the electroweak symmetry. Asymptotically, the potential is of h^4 form,

2.1 Inflation and scalar field dynamics

which would in principle be suitable for inflation but turns out to be too steep to be able to match observations of the CMB (which we will discuss in detail in Chapter 3). Hence, at first sight, it seems that the only known scalar field does not provide a realistic candidate inflaton.

However, the coupling to gravity must not necessarily be minimal. Additional, nonminimal terms are expected to arise as quantum corrections, or may be present due to the embedding into a more fundamental theory of quantum gravity. One such term in particular, which couples the Higgs directly to the Ricci scalar, effectively modifies the potential and flattens its shape at large h. Let us go through how this happens. First we write out the full action, including the gravitational terms,

$$S = \int d^4x \sqrt{-\bar{g}} \left[\frac{1}{2}(1+\xi h^2)\bar{R} - \frac{1}{2}\bar{g}^{\mu\nu}\partial_\mu h \partial_\nu h - \frac{\lambda}{4}(h^2 - v^2)^2 \right]. \quad (2.46)$$

Here the new scalar–gravity term has a coupling constant ξ, and we have called the metric $\bar{g}_{\mu\nu}$ for reasons that will become clear momentarily. The scalar–gravity coupling effectively changes the value of Newton's constant, but because currently $h \approx v \sim 10^{-16}$ is exceedingly small in Planck units, these corrections are undetectable in laboratory experiments for all values of ξ that we will consider. In fact, for small values of h there is no noticeable change at all compared to minimal coupling.

However, at large values of h (relevant during inflation) the situation can change significantly. To see this, we can perform a field redefinition, combining gravity and the scalar field such that the action for gravity returns to the standard Einstein–Hilbert form. We can also redefine the scalar field such that its kinetic action, which gets modified by the redefinition of the metric, returns to canonical form. The required redefinitions are

$$g_{\mu\nu} \equiv (1+\xi h^2)\bar{g}_{\mu\nu}, \qquad \frac{d\phi}{dh} \equiv \frac{\sqrt{1+\xi h^2 + 6\xi^2 h^2}}{1+\xi h^2}. \quad (2.47)$$

The original fields $\bar{g}_{\mu\nu}, h$ are said to be in *Jordan frame*, while the new fields $g_{\mu\nu}, \phi$ are in *Einstein frame*. The inflationary regime is supposed to occur at large field values and, as we will see, at large coupling ξ. With the approximations $\xi h \gg 1$ and $\xi \gg 1$, the relation between h and ϕ simplifies,

$$\frac{d\phi}{dh} \approx \frac{\sqrt{6}\xi h}{1+\xi h^2} \quad \to \quad \phi = \sqrt{\frac{3}{2}} \ln(1+\xi h^2), \quad (2.48)$$

and with those approximations the action may be verified to become

$$S = \int d^4x \sqrt{-g} \left[\frac{1}{2}R - \frac{1}{2}g^{\mu\nu}\partial_\mu\phi\partial_\nu\phi - \frac{\lambda}{4\xi^2}\left(1 - e^{-\sqrt{2/3}\phi}\right)^2 \right]. \quad (2.49)$$

We can see that after this reshuffling of the fields, both gravity and the scalar kinetic term have returned to their standard, canonical forms, but the potential has been modified drastically. Instead of the steep h^4 climb, at large field values $\phi \gtrsim 1$ the potential has now become very flat, and in fact it asymptotes to a constant; see Fig. 2.3. Note that the height of the potential is given both in terms of the Higgs coupling λ and the new coupling to gravity ξ. In Chapter 3, we will see that in order to match observations, the coupling is required to be

$$\xi \approx 46\,000\, \sqrt{\lambda}, \qquad (2.50)$$

which is consistent with the approximations made above. Whether such a large coupling can be achieved in quantum gravity remains a subject of debate. Details regarding the slow-roll dynamics in this model will be the subject of an exercise at the end of this chapter. Let us just point out that inflation will naturally come to an end once the field has rolled sufficiently far down the potential. From that point onward, the physics is entirely analogous to the quadratic inflation case, with reheating occurring as the field undergoes damped oscillations around the potential minimum.

Starobinsky inflation

The third model that we will present in more detail shows that inflation can also occur in the absence of a fundamental scalar field. It is based on the addition to standard gravity of a term that is quadratic in the Ricci scalar, so that the action becomes

$$S = \int d^4x \sqrt{-\bar{g}} \frac{1}{2}\left(\bar{R} + \frac{\bar{R}^2}{6M^2}\right). \qquad (2.51)$$

Here M is a mass scale and we once again denote the original metric with an overbar. In the next paragraph we will outline a theoretical justification for such an action, but this may be skipped by readers unfamiliar with renormalization.

The original Starobinsky model starts with Einstein–Hilbert gravity in the presence of massless conformally coupled matter fields, such as for example massless vector fields. (Note that the matter fields may not need to be strictly massless, as at high energies/high curvatures their masses can effectively be neglected.) The quantum fluctuations of such fields, in particular via loop corrections, give contributions to the energy–momentum tensor involving higher powers of the Riemann tensor. Their divergences are then canceled by adding counterterms that are themselves power-law expressions of the Riemann tensor, and the leading such terms are quadratic in the Riemann tensor. The action (2.51) is then seen as an effective action capturing these effects.

The field equations are derived in the usual way by varying the action (2.51) with respect to the metric. In the present case, this is a somewhat lengthy procedure, but in the end, in a flat FLRW spacetime, the equations of motion reduce to

$$\ddot{\bar{H}} - \frac{\dot{\bar{H}}^2}{2\bar{H}} + \frac{1}{2}M^2\bar{H} = -3\bar{H}\dot{\bar{H}}, \tag{2.52}$$

$$\ddot{\bar{R}} + 3\bar{H}\dot{\bar{R}} + M^2\bar{R} = 0. \tag{2.53}$$

If the linear term in the action (2.51) were absent, then one would obtain a pure de Sitter solution with constant Hubble rate and constant curvature (this can be seen by taking the $M \to 0$ limit in the equations of motion). As it stands, the linear term causes the Hubble rate to decay and inflation to eventually come to an end. An approximate solution, valid during inflation, can be found by realizing that the first two terms in Eq. (2.52) are subdominant. Upon integration one finds

$$\bar{H} \approx \bar{H}_i - \frac{M^2}{6}(t - t_i), \tag{2.54}$$

$$\bar{R} \approx 12\,\bar{H}^2 - M^2, \tag{2.55}$$

where $\bar{H}_i$ and t_i are the initial Hubble rate and initial time, respectively. These are not fixed by the model – rather, they depend on how inflation starts, which is a topic we will discuss more thoroughly in Section 3.4.

The slow-roll parameter is given by

$$\epsilon = -\frac{\dot{\bar{H}}}{\bar{H}^2} \approx \frac{M^2}{6\bar{H}^2}. \tag{2.56}$$

It grows as the Hubble rate decreases, and inflation ends when it reaches order unity, i.e., when $\bar{H} \approx M/\sqrt{6}$. At that point the curvature is of order $\bar{R} \approx M^2$. (We will see later that the mass scale must be set to $M \approx 10^{13}$ GeV in order to match observations from the CMB.) The relations above allow us to estimate the number of e-folds of inflation,

$$\mathcal{N} = \int_{t_i}^{t_f} \bar{H}\,\mathrm{d}t \approx \bar{H}_i(t_f - t_i) - \frac{M^2}{12}(t_f - t_i)^2. \tag{2.57}$$

Inflation ends at $t_f \approx t_i + 6\bar{H}/M^2$, and thus

$$\mathcal{N} \approx \frac{3\bar{H}_i^2}{M^2} \approx \frac{1}{2\epsilon(t_i)}. \tag{2.58}$$

As long as $\epsilon \lesssim \mathcal{O}(10^{-2})$, which will depend on the initial conditions, the flatness and horizon puzzles can thus be resolved.

It turns out to be possible to eliminate the $\bar{R}^2$ term in the action (2.51) by a scaling of the metric. The required transformations are

$$g_{\mu\nu} \equiv e^{\sqrt{2/3}\phi}\bar{g}_{\mu\nu}, \qquad \phi \equiv \sqrt{\frac{3}{2}}\ln\left(1 + \frac{\bar{R}}{3M^2}\right), \qquad (2.59)$$

where one should note in particular that the Ricci curvature now plays the role of an effective scalar field (which will be the inflaton). The action then becomes that of gravity, a scalar ϕ and an effective potential,

$$S = \int \mathrm{d}^4x \sqrt{-g}\left[\frac{1}{2}R - \frac{1}{2}g^{\mu\nu}\partial_\mu\phi\partial_\nu\phi - V(\phi)\right], \qquad (2.60)$$

$$V(\phi) = \frac{3M^2}{4}\left(1 - e^{-\sqrt{2/3}\phi}\right)^2. \qquad (2.61)$$

Remarkably, this is the exact same action as that of the Higgs inflation model in the large field limit (2.49) (though in the present case it is valid for all field values, so there is no strict equality between the models). To see that the number of degrees of freedom matches, and that we have not artificially created a new one, note that we have replaced the fourth-order equations of motion (2.52) and (2.53) of a higher-order gravity model with a model leading to two two-derivative equations of motion.

The shape of the potential is sketched in Fig. 2.3. At large positive field values it asymptotes to a constant, which represents the approximate de Sitter solution induced by the quadratic term in (2.51). The freedom of initial conditions in part concerns the question of where exactly on this plateau the inflaton ϕ is placed initially. In the solution (2.54) this ambiguity corresponds to choosing $\bar{H}_i$.

The inflaton then slowly rolls to smaller field values, and inflation comes to an end once the potential becomes too steep (see the exercises for further discussion). Once inflation ends, the scalar field rolls to the minimum of the potential and starts oscillating around it. If one expands the potential around the minimum, one obtains $V(\phi) \approx \frac{1}{2}M^2\phi^2$, by which argument one can see that the mass scale M in fact corresponds to the mass of the inflaton when it sits in its vacuum. In the next section we will discuss this final stage.

Let us mention an important issue regarding both Higgs and Starobinsky inflation. As we saw, and as we will see in more detail when discussing fluctuations in Chapter 3, both models are in excellent agreement with observations, as long as the coupling ξ or mass scale M take appropriate values. However, for that agreement to hold up, it is crucial in both cases that further correction terms, such as terms of the form Rh^n (in the Higgs case) or R^n (in the Starobinsky case) with $n > 2$, are subdominant. Otherwise, they modify the plateau part of the potential, thereby making it impossible to have a prolonged inflationary phase. Put differently, the success of these models depends on the first correction term being suitably large, and all higher correction terms being suitably small. This represents a significant challenge when trying to embed these models in more complete theories, such as supergravity or string theory.

2.1.6 End of inflation and reheating

During inflation the universe grows by an enormous amount, at least on the order of a factor of $e^{50} \sim 10^{22}$ in all directions. This means that, even if some amounts of radiation or matter are present when inflation begins, they become diluted to such an extent that the universe can be thought of as being empty (except for vacuum energy) once inflation is underway. When inflation comes to an end, the universe has thus become a vast, cold, empty place. *Reheating* is the process by which the inflaton subsequently decays into fermionic, and typically also bosonic, particles. These newly created particles can remain stable or disintegrate into further decay products, and interact with each other until thermal equilibrium is established. In this way the conditions for the hot big bang evolution described in Chapter 1 can be created. The inflationary paradigm thus has the potential to explain the origin of radiation and matter in our universe.

The details of the reheating mechanism depend sensitively on the particular model of particle physics that is assumed. In particular, since the energies reached are high, they depend on currently unknown physics beyond the Standard Model (even in the case of Higgs inflation). This is the first reason why we will not go into much detail here. The second reason is more important, namely that the predictions of inflationary models regarding observations of the CMB are almost entirely independent of those details (we will prove this in Chapter 3). Our aim here is therefore merely to sketch a few of the physical ideas that operate during reheating.

Perhaps the most important aspect is that of time scales. The inflaton ϕ will have a few decay channels into other particles, specified for example by terms in the Lagrangian of the form

$$\mathcal{L} \supset f\phi\chi^2 + g\phi\bar{\psi}\psi, \tag{2.62}$$

where χ is a bosonic particle, ψ a fermionic one, and f, g are coupling constants. For example, the amplitude for ϕ to decay into two χ particles will be proportional to f. The decay rate, being a probability, is proportional to the square of the amplitude – more specifically we have the rates (assuming χ and ψ to be much lighter than the inflaton)

$$\Gamma_{\phi \to \chi\chi} \sim \frac{f^2}{M}, \qquad \Gamma_{\phi \to \bar{\psi}\psi} \sim g^2 M, \qquad (2.63)$$

where M is the mass of the inflaton, which enters the decay rates differently for bosons and fermions (the precise rates are calculated in quantum field theory textbooks). The point is that these decay rates, being proportional to the square of the coupling constants, are small compared to the Hubble rate during inflation,

$$\Gamma \ll H_{\text{inflation}}. \qquad (2.64)$$

This implies that during inflation the universe expands so fast that the inflaton effectively does not have time to decay. However, at the end of inflation, when the accelerated expansion ends and the Hubble rate has decreased sufficiently,

$$\Gamma \sim H_{\text{reheating}}, \qquad (2.65)$$

decay processes start becoming important. The details obviously depend on the specifics of the particle physics model (and additional effects, such as Bose–Einstein condensation, can come into play), but typically copious numbers of particles are produced at this point.

This can be very roughly estimated as follows: Around the potential minimum, the potential is that of a mass term $V(\phi) \approx \frac{1}{2} M^2 \phi^2$ and the inflaton behaves like a harmonic oscillator, with equal time-averaged kinetic and potential energies. Hence it behaves like a fluid with zero average pressure, since $p = \frac{1}{2}\dot{\phi}^2 - V(\phi)$. It follows that on average the universe expands just as during matter domination, $H \approx 2/(3t)$. More precisely, one can verify that up to corrections of order $1/(Mt)$, the equations of motion (2.17) and (2.19) are solved by

$$a \propto t^{\frac{2}{3}}, \qquad \phi = \frac{2\sqrt{2}}{\sqrt{3} M t} \cos(Mt), \qquad (2.66)$$

which confirms that ϕ undergoes oscillations with an increasingly damped amplitude. The number density of particles may now be estimated as

$$n_\phi = \frac{\rho}{M} = \frac{1}{2M}(\dot{\phi}^2 + M^2 \phi^2) = \frac{4M}{3(Mt)^2}. \qquad (2.67)$$

If $Mt \gtrsim 1$ and $M \sim 10^{13}$ GeV, then we have up to about $n_\phi \sim 10^{-6}$ particles per Planck volume. This translates into a staggering 10^{98} particles per cubic meter.

At this point, the particles are not in a thermal distribution yet, but the repeated collisions between particles quickly lead to an equilibrium. The range of possible temperatures resulting from reheating is large, but typically the end result is on the order of the energy density during inflation – i.e., it is given to a first approximation by the height of the potential. Hence a value of $\mathcal{T} \sim 10^{13}$ GeV $\sim 10^{17}$ K is commonly quoted.

We end with two comments. The first is that the term *re*-heating is a little misleading, as there is currently no reason to assume that the universe was hot at any previous time. And secondly we note that because of the high temperatures reached at reheating, baryogenesis may occur during this epoch, thus allowing another one of the big bang puzzles to be potentially resolved.

2.2 An alternative: the ekpyrotic universe

Inflation is by far the most popular theory for explaining the large-scale flatness of the universe. However, as we will discuss in Chapter 3, the observational evidence for inflation is still indirect. In other words, to date there exists no observation that singles out an early accelerated expansion phase "beyond a reasonable doubt." If only for this reason, we may wonder whether other explanations for the smoothness of the universe might exist. So far, only one other dynamical mechanism is known to achieve this – named *ekpyrosis* – and we will review it here.

2.2.1 The ekpyrotic phase

Let us go back to the Friedmann equation in the presence of different sources, represented here by their energy densities,

$$3H^2 = \frac{-3K}{a^2} + \frac{\rho_{\rm m}}{a^3} + \frac{\rho_{\rm r}}{a^4} + \frac{\rho_\sigma}{a^6} + \cdots + \frac{\rho_\phi}{a^{3(1+w_\phi)}}. \quad (2.68)$$

In addition to homogeneous curvature, pressure-free matter and radiation, we have included the energy density of anisotropies in the curvature of the universe (σ) and a scalar field (ϕ) with equation of state w_ϕ. The scaling of the anisotropy term can be calculated as follows: Consider a metric of the (Bianchi I) form

$$\mathrm{d}s^2 = -\mathrm{d}t^2 + a(t)^2 \sum_i e^{2\beta_i}(\mathrm{d}x^i)^2, \quad (2.69)$$

with $\sum_i \beta_i = 0$, so that a denotes the average scale factor, while the β_i parametrize anisotropies in the x, y, z spatial directions. Then the Friedmann equation picks up a new term

$$3H^2 = \cdots + \frac{1}{2}\sum_i \dot{\beta}_i^2, \qquad (2.70)$$

while the ij Einstein equations give

$$\ddot{\beta} + 3H\dot{\beta} = 0 \quad \to \quad \dot{\beta} \propto \frac{1}{a^3}. \qquad (2.71)$$

Thus the new term in the Friedmann equation scales as $\sum_i \dot{\beta}_i^2 \propto 1/a^6$.

In an expanding universe, as the scale factor a grows, matter components with a slower falloff of their energy density come to dominate. If there is an inflationary scalar field, then eventually the inflaton, whose energy density is roughly constant, dominates the cosmic evolution and determines the (roughly constant) Hubble parameter while causing the scale factor to grow exponentially, $a \propto e^{Ht}$. During inflation, the relative energy densities in the curvature $-K/(a^2 H^2)$ and in the anisotropies $\rho_\sigma/(3a^6 H^2)$ fall off quickly, and the universe is rendered exponentially flat; as we calculated earlier, the flatness puzzle is then resolved as long as the scale factor grows by at least 50 e-folds.

Interestingly, the same problem can be solved by having a contracting phase before the standard hot big bang phase of the universe. In a contracting phase, the argument presented in the previous paragraph is reversed. Now terms with a quicker falloff in the scale factor a come to dominate. In particular, the relative energy density in the homogeneous curvature $-K/(a^2 H^2)$ quickly becomes subdominant, and thus flatness is essentially automatic. However, this threatens being spoiled by the rapid growth of anisotropies, which grow in proportion to a^{-6}. Thus the anisotropies are the real danger in a contracting universe. But the point is that the energy density in a (homogeneous, isotropic) scalar field can grow even faster, as long as

$$w_\phi > 1 \quad \leftrightarrow \quad \epsilon > 3 \quad \text{(condition for ekpyrosis)}. \qquad (2.72)$$

In such a case, the relative energy densities of homogeneous and anisotropic curvature are both suppressed compared to the energy density in the scalar field.

Ekpyrosis is a conjectured contracting phase with an ultrastiff equation of state $w_\phi > 1$ preceding the hot big bang evolution.

How can such a large equation of state be implemented? Suppose that,

2.2 An alternative: the ekpyrotic universe

instead of a flat, positive potential the scalar field ϕ has a steep, negative potential. As a concrete example, let us consider a negative exponential

$$V(\phi) = -V_0 e^{-c\phi}, \qquad (2.73)$$

where V_0 and c are positive constants. Then, in analogy with power-law inflation (2.26), the equations of motion may be solved by a scaling solution

$$a \propto (-t)^{\frac{1}{\epsilon}}, \quad \phi = \sqrt{\frac{2}{\epsilon}} \ln\left[-\left(\frac{V_0 \epsilon^2}{\epsilon - 3}\right)^{\frac{1}{2}} t\right], \quad V = -\frac{1}{\epsilon t^2}, \quad \epsilon = \frac{c^2}{2}. \qquad (2.74)$$

Here the time coordinate is taken to be negative and is running towards 0^-. The equation of state $w > 1, \epsilon > 3$ can then simply be achieved if the potential is sufficiently steep, $c > \sqrt{6}$. The scalar field is quickly rolling down its potential, yet notice that the universe is contracting very slowly, due to the high pressure that counteracts gravitational collapse. If nothing else happens, then as $t \to 0^-$ both the kinetic and potential energies blow up, and a big crunch singularity is reached. This tells us that the ekpyrotic phase must come to an end (e.g., due to the potential reaching a minimum and turning around) and a *bounce* must occur, linking the ekpyrotic contracting phase to the hot big bang expanding phase. We will elaborate on this topic in the following two subsections below.

How long must the ekpyrotic phase last in order to solve the flatness puzzle? As shown in Section 1.3.2, it is the comoving horizon $1/(aH)$ that must shrink by a factor of at least e^{50} if we assume an original curvature contribution to the Friedmann equation of order unity. From (2.74) we infer

$$V \propto |aH|^{\frac{2\epsilon}{\epsilon - 1}}, \quad t = -(-\epsilon V)^{-\frac{1}{2}}, \qquad (2.75)$$

which imply that the potential V must grow in magnitude by a factor of at least $e^{100\epsilon/(\epsilon-1)} \gtrsim 10^{48}$, assuming $\epsilon \sim \mathcal{O}(10)$. If we assume a potential minimum at the grand unified scale, $|V_{\min}| \sim (10^{-2} M_P)^4$, then the end of the ekpyrotic phase occurs at time $t_{\text{ek-end}} \sim -10^3 t_P$, where M_P and t_P denote the Planck mass and time, respectively. This implies that the ekpyrotic phase must have started at the latest at time

$$|t_{\text{ek-beg}}| \gtrsim 10^{27} t_P \approx 10^{-16} s, \qquad (2.76)$$

a short time in cosmology, although quite long by high energy physics standards, and many orders of magnitude longer than a typical inflationary phase. In the cyclic picture of the universe, which will be discussed below, the potential V interpolates between the grand unified and the dark energy scales. In that case one finds

$$|t_{\text{ek-beg}}| = \sqrt{\frac{V_{\text{ek-end}}}{V_{\text{ek-beg}}}}|t_{\text{ek-end}}| \approx \sqrt{10^{114}}10^3 \, t_P \approx 10^{17} \, s \, . \qquad (2.77)$$

In this case, the ekpyrotic phase lasts on the order of 10 billion years, i.e., the time scale associated with dark energy. During this entire time, the universe only contracts by a rather modest factor of about a million.

Just like inflation, ekpyrosis is also an attractor. For the simple case of spatially independent (but time-dependent) shifts in the initial conditions, this can be shown by repeating the analysis performed around Eqs. (2.28) and (2.29). Just as in that case, the relevant solution for the perturbations $\delta a(t)$, $\delta\phi(t)$ is proportional to $|t|^{1-3/\epsilon}$. However, now the time coordinate is decreasing as ekpyrosis progresses, and the equation of state satisfies $\epsilon > 3$. Thus, once again, perturbations are decaying and an attractor solution is approached.

Before embedding the ekpyrotic phase into a more complete cosmological model, let us comment on the origin of the name: Ekpyrosis means "out of fire" in Greek and was the name of a cosmological model developed by the Stoic philosophers, in which the universe periodically collapses and is consumed by fire, before reemerging from the ashes. The modern version of this model is presented next.

2.2.2 Cyclic scenario and cosmic bounces

The cyclic model is an ambitious attempt at providing a complete history of the universe, incorporating both the ekpyrotic mechanism and dark energy in an essential way. We will present a brief overview.

The cyclic model relies on having a scalar field potential of the general shape depicted in Fig. 2.4. Since the evolution is periodic, we may start at any time, for instance during the current dark energy phase. The scalar field finds itself on the dark energy plateau (on the right in the figure) and slowly rolls down the potential, thereby providing a period of dark energy domination. The dark energy is slowly decaying and, at some point in the future, the potential energy becomes negative and the universe reverts from expansion to contraction. We then enter an ekpyrotic phase, which locally flattens and homogenizes the universe. The ekpyrotic phase terminates when the potential reaches a minimum and turns around. At that point a brief period of kinetic energy domination follows, during which the scalar field rapidly shoots off towards large negative values.

Here we reach the phase of greatest uncertainty in the model: The scalar

2.2 An alternative: the ekpyrotic universe

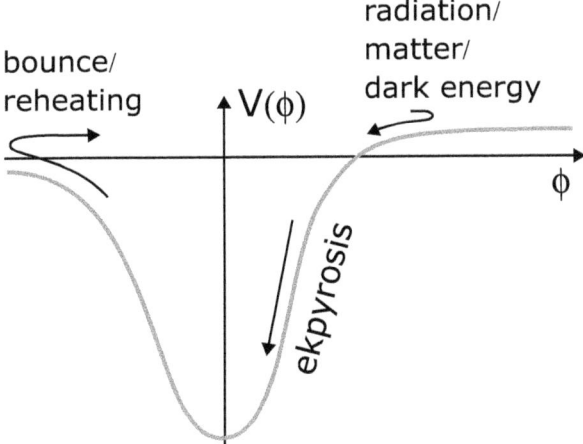

Figure 2.4 Cartoon of the potential in the cyclic model of the universe, incorporating an ekpyrotic phase.

field has to turn around while the scale factor undergoes a bounce – i.e., a transition from contraction to expansion. The bounce thus replaces the big bang singularity, and it is also at this stage that reheating is supposed to occur, filling the universe with radiation and pressure-free matter. The challenges in modeling a bounce will be elaborated on below. During the bounce, the scalar field acquires a small boost and quickly rolls back across the potential well and onto the positive energy plateau, where the scalar gets quasi-stabilized. Now the universe undergoes the usual periods of radiation and matter domination, while the scalar field starts rolling back slowly. Eventually, the resulting dark energy comes to dominate the energy density of the universe, and the whole cycle starts again. In this way, physical processes in the universe today provide the initial conditions for the next cycle.

What should be noted is the economy of ingredients and the seamless interplay between the early and the late universe. Further note that in this model, the universe is also flattened during the dark energy phase, which implies that the ekpyrotic phase can be correspondingly shorter. Dark energy, which may be regarded as a low-energy inflationary phase, moreover contributes to stabilizing the cycles and thus plays a crucial role, unlike in standard big bang cosmology.

An important aspect is that the cyclicity only applies to local quantities, such as the Hubble rate. Other quantities, in particular the scale factor of the universe, need not evolve cyclically: The universe expands by large amounts during radiation, matter, and dark energy domination, while it only contracts a little during the ekpyrotic phase. Thus, over the course of each cycle, the universe grows by a huge factor. This is a beneficial feature, as it prevents the buildup of entropy.

It is evident that one may consider many extensions or variations of this model – for example, one might consider that coupling constants could evolve from one cycle to the next, and envisage a cosmic selection scenario in which their present values would be reached dynamically. An example will be considered in Exercise 9.4. However, there is one crucial issue that plagues all ekpyrotic/cyclic models, and that is the nature of the bounce.

To see why it is hard to model a bounce, consider the rate of change of the Hubble rate, obtained by combining (1.36) and (1.37),

$$\dot{H} = \frac{K}{a^2} - \frac{1}{2}(\rho + p). \tag{2.78}$$

At the bounce, $H = 0$ and $\dot{H} > 0$. The ekpyrotic phase drives the homogeneous curvature to zero, so that the first term on the right-hand side plays no role. Then we see that having a bounce is synonymous with requiring a violation of the null energy condition, as one would need $\rho + p < 0$. This is a very obstinate obstacle, as violations of the null energy condition are associated with instabilities, such as ghosts (fields with wrong-sign kinetic terms). By definition, a standard scalar field cannot violate the null energy condition ($\rho + p = \dot{\phi}^2$), for instance. In fact, no form of matter appearing in nature is known to violate this bound. The best that has been achieved so far is to construct models with higher-derivative kinetic terms, which contain bounce solutions that are perturbatively stable, if the Lagrangian is suitably tuned. The trouble is that these theories contain other solutions that contain instabilities, and this makes one doubt that they could be nonperturbatively stable at the quantum level.

We may summarize our discussion by saying that the ekpyrotic phase provides a dynamical alternative to inflation for smoothing the universe, but that it remains unclear at present whether a large contracting region of the universe can evolve into an expanding region according to the laws of quantum gravity. For this reason, we will mainly concentrate on inflationary models in this book.

Exercises

2.1 Try to get a feeling for how extreme inflation is in comparison with terrestrial experience: Assume that the inflationary potential has an energy scale of 10^{13} GeV, and that inflation operates for 50 e-folds.

Then estimate the duration of inflation and the growth of the universe during inflation. Also calculate the size that the currently observable part of the universe had at the beginning of inflation.

2.2 For the effective potential (2.61) of Starobinsky and Higgs inflation, and in the slow-roll limit, calculate the slow-roll parameters, the field value where inflation ends, and the field range required for 50 e-folds of inflation. Also approximate the slow-roll parameters in terms of the number of e-folds.

2.3 Quadratic inflation with $V(\phi) = \frac{1}{2}M^2\phi^2$ can only be solved approximately. Show that, when a negative cosmological constant of suitable size is added, there is an analytic flat FLRW solution in which the scalar field rolls with constant velocity $\dot{\phi}(t) = c$. Calculate the stability of the solution to time-dependent perturbations and interpret it.

2.4 Show that the Hubble rate evolves cyclically in the cyclic model (as a first approximation, you may ignore the matter and bounce phases, because the matter phase is short and the bounce phase effectively changes $-|H|$ into $+|H|$). Also calculate how large the observable universe was one cycle ago.

3
Observing perturbative quantum gravity: fluctuations

It is difficult to directly probe the background evolution of the very early universe, essentially because we only see out to the surface of last scattering, which is (optically) opaque to what happened before. Eventually, neutrinos and gravitational waves will allow us to probe the deeper past, but it will take time before such observations will become feasible. That said, it turns out that there is a huge amount of information in the details, i.e., in the small temperature fluctuations, of the CMB sky.

In this chapter, we will discuss the remarkable insight that these fluctuations may be a result of perturbative quantum gravity. This is perhaps our best experimental evidence for quantum gravity, which, rather than being a far-out speculation, may already have caused effects visible in all directions in the sky.

We will start by describing the observations of temperature fluctuations, then explain why these cannot plausibly be explained by classical physics, and finally work through the (somewhat subtle) calculation of quantum fluctuations of gravity and scalar matter. At various later stages in the book, we will return to these calculations from different points of view.

3.1 Observed fluctuations in the CMB

In Chapter 1, we discussed the background blackbody nature of the cosmic background radiation, and the dipole arising from our intrinsic motion across the cosmos. However, there are additional small anisotropies encoded in the CMB, and these provide a wealth of information about the early universe. It is useful to characterize these anisotropies in terms of the higher multipoles of the temperature field. Since the sky appears as a sphere surrounding us, it is appropriate to take these multipoles as spherical harmonics Y_{lm},

$$\frac{\delta \mathcal{T}(\hat{n})}{\mathcal{T}} = \sum_{l=1}^{\infty} \sum_{m=-l}^{l} a_{lm} Y_{lm}(\hat{n}), \tag{3.1}$$

where $\hat{n}$ denotes the direction in the sky. In principle $\delta\mathcal{T}(\hat{n})$ and the a_{lm} are spacetime dependent, but over the range of times and places that humans are currently able to observe the CMB they are essentially constant. The spherical harmonics are the analogs of Fourier transforms on the sphere (they are also complex functions in general), and they obey the orthonormality condition

$$\int d\Omega \, Y_{lm}(\hat{n}) \, Y^*_{l'm'}(\hat{n}) = \delta_{ll'} \delta_{mm'}, \tag{3.2}$$

where Ω is the solid angle on the projected sky. One may think of the parameter l as the wavenumber ($l=1$ is the dipole, $l=2$ the quadrupole, etc.) specifying the angular size of the fluctuation on the sky. The different m then specify independent fluctuation modes of that size – see Appendix C for examples.

The a_{lm} have zero average by definition, but they have a nonzero variance that contains statistical information encoded in the CMB,

$$\langle a_{lm} \rangle = 0, \qquad \langle a_{lm} a^*_{l'm'} \rangle = C^{\mathrm{TT}}_l \delta_{ll'} \delta_{mm'}. \tag{3.3}$$

One may also invert the last relation,

$$C^{\mathrm{TT}}_l = \frac{1}{2l+1} \sum_m \langle a^*_{lm} a_{lm} \rangle. \tag{3.4}$$

It is these rotationally invariant C^{TT}_l coefficients, or in fact the combination $D^{\mathrm{TT}}_l = l(l+1)C^{\mathrm{TT}}_l/2\pi$, that are often plotted in graphical representations of the statistical properties of the cosmic background radiation; see the left panel in Fig. 3.1. The superscript TT here means that we are looking at correlations between two temperature fluctuations. Other quantities of interest will be discussed below.

The graph shows a roughly flat distribution at small wavenumbers, a notable peak at $l \approx 200$, and further peaks and troughs at larger l values, with the overall amplitude decaying at large wavenumber. It turns out that this pattern is very far from random. In fact, even though we cannot see beyond the surface of last scattering, the physics at the times leading up to last scattering is very well understood (serious uncertainties only arise once we extrapolate to energies higher than TeV scales, cf. our discussion in Chapter 1). This means that we can evolve the observed fluctuations some

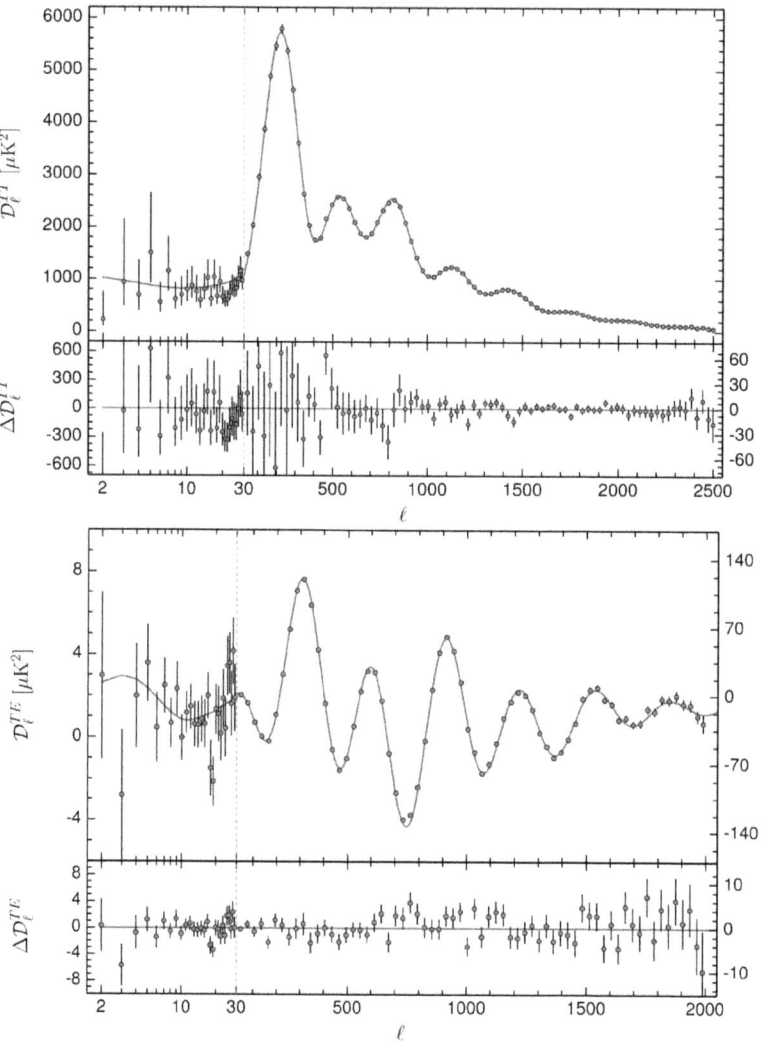

Figure 3.1 *Upper panel:* This graph shows the angular power spectrum of the CMB radiation – i.e., it shows the amplitude of the temperature fluctuations as a function of the multipole moment l. To be precise: What is depicted on the vertical axis is the combination $D_l^{\text{TT}} = l(l+1)C_l^{\text{TT}}/2\pi$. *Lower panel:* the corresponding temperature–E-mode cross-correlation spectrum. Figures reproduced with permission from Planck Collaboration; Akrami et al. (2020b). *Credit: ESA/Planck Collaboration.*

way back in time, and when doing this something remarkable happens. We will only describe this heuristically here, both because precise calculations are performed numerically and because the focus of this book lies on earlier cosmological stages.

Recall from Section 1.3.3, and see in particular the small circles in Fig. 1.5, that the horizon at recombination currently subtends an angle of a little over 1 degree in the sky. This is comparable to the angular size of the first peak at $l \approx 200$ seen in the angular power spectrum in the left panel of Fig. 3.1.[1] Any fluctuations that are larger in wavelength (i.e., have small wavenumber l) did not evolve in the time leading up to recombination, simply by causality. In particular, the flat region at small $l \lesssim 30$ gives us a direct view of the "primordial" perturbations. Note that the uncertainties become large at very small l, because at such small l only a few modes fit on the entire sky, and hence statistical variance is significant for these long-wavelength modes (the variance scales as $\Delta C_l/C_l \sim 1/\sqrt{l}$). This intrinsic limitation is known as *cosmic variance*, but it only affects the very largest modes on the sky significantly.

Given that in the period before recombination the horizon was steadily growing, modes with increasingly small wavelength (large wavenumber) had more and more time to evolve prior to last scattering. During that period, the universe was filled with a radiation-matter plasma. Small fluctuations may then be thought of as acoustic waves – gravity causes the perturbations to collapse, while pressure causes them to expand – and they behave like a (forced) harmonic oscillator. In other words, their amplitude alternately grows and shrinks. The first peak is caused by the modes that have grown to maximal amplitude for the first time just when recombination happens. Then there will be modes that started oscillating a little earlier and have just reached their first minimum of the amplitude at the time of recombination – these modes, at slightly larger wavenumber l, give rise to the first trough. At larger l this pattern then repeats for modes that have reached subsequent maxima and minima, leading to the pattern of alternating peaks and troughs seen in Fig. 3.1.

Three comments: First, all fluctuation modes, once the horizon has become larger than them and they start evolving, must start their evolution with the *same phase* in order for this pattern of peaks and troughs to emerge. If, for the same wavenumber l, some modes would start at the peak and some others at minimum amplitude and some more at intermediate values, then they would not all reach a peak (say) simultaneously at recombination. Thus we learn that all perturbation modes with the same l must have started their evolution with the same phase. As a second comment let us point out that, using known plasma physics, one can calculate the size of the oscillations

---

[1] Up to a given l_* the number of a_{lm} is $\sum_{l=0}^{l_*}(2l+1) = (l_*+1)^2$. For $l_* = 200$ we get a total of 40 401. A full sky corresponds to about 41 253 degrees squared, hence $l = 200$ corresponds to angular scales of approximately 1 degree on the sky.

62 *Observing perturbative quantum gravity: fluctuations*

that are at the first peak, since this size depends on the speed of sound. In other words, the actual, physical size of these fluctuations at recombination provides a standard ruler. Thus, by measuring the corresponding angular size on the sky, we can find out (since we know the redshift) what the geometry of the universe is in between last scattering and us. Such measurements allow one to find out that the spatial geometry of the universe is flat to high precision. The sizes and angular locations of higher peaks give us information about other cosmological parameters, such as the amount of baryonic matter in the universe. And third, at very high l the numerous scattering interactions between photons and electrons (right around the time of recombination) tend to wash out temperature differences – this explains why the angular power spectrum falls off rapidly at large l.

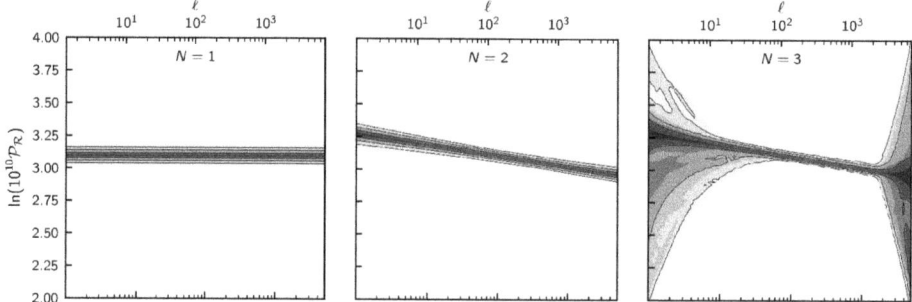

Figure 3.2 Reconstructions of the power spectrum of curvature fluctuations at pre-recombination times, fitted by curves with 1, 2 or 3 anchor points, and shown as a function of angular scales. As one can clearly see, over the range of modes where the statistical significance is high, the spectrum is nearly flat. Figure reproduced with permission from Planck Collaboration; Akrami et al. (2020b). *Credit: ESA/Planck Collaboration.*

Now that we understand the physics, at least qualitatively, we can take a look at the result of evolving the angular power spectrum back to pre-recombination times, more precisely back to when even the largest l modes shown were still larger than the horizon at the time. This is shown in Fig. 3.2 (the figure shows the spectrum of curvature perturbations, which are directly proportional to temperature fluctuations), where the resulting primordial spectrum is fitted with curves anchored at 1, 2, or 3 points. The spread between the curves reflects the uncertainty in observations. What is striking is that we now see that over the entire range of scales where the uncertainty in the reconstruction is small, the primordial spectrum was nearly flat. Hence the flatness seen in the plateau at small l in fact extended to much larger wavenumbers at early times. There is a slight *red tilt* to the spectrum, by which we mean that there is a little more power on large scales. We will see the significance of this feature later in this chapter.

Of course, the CMB is a projection onto the sky, and as such it is a 2-dimensional map. When we will calculate the behavior of early universe fluctuations, they will evolve in 3 dimensions, so in comparing theory and observations one must take into account this projection – see Fig. 3.3. It may be useful at this point to introduce the actual quantities that observational cosmologists typically use.

Figure 3.3 A 3-dimensional perturbation with wavenumber k leads to oscillating perturbations that freeze out at recombination. The observer in the center, a few billion years later, observes the projection thereof onto the 2-dimensional spherical sky, where the perturbation now has angular wavenumber l. The two concentric circles mark the beginning and end of recombination, which occurs almost instantaneously. The angle subtended on the sky depends on the geometry, and hence composition, of the universe.

It turns out that at very early times, the most useful quantity to characterize the departure from a purely flat FLRW universe is the *curvature perturbation* $\mathcal{R}(x^\mu)$. We will define it rigorously in Section 3.2.1, but as the name suggests, it can be thought of as a location- and time-dependent perturbation in the curvature of the universe. Thus it is a meaningful quantity even at times when there might not have been any radiation present in the universe. However, when radiation is present, then a slight perturbation in curvature corresponds to a small change in the local expansion history, and thus also to a local fluctuation in the temperature (given that the radiation cools as the universe expands). Hence, a primordial curvature perturbation induces a temperature perturbation in the CMB.[2] We can decompose the curvature perturbation in Fourier space as

$$\mathcal{R}_\mathbf{k}(t) = \int d^3x \, \mathcal{R}(x^\mu) \, e^{-i\mathbf{k}\cdot\mathbf{x}}, \qquad (3.5)$$

and then define its power spectrum $P_\mathcal{R}$ via

$$\langle \mathcal{R}_\mathbf{k} \mathcal{R}_{\mathbf{k}'} \rangle \equiv (2\pi)^3 \delta(\mathbf{k} + \mathbf{k}') P_\mathcal{R}(k). \qquad (3.6)$$

The spatial isotropy of the background means that the power spectrum depends only on the magnitude $k = |\mathbf{k}|$. The relationship between the power spectrum of the curvature perturbations and the angular power spectrum of the CMB temperature fluctuations is then expressed via the definition

$$C_l^{\mathrm{TT}} \equiv \int \frac{d^3k}{(2\pi)^3} \, P_\mathcal{R}(k) \, \mathcal{T}_l^2(k), \qquad (3.7)$$

where $\mathcal{T}_l(k)$ is the *transfer function*, which includes both the evolution of the fluctuations prior to recombination (i.e., the acoustic oscillations), as well as the transport and projection effects in between recombination and now. The important point is that the transfer function depends entirely on known and well-tested plasma physics. Its detailed form is usually computed numerically, which involves solving the full Boltzmann equations describing the various matter components in the early universe and their interactions. This knowledge of the transfer function is used in reconstructing the primordial spectrum, as shown in Fig. 3.2.

It is very useful to define the *variance* $\Delta_\mathcal{R}^2$ and the *spectral index* n_s of the curvature perturbations as

$$\Delta_\mathcal{R}^2 \equiv \frac{k^3}{2\pi^2} P_\mathcal{R}, \qquad n_s - 1 \equiv \frac{d \ln \Delta_\mathcal{R}^2}{d \ln k}. \qquad (3.8)$$

A value of $n_s = 1$ corresponds to a *scale-invariant* spectrum – this corresponds to having equal power in each logarithmic interval of scales k, as can be seen by performing the Fourier integral to the real space correlation function

$$\langle \mathcal{R}^2(\mathbf{x}) \rangle = \int \frac{d^3k}{(2\pi)^3} \, P_\mathcal{R}(k) = \int d \ln k \, \Delta_\mathcal{R}^2, \qquad (3.9)$$

keeping in mind that $d^3k = 4\pi k^2 dk$. Because it fits the data well, one often

---

[2] On large scales one has $\delta \mathcal{T}/\mathcal{T} = \mathcal{R}/3$. There is a subtlety here: When the universe expands a little less, then this gives rise to an overdense region, which is thus hotter than average at recombination. However, to reach us today, the fluctuation has to climb out of its gravitational potential well, and this effect overcompensates, so that the fluctuation eventually appears as a cold spot on the CMB sky.

assumes a power-law form

$$\Delta_{\mathcal{R}}^2 = A_{\mathcal{R}} \left(\frac{k}{\tilde{k}}\right)^{n_s - 1}, \qquad (3.10)$$

where $A_{\mathcal{R}}$ is the *amplitude* and $\tilde{k}$ is a reference scale called the *pivot scale*. The data from the Planck satellite then provides the following measurements at the pivot scale $\tilde{k} = 0.05\,\mathrm{Mpc}^{-1}$ (here we are assuming that the scale factor is normalized such that its present-day value is $a_0 = 1$)

$$A_{\mathcal{R}} = (2.09 \pm 0.04) \times 10^{-9}, \qquad (3.11)$$
$$n_s = 0.965 \pm 0.005, \qquad (3.12)$$

where the errors are quoted at 1σ – see Planck Collaboration; Akrami et al. (2020b). Thus the spectrum is found to be very close to scale-invariant, but slightly *red*, meaning that there is more power on larger scales (smaller k). A major goal of early universe cosmology is to find an explanation for the existence and the properties of these primordial curvature perturbations.

We should also discuss (at least heuristically) another aspect of the CMB radiation, namely its polarization. The polarization is produced right at last scattering, due to Thomson scattering of the light off electrons. The electrons have a velocity $\mathbf{v_e}$ that, due to the tight coupling to photons, leads to a dipole in the photon distribution. Moreover, gradients in the velocity field lead to a quadrupole, which in turn generates the polarization. This polarization is usefully described via an intensity matrix I_{ij} in the plane perpendicular to propagation, with

$$I_{ij} = \begin{pmatrix} T + Q & U \\ U & T - Q \end{pmatrix}. \qquad (3.13)$$

Here T denotes the temperature anisotropy, while Q and U are the Stokes parameters, and they can be decomposed into spin-2 spherical harmonics

$$(Q \pm iU)(\hat{n}) = \sum_{l,m} a_{\pm 2, lm}\, {}_{\pm 2}Y_{lm}(\hat{n}). \qquad (3.14)$$

Particular linear combinations of these coefficients are then called E-mode polarization (symmetric under parity transformations) and B-mode polarization (antisymmetric under parity transformations),

$$a_{E,lm} = -\frac{1}{2}(a_{2,lm} + a_{-2,lm}), \qquad a_{B,lm} = -\frac{1}{2i}(a_{2,lm} - a_{-2,lm}). \qquad (3.15)$$

On top of the temperature–temperature cross-correlation described earlier in this section, we therefore have more possibilities, as we can consider all of the following angular 2-point functions

$$C_l^{XY} = \frac{1}{2l+1} \sum_m \langle a_{X,lm}^* a_{Y,lm} \rangle, \quad (3.16)$$

where now $X, Y = T, E, B$ (although the TB and EB combinations vanish due to symmetry/antisymmetry). These spectra are useful, because scalar perturbation modes produce only E-modes, while tensor perturbations (gravitational waves) produce both E-modes and B-modes. Thus, a detection of *primordial B-modes* in the CMB would be strong evidence for the presence of gravitational waves in the early universe.

By analogy with scalar perturbations, for tensor perturbations the amplitude $\mathcal{A}_t$ and the spectral index n_t are conventionally defined in terms of the variance Δ_h^2 and power spectrum P_h (which will be defined precisely later in this chapter) via

$$\Delta_h^2(k) \equiv \frac{k^3}{2\pi^2} P_h = \mathcal{A}_t \left(\frac{k}{k_\star} \right)^{n_t}, \quad (3.17)$$

where $n_t = 0$ corresponds to a scale-invariant spectrum (note the difference with the corresponding definition for scalar modes, where $n_s = 1$ characterizes a scale-invariant spectrum). It is conventional to also define the *tensor-to-scalar ratio* r as the ratio of the amplitude of tensor fluctuations to that of scalar fluctuations, at a specified pivot scale,

$$r(k_\star) \equiv \frac{\mathcal{A}_t}{\mathcal{A}_\mathcal{R}}. \quad (3.18)$$

The nondetection of a primordial BB cross-correlation by Planck and the Background Imaging of Cosmic Extragalactic Polarization (BICEP) observations puts an upper bound on r of

$$r < 0.056 \quad (\text{at } 95\% \text{ confidence level}), \quad (3.19)$$

at the pivot scale $k_\star = 0.002 \, \text{Mpc}^{-1}$ (Planck Collaboration; Akrami et al. (2020b)). Thus we already know that gravitational waves, at least in the span of wavelengths relevant to CMB physics, could not have played a significant role in the times leading up to recombination.

Of special interest is the temperature–E-mode cross-correlation spectrum, shown in the lower panel of Fig. 3.1. In the figure, one can make out a similar pattern of peaks and troughs to the pure temperature auto-correlation. But there is an additional feature, which is surprising: There is a trough at $l \approx 100$, i.e., on scales that were larger than the horizon at recombination.

Its origin can be understood as follows: The temperature fluctuation corresponds to a fluctuation in the energy density of the photon-baryon fluid, while the E-mode polarization is sourced by the electron velocity just before last scattering. The equation of continuity $\dot{\rho}+\nabla\cdot(\rho\mathbf{v_e}) = 0$ then implies that the electron velocity behaves essentially as the time derivative of the density, and thus around the first peak in the temperature, the E-mode is vanishingly small, and the cross-correlation between the two is especially small. On slightly smaller scales, there is an anticorrelation, causing the trough between $l \approx 100$ and $l \approx 200$. Thus this trough provides direct evidence that the primordial perturbations were already in place before recombination, on superhorizon scales. No causal physics could have generated such a correlation during the hot big bang phase leading up to recombination, since during that phase the comoving horizon was steadily growing, and thus smaller than at recombination.

If we want a causal explanation for their origin, then there must have been an earlier period, preceding the hot big bang evolution, during which the comoving horizon was larger than at recombination. As we saw in Chapter 2, inflation (or ekpyrosis) could have provided just such a mechanism. We will see below that inflation indeed has the capacity of generating precisely the kind of primordial spectra we have just described, by amplifying small quantum fluctuations in the inflationary expansion. Before turning to that calculation, we will briefly reinforce the point that classical statistical fluctuations cannot be considered as the origin of the primordial fluctuations.

3.1.1 Historical aside: the need for seeds

There is a lot of stuff in the universe: stars, dust, galaxies, etc. Given that all this matter is electrically neutral, the only force that affects it significantly on large scales is gravity. This leads to the idea that the bound structures in the universe have formed via gravitational collapse. One may then wonder whether all the large-scale structure in the universe may have formed via gravitational collapse of random, statistical fluctuations in the matter content of the universe – i.e., whether the structure that we see is an inevitable consequence of having (classical) matter and gravity. To test this idea, we must look at how density perturbations evolve under the influence of gravity.

We will first analyze the simple case of matter in a static universe, and we will approximate gravity by its Newtonian version, which, on small scales and for sources not too large, is a good approximation. Assuming a perfect fluid description, the matter is described by its energy density $\rho(\mathbf{x},t)$, pressure $p(\mathbf{x},t)$, and velocity 3-vector $\mathbf{v}(\mathbf{x},t)$. We will denote the gravitational potential by Φ, and it is determined via the Poisson equation

$$\nabla^2 \Phi = 4\pi G \rho, \tag{3.20}$$

where we have reinstated Newton's constant. We also have the equation of continuity (matter conservation)

$$\frac{\partial \rho}{\partial t} = -\nabla \cdot (\rho \mathbf{v}), \tag{3.21}$$

and the Euler equation (this is the equation for the force on a small matter element, due to pressure and gravity)

$$\frac{d\mathbf{v}}{dt} = \frac{\partial \mathbf{v}}{\partial t} + (\mathbf{v} \cdot \nabla)\mathbf{v} = -\frac{\nabla p}{\rho} - \nabla \Phi. \tag{3.22}$$

Now we can look at perturbations. We will assume that the perturbations are adiabatic, i.e., that the pressure perturbation depends only on the energy density perturbation (and not on additional entropy perturbations),

$$\delta p = c_s^2 \delta \rho, \qquad c_s^2 \equiv \frac{\partial p}{\partial \rho}|_S, \tag{3.23}$$

where c_s is the speed of sound. Then by taking the divergence of the perturbed Euler equation, assuming a static universe (i.e., constant background energy density) and substituting the perturbed continuity and Poisson equations, we get an equation purely for the perturbed density $\delta \rho$,

$$\frac{\partial^2 \delta \rho}{\partial t^2} - c_s^2 \nabla^2 \delta \rho - 4\pi G \rho \delta \rho = 0. \tag{3.24}$$

Since the equation is linear, it is convenient to solve it in Fourier space,

$$\delta \rho(\mathbf{x}, t) = \int \frac{d^3 k}{(2\pi)^3} e^{i\mathbf{k} \cdot \mathbf{x}} \delta \rho_\mathbf{k}, \tag{3.25}$$

where it becomes

$$\frac{\partial^2 \delta \rho_\mathbf{k}}{\partial t^2} + \left(c_s^2 k^2 - 4\pi G \rho \right) \delta \rho_\mathbf{k} = 0. \tag{3.26}$$

The behavior of the solutions crucially depends on the wavelength $\lambda = 2\pi/k$. Let us define the *Jeans length* as

$$\lambda_J \equiv c_s \sqrt{\frac{\pi}{G\rho}}. \tag{3.27}$$

For short wavelengths $\lambda < \lambda_J$, the pressure dominates and the solutions

3.1 Observed fluctuations in the CMB

are oscillatory. But for wavelengths longer than the Jeans length, gravity dominates and we have a solution of the form

$$\delta\rho_{\mathbf{k}} \sim e^{\pm\omega t}, \qquad \omega = \sqrt{4\pi G\rho\left(1 - \frac{\lambda_J^2}{\lambda^2}\right)}. \qquad (3.28)$$

Thus in general a small long-wavelength perturbation will grow exponentially fast. If we have a large number N of particles, then statistical physics tells us to expect fluctuations on the order of $\delta\rho/\rho \sim N^{-1/2}$. For a (large) star, we have $N \approx 10^{60}$, so the question is whether a fluctuation of size $\delta\rho/\rho \approx 10^{-30}$ could collapse into an actual star under gravity. The time scale for collapse is $t_{\text{collapse}} \approx 1/\sqrt{4\pi G\rho}$. In the early universe, say around the time that the first atoms formed, the universe was a thousand times smaller in length than now, so that the density was about $\rho \approx 10^{-17}$ kg·m^{-3} (the current density is about one hydrogen atom per cubic meter), leading to a collapse time scale of about a million years. Thus an initial perturbation of $10^{-30} \approx e^{-70}$ could have grown to order unity in about 70 million years. At first, this sounds very promising, but there is one crucial effect that we have neglected: the expansion of the universe.

In an expanding universe, the velocities will obey the Hubble law,

$$\mathbf{v} = H(t) \cdot \mathbf{x}, \qquad (3.29)$$

where H is the Hubble rate. The equation of continuity then implies that the background energy density evolves as

$$\dot\rho = -3H\rho, \qquad (3.30)$$

which agrees with the relativistic equation for pressure-free matter. Combining the Euler and Poisson equations then also leads to the Friedmann equation

$$3\dot H + 3H^2 = -4\pi G\rho. \qquad (3.31)$$

If we now repeat the calculations performed above, we find the following Fourier space equation for the density contrast $\delta \equiv \frac{\delta\rho}{\rho}$,

$$\frac{\partial^2 \delta_{\mathbf{k}}}{\partial t^2} + 2H\frac{\partial \delta_{\mathbf{k}}}{\partial t} + \left(c_s^2 \frac{k^2}{a^2} - 4\pi G\rho\right)\delta_{\mathbf{k}} = 0, \qquad (3.32)$$

where k is now a co-moving wavenumber (the physical wavenumber being k/a). During matter domination, the scale factor of the universe grows as $a(t) \propto t^{2/3}$, so that the expansion rate is $H = \dot a/a = 2/(3t)$. Also, the energy density is given by the Friedmann equation $3H^2 = 8\pi G\rho$, so that

$\rho = 1/(6\pi G t^2)$. For long-wavelength perturbations, i.e., small wavenumber k, we have the approximate equation

$$\frac{\partial^2 \delta_{\mathbf{k}}}{\partial t^2} + \frac{4}{3t}\frac{\partial \delta_{\mathbf{k}}}{\partial t} - \frac{2}{3t^2}\delta_{\mathbf{k}} = 0, \qquad (3.33)$$

which is solved by

$$\delta = c_1 t^{2/3} + \frac{c_2}{t}, \qquad (3.34)$$

with integration constants c_1, c_2. (This result agrees precisely with more involved, fully relativistic calculations.) Now we see that the growing mode evolves as a power law, and in fact the growing mode has $\delta \propto a(t)$. Since the universe grew by a factor of about 1 000 since the first atoms formed, perturbations only grew by this factor also – i.e., a perturbation of order unity now was of order 10^{-3} at the beginning of the matter phase. This is an enormously slower growth than what one obtains by neglecting the expansion of the universe. In an expanding universe it is thus unreasonable to think that the perturbations grew out of standard statistical fluctuations. The situation does not improve by going further back in time, because during the preceding radiation-dominated era there were also dissipation effects at work. Hence we conclude that something must have caused fluctuations to already exist in the very early universe – i.e., we need some new physics to explain the existence of primordial density perturbations. This new physics can be provided by quantum fluctuations during inflation, as we will discuss in what follows.

3.2 Cosmological perturbation theory

The proper treatment of cosmological fluctuations is technically somewhat involved. Before presenting it in detail, it may be useful to understand why, at the heuristic level, inflationary perturbations might play such an important role. Consider then fluctuations in the scalar field,

$$\phi(t, \vec{x}) = \bar{\phi}(t) + \delta\phi(t, \vec{x}), \qquad (3.35)$$

where $\bar{\phi}$ denotes the unperturbed, homogeneous solution, and $\delta\phi$ is a (local) perturbation that depends on both time and space. Since the scalar field drives inflation, small fluctuations lead to slightly different expansion histories in different regions (reheating will happen at different times in different regions), thus some regions expand a little more and are a little colder, while other regions expand a little less and are still hotter. Hence such fluctuations in the scalar field can lead to primordial temperature fluctuations. But why are there fluctuations in the scalar field in the first place? As we will see, the origin for these fluctuations comes from quantum fluctuations, which are necessarily present because of the uncertainty principle. Note that under more ordinary circumstances, say in your living room, quantum fluctuations do not get amplified into classical perturbations affecting the expansion of spacetime. This is something rather special, which inflation manages to achieve. To see whether these fluctuations also match the statistical distribution of fluctuations seen in the CMB, we must do a detailed calculation.

To perform this calculation, we must first introduce the formalism for classical perturbations before turning to a quantum treatment. The main complication arises because in general relativity we have the freedom to change coordinates, so that one must clarify what constitutes a physical fluctuation and what represents just a different choice of coordinates. In this section, we will assume that the fluctuations we are dealing with are small, so that we can work to linear order in these fluctuations.

We start by looking at perturbations of the metric $g_{\mu\nu} = \bar{g}_{\mu\nu} + \delta g_{\mu\nu}$, where $\bar{g}_{\mu\nu}$ denotes the background spacetime. The most general perturbed Robertson–Walker metric can be written as

$$\mathrm{d}s^2 = -(1+2A)\mathrm{d}t^2 + 2a(t)(B_{,i} + G_i)\mathrm{d}x^i\mathrm{d}t + a^2(t)[\delta_{ij} + h_{ij}]\mathrm{d}x^i\mathrm{d}x^j, \quad (3.36)$$

with

$$h_{ij} = 2\psi\delta_{ij} + 2\partial_i\partial_j E + \partial_i F_j + \partial_j F_i + \gamma_{ij}, \quad (3.37)$$

and where we impose the additional conditions

$$\partial^i F_i = 0, \quad \partial^i G_i = 0, \quad \gamma^i_i = 0, \quad \partial^i \gamma_{ij} = 0. \quad (3.38)$$

The perturbations can be separated into three categories, according to their transformation properties with respect to spatial, three-dimensional, rotations (encoded in their spatial $i, j, \ldots$ index structure):

- scalar – no index: A, ψ, B, E
- vector – one i index, divergence-free: F_i, G_i
- tensor – symmetric ij indices, transverse and traceless: γ_{ij}

The conditions (3.38) ensure that there is no scalar component present in the vector perturbations (this is the divergence-free condition $\partial^i F_i = 0$, $\partial^i G_i = 0$, which, due to the summation over indices, is a scalar condition), and no scalar nor vector in the tensor perturbations (these are the conditions of tracelessness $\gamma^i_i = 0$ and transversality $\partial^i \gamma_{ij} = 0$, respectively). Vector perturbations quickly decay in all the models we will consider in this book, hence we will ignore them and focus only on scalar and tensor modes.

3.2.1 Dealing with gauge transformations

We consider a small local change in the coordinates,

$$x^\mu \to x'^\mu = x^\mu + \xi^\mu, \tag{3.39}$$

where the vector ξ^μ can be decomposed as $\xi^\mu = (\xi^0, \xi^i)$ with $\xi^i = \xi_T^i + \partial^i \xi$. Here ξ is a scalar and we impose $\partial_i \xi_T^i = 0$ such that ξ_T^i is a divergence-free 3-vector. Thus ξ^0 and ξ are the two scalar transformation parameters, and since we disregard vector perturbations here we will not need to consider ξ_T^i.

In general, under such a coordinate transformation the fields and their perturbations change. It turns out to be convenient to look at *gauge transformations*, where the background fields remain unperturbed and the entire change is accounted for by the field perturbations. Thus, for scalar and tensor quantities we write

$$s(x^\mu) \to s'(x^\mu) = s(x^\mu) + \Delta s(x^\mu), \tag{3.40}$$
$$t_{\rho\sigma}(x^\mu) \to t'_{\rho\sigma}(x^\mu) = t_{\rho\sigma}(x^\mu) + \Delta t_{\rho\sigma}(x^\mu). \tag{3.41}$$

Tensor quantities are defined by $t'_{\rho\sigma}(x'^\mu) = t_{\lambda\kappa}(x^\mu)\frac{\partial x^\lambda}{\partial x'^\rho}\frac{\partial x^\kappa}{\partial x'^\sigma}$, with analogous definitions for tensors with different numbers of indices (note that the definition guarantees the invariance of the spacetime interval $ds^2 = g_{\mu\nu}dx^\mu dx^\nu$). It follows that scalars and tensors transform as

$$\Delta s(x) = s'(x) - s(x) = s'(x') - \xi^\mu \partial_\mu s(x) - s(x) = -\xi^\mu \partial_\mu s(x), \tag{3.42}$$
$$\begin{aligned}\Delta t_{\rho\sigma}(x) &= t'_{\rho\sigma}(x) - t_{\rho\sigma}(x) \\ &= t'_{\rho\sigma}(x') - \xi^\mu(x)\partial_\mu t_{\rho\sigma}(x) - t_{\rho\sigma}(x) \\ &= -\xi^\mu(x)\partial_\mu t_{\rho\sigma}(x) - t_{\lambda\sigma}(x)\partial_\rho\left(\xi^\lambda(x)\right) - t_{\lambda\rho}(x)\partial_\sigma\left(\xi^\lambda(x)\right)\end{aligned} \tag{3.43}$$

under gauge transformations. Note the extra "transport" terms, $-\xi^\mu \partial_\mu s(x)$ and $-\xi^\mu(x)\partial_\mu t_{\rho\sigma}(x)$, which pull the transformed scalar and tensor quantities back to the x^μ coordinate position, so that we can compare them with their original selves. This allows us to calculate how the various components of the metric (3.36) transform, namely

$$A \to A + \dot\xi_0, \tag{3.44}$$
$$B \to B + \frac{1}{a}(-\xi_0 - \dot\xi + 2H\xi), \tag{3.45}$$
$$\psi \to \psi + H\xi_0, \tag{3.46}$$

3.2 Cosmological perturbation theory

$$E \to E - \frac{1}{a^2}\xi, \qquad (3.47)$$

$$\gamma_{ij} \to \gamma_{ij}. \qquad (3.48)$$

Note that the tensors γ_{ij} are left unchanged. It is thus simpler to work with tensor perturbations than with scalar perturbations. Nevertheless, we will first treat the scalar perturbations in detail and return to the tensor perturbations later in this Chapter, whose treatment will then be rather straightforward.

Even though the scalar metric perturbations transform under gauge transformations, one can find *gauge-invariant* quantities that do not. Noting that the combination $B - a\dot{E}$ only changes in proportion to ξ_0, one can form the well-known Bardeen potentials

$$\Phi = A + \frac{d}{dt}[a(B - a\dot{E})], \qquad (3.49)$$

$$\Psi = \psi - aH(B - a\dot{E}), \qquad (3.50)$$

which do not transform (you may check that the various transformations of the single perturbation variables cancel each other out).

The fact that the individual perturbations change under a general coordinate transformation, as shown in Eqs. (3.44) – (3.47), is problematic: It means that these perturbations do not give us physically meaningful quantities.

If we want to make sure that we are talking about physical quantities only, there are two options:

1. Work with gauge-invariant quantities.
2. Fix a gauge, i.e., fix the initially arbitrary ξ^0, ξ functions in such a way that the perturbations become unambiguous.

Both methods yield sensible results, and it is typically a matter of convenience which method one uses. It also often happens that, when working in a particular fixed gauge, one arrives at equations that happen to involve only gauge-invariant quantities. In that case these equations are then automatically valid in all gauges.

A useful example of a gauge choice is *comoving gauge* in which, even in the perturbed spacetime, one is moving along with the background matter field. For the case of scalar field driven inflation, in the perturbed spacetime surfaces of constant time are also surfaces of constant scalar field. This can be achieved by choosing coordinates that are such that the scalar field effectively remains unperturbed. Eq. (3.42) implies that a scalar field perturbation transforms under a change of time slicing ($t \to t + \xi^0 \equiv t + \delta t$)

as
$$\delta\phi \to \delta\phi - \dot{\phi}\delta t\,. \tag{3.51}$$

In order to stay in comoving gauge,
$$\delta\phi_{\text{com}} = 0\,, \tag{3.52}$$

we can see that we need to restrict coordinate changes such that
$$\delta t = \frac{\delta\phi}{\dot{\phi}}\,. \tag{3.53}$$

Thus, at least as long as $\dot{\phi} \neq 0$, it is possible to find coordinates such that the scalar field remains unperturbed, while all the perturbations reside in the metric. We will make use of this gauge below (and define it more accurately) to calculate the action and equation of motion of a very important quantity, namely the comoving curvature perturbation. Keeping in mind that
$$\psi \to \psi - H\delta t\,, \tag{3.54}$$

we can see that
$$\mathcal{R} \equiv \psi - H\frac{\delta\phi}{\dot{\phi}} \tag{3.55}$$

has an invariant meaning. This is the *comoving curvature perturbation*. With a more general matter content it is defined via
$$\mathcal{R} \equiv \psi + \frac{2H}{\rho + p}\delta q\,, \tag{3.56}$$

where δq denotes the perturbation of the off-diagonal part of the stress–energy tensor, $\delta T_i^0 = \partial_i \delta q$. In comoving gauge we simply have $\mathcal{R}_{\text{com}} = \psi$ and thus in that gauge the comoving curvature perturbation specifies the fluctuations in the scale factor (on constant-ϕ surfaces). But a change in the scale factor induces a change in the curvature, hence the name. The comoving curvature perturbation is important because under certain conditions it is conserved on very large scales, and moreover it is directly related to what we observe in the CMB. We will discuss this variable in detail in Section 3.2.3.

Another closely related perturbation variable is the *curvature perturbation on uniform-density hypersurfaces*. It is defined via
$$\zeta \equiv \psi - \frac{H}{\dot{\rho}}\delta\rho, \tag{3.57}$$

where $\rho = \tfrac{1}{2}\dot{\phi}^2 + V$ denotes the energy density. Note that this quantity is gauge invariant since under a gauge transformation $\delta\rho \to \delta\rho - \dot\rho\delta t$. On hypersurfaces of constant energy density we have $\delta\rho = 0$, and then ζ corresponds to a local change in the scale factor – i.e., once again it corresponds to a perturbation in the homogeneous curvature of the universe. Note that in a spatially flat universe, by virtue of the Friedmann equation $3H^2 = \rho$, surfaces of constant energy density are also surfaces of constant Hubble rate.

In situations where spatial gradients can be neglected, in particular for very long-wavelength perturbations, we can directly relate ζ and $\mathcal{R}$: employing the scalar field equation of motion with spatial gradients dropped, $\ddot\phi + 3H\dot\phi + V_{,\phi} = 0$, we can rewrite $\delta\rho = \dot\phi\delta\dot\phi + V_{,\phi}\delta\phi = (\ddot\phi + V_{,\phi})\delta\phi = -3H\dot\phi\delta\phi$. Together with the equation of continuity $\dot\rho = -3H(\rho + p)$, we then obtain $\delta\rho/\dot\rho = \delta\phi/\dot\phi$ so that on large scales we simply have

$$\mathcal{R} \approx \zeta, \qquad \text{when} \quad \partial_i^2 \ll \partial_0^2. \tag{3.58}$$

Thus on large scales the two curvature perturbations may be used interchangeably.

There exist many more gauge choices, which are useful in different situations. A further example is *flat gauge*, in which one sets the homogeneous perturbation of the scale factor to zero and instead focuses on the scalar field perturbations. Which gauge to choose is in principle arbitrary, but depending on the circumstances and the associated physical intuition, one choice can make a calculation vastly simpler and physically more transparent than another choice. There exists no general rule here: Initially trial and error, and later experience, will guide one's choice.

3.2.2 The ADM formalism

The cosmological background solutions that we are interested in effectively split spacetime into space and time, as the background fields (the metric and the scalar field) depend on time alone. Thus surfaces of constant scalar field correspond to equal-time slices through the spacetime. In calculating the perturbations around these cosmological spacetimes, it is useful to split the metric in a similar fashion, as first described by Arnowitt, Deser, and Misner (ADM). This formalism is of general use in cosmology and will prove crucial in Chapters 4 and 5.

The ADM formalism starts by rewriting the metric in the form

$$\mathrm{d}s^2 = -N^2 \mathrm{d}t^2 + h_{ij}(\mathrm{d}x^i + N^i \mathrm{d}t)(\mathrm{d}x^j + N^j \mathrm{d}t). \tag{3.59}$$

We may understand its structure as follows. Consider two spatial hypersurfaces that are infinitesimally close together – see Fig. 3.4. Going up from

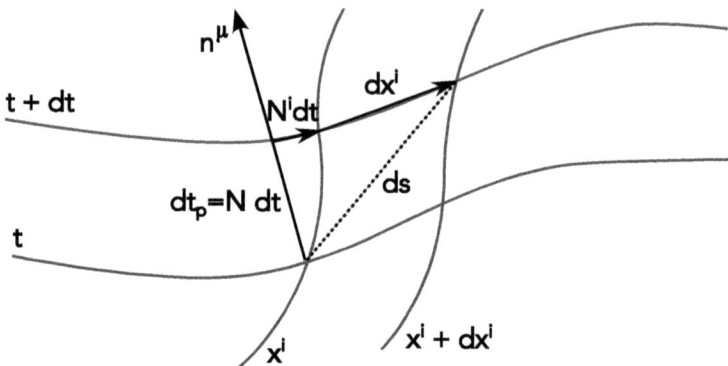

Figure 3.4 The ADM decomposition of the metric, describing the spacetime interval between the two hypersurfaces at t and $t + \mathrm{d}t$.

the bottom to the top hypersurface takes a certain amount of proper time, which we define to be $\mathrm{d}s^2 = -\mathrm{d}t_p^2 = -N^2\mathrm{d}t^2$. Here $N(t, x^i)$ is the *lapse function*, so-called precisely because it tells us how much proper time elapsed between being at $x^0 = t$ and $x^0 = t + \mathrm{d}t$, going up along the unit normal n^μ to the surface at constant t. This vector satisfies $n^\mu n_\mu = -1$ and has a simple covariant form, $n_\mu = (-N, 0, 0, 0)$. The metric on a hypersurface is given by h_{ij} and it allows us to measure distances inside that hypersurface. Now, the spatial coordinates can shift when going from the bottom to the top hypersurface. Starting at x^i on the hypersurface at $x^0 = t$ and going up along the normal, we will not end up at x^i in general, but rather at a shifted location $x^i - N^i\mathrm{d}t$. For this reason $N^i(t, x^i)$ is called the *shift vector*. When $N^i = 0$, we have comoving coordinates. The full ADM metric (3.59) is then the spacetime interval (Pythagorean theorem in a pseudo-Riemannian manifold) resulting from these displacements.

At this point the metric is still completely general, and we can re-express the action,

$$S = \int \mathrm{d}x^4 \sqrt{-g}[\frac{1}{2}R - \frac{1}{2}g^{\mu\nu}\partial_\mu\phi\partial_\nu\phi - V(\phi)], \tag{3.60}$$

in terms of these variables. For the remainder of this section, we will also keep ϕ general, i.e., not restricting it to a function of time alone. Note that $\sqrt{-g} = N\sqrt{h}$. We can relate the 4-dimensional Ricci scalar R to the 3-dimensional Ricci scalar $R^{(3)}$ via the Gauss–Codazzi equation (which we will not derive here)

$$R = R^{(3)} + K^{ij}K_{ij} - K^2, \tag{3.61}$$

where K_{ij} is the extrinsic curvature, i.e., it is the curvature of the 3-dimensional hypersurfaces of constant time, as embedded in the 4-dimensional spacetime. In other words, the extrinsic curvature describes how curved the constant-time slices look from the 4-dimensional point of view. To describe this, one makes use of the unit normal vector n^μ to the constant-t surfaces. The extrinsic curvature then expresses how this vector changes as it is transported along a constant-t surface,

$$K_{ij} \equiv -\nabla_i n_j = \Gamma^0_{ij} n_0 = \frac{1}{2N}\left(-\dot{h}_{ij} + D_i N_j + D_j N_i\right), \qquad (3.62)$$

$$K \equiv K^i_i = h^{ij} K_{ij}, \qquad (3.63)$$

where D_i is the covariant derivative for the spatial metric h_{ij}. In terms of the rescaled field

$$E_{ij} = N K_{ij}, \qquad (3.64)$$

the action then becomes

$$S = \frac{1}{2}\int d^4x \sqrt{h} N \left(R^{(3)} - h^{ij}\partial_i\phi\partial_j\phi - 2V\right) \\ + \frac{1}{2}\int d^4x \sqrt{h} N^{-1}\left(E_{ij}E^{ij} - E^2 + (\dot{\phi} - N^i\partial_i\phi)^2\right). \qquad (3.65)$$

In the ADM formalism, the 3-metric h_{ij} is considered to be the dynamical field. The momentum conjugate to h_{ij} is essentially the extrinsic curvature,

$$\pi^{ij} = \frac{\partial \mathcal{L}}{\partial \dot{h}_{ij}} = -\sqrt{h}\left(K^{ij} - K h^{ij}\right). \qquad (3.66)$$

The lapse and shift appear without time derivatives, i.e., as Lagrange multipliers. Their variation leads to constraints,

$$R^{(3)} - h^{ij}\partial_i\phi\partial_j\phi - 2V = \frac{1}{N^2}\left(E_{ij}E^{ij} - E^2 + (\dot{\phi} - N^i\partial_i\phi)^2\right), \qquad (3.67)$$

$$D_i\left[\frac{1}{N}(E^i_j - \delta^i_j E)\right] = \frac{\dot{\phi} - N^i\partial_i\phi}{N}\partial_j\phi. \qquad (3.68)$$

These constraints, equivalent to the 0μ Einstein equations, will play an important role in the treatment of cosmological perturbations below, and an even more prominent role in the canonical quantization of gravity.

The remaining Einstein equations are obtained by varying the action with respect to the 3-metric. They are somewhat cumbersome in the $1+3$ split, and in practice it is typically easier (and equivalent) to work with the standard 4-dimensional space-space Einstein equations (2.15), as well as the scalar equation of motion (2.16). For now, this is all we need; we will develop the ADM formalism further in Chapter 4.

3.2.3 The comoving curvature perturbation

Our first goal here is to calculate the action for the comoving curvature perturbation (3.55). Thus we will now revert to background fields $a(t), \phi(t)$ being functions of time only, and choose comoving gauge, in which the inflaton fluctuation vanishes

$$\delta\phi = 0. \tag{3.69}$$

The scalar degree of freedom is represented by the comoving curvature perturbation $\mathcal{R}(x^\mu)$. The metric on spatial hypersurfaces is then given by

$$h_{ij} = a^2[(1 + 2\mathcal{R})\delta_{ij} + \gamma_{ij}]. \tag{3.70}$$

Note that this gauge is fully fixed: ξ^0 is chosen to remove $\delta\phi$ and ξ to remove E, while A is absorbed into the lapse function and B into the shift. First, we drop the tensor fluctuations γ_{ij} – at linear order, they evolve independently, and thus we can deal with them separately later on. Note that in comoving gauge all the perturbations are in the metric, and time is moving along with the background scalar field.

In the ADM formalism, we saw that h_{ij} is the dynamical field while the lapse and shift functions lead to constraints. To linear order the constraints (3.67) and (3.68) are solved by

$$N = 1 + \frac{\dot{\mathcal{R}}}{H}, \tag{3.71}$$

$$N_i = -\frac{\partial_i \mathcal{R}}{H} + a^2 \frac{\dot{\phi}^2}{2H^2} \partial_i \left((\partial_j \partial^j)^{-1}\dot{\mathcal{R}}\right), \tag{3.72}$$

where $(\partial_j \partial^j)^{-1}$ is defined such that $(\partial_j \partial^j)^{-1}(\partial_k \partial^k)\mathcal{R} = \mathcal{R}$. In verifying this solution it is useful to know the linear order relation $R^{(3)} = -4\partial_j \partial^j \mathcal{R} = 4k^2\mathcal{R}$, which incidentally shows that the physical meaning of the gauge-invariant quantity $\mathcal{R}$ is indeed that it describes the local change in the *intrinsic* curvature of the spatial hypersurfaces.

We can now plug these expressions into the action (3.65) and write it out explicitly up to second order. Since we are interested in the linear equations of motion, we must work out the action up to second order, as the equation of motion is obtained after varying with respect to the linear variable. We may also simplify the action by making use of the background equations of motion since, at the perturbed level, background quantities are considered as external fixed functions. The calculation of the quadratic action, which you are strongly urged to do by yourself, is fairly long. After several integrations by parts one finally obtains a surprisingly simple result,

$$S_2 = \int d^3x dt\, \epsilon \left[a^3 \dot{\mathcal{R}}^2 - a(\partial_i \mathcal{R})^2 \right]. \tag{3.73}$$

Note the overall factor of the slow-roll parameter ϵ, which suppresses the fluctuations in the slow-roll regime. The rest of the action may be recognized as that of a massless scalar field in an FLRW background.

The equation of motion resulting from (3.73) is

$$\ddot{\mathcal{R}} + \left(3H + \frac{\dot{\epsilon}}{\epsilon} \right) \dot{\mathcal{R}} - \frac{1}{a^2} \partial_i \partial^i \mathcal{R} = 0. \tag{3.74}$$

Interestingly, the potential of the theory does not appear explicitly. However, the evolution of the scale factor a is of course determined via the potential, and in this indirect way the potential determines the evolution of $\mathcal{R}$.

The great usefulness of the comoving curvature perturbation is not only due to the simplicity of its action and equation of motion, but due to its properties on large scales. If we consider a fluctuation with a very long wavelength $a/k \gg 1/H$ (a wavelength larger than the horizon), then the gradient term in the equation of motion (3.74) may be neglected. In this case the two solutions to the equation are

$$\mathcal{R} = \text{constant}, \qquad \mathcal{R} \propto \frac{1}{a^3 \epsilon}. \tag{3.75}$$

The second solution decays rapidly during inflation. The first solution then quickly dominates, and keeps the curvature perturbation *conserved* for as long as the perturbation is super-Hubble. This property allows one to track long-wavelength perturbations over comparatively long time scales, irrespective of what happens on smaller scales. It is this property that will allow us to connect certain processes in the early universe to present-day observations without having to know the details of all of the small-scale physics that has occurred in the meantime.

There is one caveat to this statement, which we ought to mention and which plays an important role in ekpyrotic models, and also in some multi-field inflationary models: if there is more than one matter component, for instance if there is more than one scalar field, then there exist so-called *entropy* perturbations as well. These are perturbations in the relative energy density of the various matter components, but such that the total energy density (and thus the Hubble rate) remains unperturbed. In the presence of entropy perturbations the curvature perturbation can evolve, even on large scales. That being said, it turns out that the models favored by observations only have a single field driving inflation.

On very small scales, the gradient term is important, but the damping term $\left(3H + \frac{\dot{\epsilon}}{\epsilon}\right)\dot{\mathcal{R}}$ in the equation of motion for the curvature perturbation can be neglected. The solutions are thus oscillatory. Since small perturbations do not grow compared to the background solution, we see that the inflationary solution is classically stable. The inclusion of quantum effects, especially of large quantum fluctuations, will require a reassessment of the question of stability.

Before continuing with our analysis of the curvature perturbation at linear order (i.e., quadratic order in the action), let us mention that there is in principle no reason to stop at that order. One can straightforwardly (though lengthily) expand the full action (3.65) up to any desired order in $\mathcal{R}$,

$$S = \int \mathrm{d}^4 x \mathcal{L}[\mathcal{R}(x)] = \int \mathrm{d}^4 x \left(\mathcal{L}^{(2)}[\mathcal{R}(x)] + \mathcal{L}^{(3)}[\mathcal{R}(x)] + ... \right). \quad (3.76)$$

We just derived $\mathcal{L}^{(2)}$, which describes the free propagation of the field. At the next order, $\mathcal{L}^{(3)}$ describes self-interactions of the curvature perturbation. In inflationary models, this cubic action is suppressed by an additional factor of the slow-roll parameters, and in fact higher orders are suppressed even more. When describing the theory quantum mechanically (as we will do shortly), the path integral for the free theory is a Gaussian integral, and for this reason one says that the perturbations are highly Gaussian. The term $\mathcal{L}^{(3)}$ represents the leading *non-Gaussian* corrections – we will look at an example in Section 3.2.8. Such corrections have not been observed yet, in fact observations by the Planck satellite have put rather stringent upper bounds on their size. As the observations are continually improving, non-Gaussian corrections will eventually be observed, though it remains hard to predict the associated timescale.

A final comment: If we look back at the definition of the comoving curvature perturbation in Eq. (3.55), we see that it combines both fluctuations in the metric and fluctuations in the scalar field. This means that the perturbations of the scalar field ϕ involve gravity in an essential way, due to the gauge transformations inherent in the system (and in comoving gauge, the entire fluctuation resides in the metric). When quantizing these fluctuations, as we will do next, one is therefore quantizing fluctuations of both matter and gravity, in a specific combination. This will therefore be our first quantum gravity calculation.

3.2.4 Quantization

We are now ready for our first calculation of quantum effects. The strategy we will adopt is to treat the inflationary background as classical and fixed, while quantizing small fluctuations around this background. This approach will of course only remain consistent as long as the fluctuations remain small, and we will have to verify this assumption at the end of the calculation. For now, one might simply think of the current procedure as a sort of Born–Oppenheimer approximation, which is also useful in other branches of physics (for instance in the quantization of atoms, one can often treat the heavy nucleus as classical, with quantized electrons surrounding it).

For the quantization of the perturbations, it is convenient to define the Mukhanov–Sasaki variable

$$v \equiv z\mathcal{R} \quad, \quad z^2 \equiv a^2 \frac{\dot\phi^2}{H^2} = 2a^2 \epsilon \,. \tag{3.77}$$

Switching to conformal time $\mathrm{d}t = a\mathrm{d}\tau$ (and with conformal time derivatives denoted by primes), the action (3.73) now becomes canonically normalized,

$$S_2 = \int \mathrm{d}^3x \mathrm{d}\tau \, \mathcal{L}^{(2)}(v) = \frac{1}{2} \int \mathrm{d}^3x \mathrm{d}\tau \left[(v')^2 - (\partial_i v)^2 + \frac{z''}{z} v^2 \right] \,. \tag{3.78}$$

The momentum conjugate to the field v is

$$\pi = \frac{\partial \mathcal{L}^{(2)}}{\partial v'} = v' \,. \tag{3.79}$$

We can quantize this system, which effectively represents a scalar field with a time-dependent mass, in analogy with ordinary quantum mechanics,[3] by promoting the fields to operators and imposing the commutation relations

$$[\hat{v}(\tau, \mathbf{x}), \hat{\pi}(\tau, \mathbf{y})] = i\delta(\mathbf{x} - \mathbf{y}) \,, \tag{3.80}$$

as well as the trivial commutators

$$[\hat{v}(\tau, \mathbf{x}), \hat{v}(\tau, \mathbf{y})] = [\hat{\pi}(\tau, \mathbf{x}), \hat{\pi}(\tau, \mathbf{y})] = 0 \,. \tag{3.81}$$

Note that the commutation relations are imposed on spatial hypersurfaces of constant time – in quantum theory time plays a distinguished role compared to space, a feature we will repeatedly come back to over the course of this book.

---

[3] What is presented here is the canonical quantization. We will later encounter equivalent quantizations in the Schrödinger picture, for instance in Section 3.2.6, and using the path integral formalism, for instance in Section 7.1.2.

The action is quadratic, and will thus lead to a linear equation of motion. This implies that it is useful to expand the perturbations into Fourier modes

$$v(\tau, \mathbf{x}) = \int \frac{d^3k}{(2\pi)^3} v_\mathbf{k}(\tau) e^{i\mathbf{k}\mathbf{x}}. \tag{3.82}$$

Here $\mathbf{x}$ are comoving coordinates, and correspondingly $\mathbf{k}$ are comoving wave numbers (corresponding to fluctuations with a physical wavelength a/k). The equation of motion for each Fourier mode is then

$$v_\mathbf{k}'' + (k^2 - \frac{z''}{z}) v_\mathbf{k} = 0. \tag{3.83}$$

The linearity of the equation implies that each mode evolves independently, and there is no mode mixing. Hence we can also regard the cosmological fluctuations as a system of independent harmonic oscillators, with a time-dependent mass.

When the field is promoted to an operator, in effect all of its Fourier modes are promoted too,

$$\hat{v}(\tau, \mathbf{x}) = \int \frac{d^3k}{(2\pi)^3} \hat{v}_\mathbf{k}(\tau) e^{i\mathbf{k}\mathbf{x}}. \tag{3.84}$$

We can proceed by writing each Fourier mode operator as a linear combination of annihilation and creation operators

$$\hat{v}_\mathbf{k} = v_k(\tau) \hat{a}_\mathbf{k} + v_k^\star(\tau) \hat{a}^\dagger_{-\mathbf{k}}. \tag{3.85}$$

Here the $v_k(\tau)$ are time-dependent, complex solutions of the equations of motion (3.83), which, because of the spatial isotropy of the background, depend only on the modulus $k = |\mathbf{k}|$. The definition above implies the relation $\hat{v}_{-\mathbf{k}} = \hat{v}_\mathbf{k}^\dagger$, which ensures that the comoving curvature perturbation is real valued, since the Fourier sum (3.84) then contains sums of terms and their Hermitian conjugates.

The annihilation/creation operators can be taken to satisfy the nontrivial commutation relation

$$[\hat{a}_\mathbf{k}, \hat{a}^\dagger_{\mathbf{k}'}] = (2\pi)^3 \delta(\mathbf{k} - \mathbf{k}'), \tag{3.86}$$

with the other commutators vanishing,

$$[\hat{a}_\mathbf{k}, \hat{a}_{\mathbf{k}'}] = 0, \qquad [\hat{a}^\dagger_\mathbf{k}, \hat{a}^\dagger_{\mathbf{k}'}] = 0. \tag{3.87}$$

Inserting the expansions (3.84)–(3.85) with the commutators (3.86)–(3.87) into the quantization conditions (3.80)–(3.81), one finds that this is all consistent as long as the mode functions satisfy the normalization condition (called the *Wronskian*),

$$v_k v_k^{*\prime} - v_k^\star v_k' \equiv i. \tag{3.88}$$

This constraint, which is conserved in time by virtue of the equation of motion (3.83), provides one boundary condition for the mode functions $v_k(\tau)$. Note that we could equally well have derived the Wronskian from the Klein–Gordon inner product – see Section 6.2 – and then the commutation relations (3.86)–(3.87) would have followed.

A second boundary condition, which fixes the mode functions completely, comes from physical considerations. We will discuss these now by looking at the actual solutions in the slow-roll case.

3.2.5 Slow-roll solutions

A realistic inflationary phase must last sufficiently long, which is typically achieved by having the scalar field roll very slowly in the potential. To solve Eq. (3.83), we need to calculate the quantity z''/z, where $z^2 = 2a^2\epsilon$. Using the definitions of the slow-roll parameters and converting to conformal time, we have to linear order in slow-roll

$$\frac{z'}{z} = a\frac{\dot z}{z} = aH\left(1 - \frac{1}{2}\eta\right), \tag{3.89}$$

$$\frac{z''}{z} = (aH)^2\left(2 - \epsilon - \frac{3}{2}\eta\right), \tag{3.90}$$

$$\mathcal{H} \equiv \frac{a'}{a} = aH = -\frac{1}{\tau}(1+\epsilon). \tag{3.91}$$

The equation of motion (3.83) then becomes

$$v_k'' + \left(k^2 - \frac{\alpha^2 - \frac{1}{4}}{\tau^2}\right)v_k = 0, \quad \text{where} \quad \alpha = \frac{3}{2} + \epsilon - \frac{3}{2}\eta. \tag{3.92}$$

It is important to understand this equation qualitatively at first: At early times and/or on small scales, more precisely when $|k\tau| \gg 1$, the very last term may be neglected, so that we simply find the mode equation in Minkowski space,

$$v_k'' + k^2 v_k = 0, \quad |k\tau| \gg 1. \tag{3.93}$$

That is the equation of motion for a harmonic oscillator, with solutions proportional to $e^{\pm ik\tau}$. The standard Minkowski vacuum, i.e., the positive-frequency vacuum solution that is also normalized so as to satisfy (3.88), is then given by

$$v_k = \frac{1}{\sqrt{2k}} e^{-ik\tau}. \tag{3.94}$$

Note that in the small-scale limit the curvature of spacetime is not felt, and thus, as a physical requirement, we can impose that the mode functions should limit to the Minkowski vacuum solution in this small-scale limit, that is to say we require

$$\lim_{|k\tau|\to\infty} v_k = \frac{1}{\sqrt{2k}} e^{-ik\tau}. \tag{3.95}$$

A solution satisfying this requirement is referred to as the *Bunch–Davies vacuum*. This is the initial condition that fully specifies the solution.

At late times and/or on large scales, more precisely when $|k\tau| \ll 1$, the k^2 term in (3.92) can be neglected and the equation changes to a harmonic oscillator with time-dependent tachyonic mass, to zeroth order in slow-roll

$$v_k'' - \frac{2}{\tau^2} v_k = 0, \quad |k\tau| \ll 1. \tag{3.96}$$

Now the solution has an entirely different behavior, as it switches from oscillating to growing, the growing mode solution approximately being $v_k \propto z \propto 1/\tau$. Thus we learn that the fluctuations get amplified as soon as they *exit the horizon*, i.e., from the moment when $|k\tau| \approx 1$. Remember that conformal time τ also measures the comoving particle horizon – see Eq. (1.73) and below – and keep in mind that k is the comoving wavenumber. The condition $|k\tau| = 1$ thus designates the moment when the physical wavelength of the perturbation a/k, which gets stretched during inflation, becomes equal to the physical horizon size $1/H$, which remains roughly constant, so that another way to write this statement is to say that horizon exit occurs when $k = aH$.

The equation of motion (3.92) may be recognized as the Bessel equation, whose solutions involve the Bessel functions with index α. The general solution is a linear combination of two independent solutions, which we take here to involve the so-called Hankel functions $H_\alpha^{(1,2)}$ of the first and second kind,

$$v_k = \sqrt{-k\tau}\left(c_1 H_\alpha^{(1)}(-k\tau) + c_2 H_\alpha^{(2)}(-k\tau)\right). \tag{3.97}$$

Appendix C lists a few useful properties of these functions, in particular their asymptotic expressions. Using (C.22) we can see that choosing the Bunch–Davies vacuum selects the solution proportional to the Hankel function of the first kind,

$$v_k(\tau) = \sqrt{-\tau}\sqrt{\frac{\pi}{4}}H_\alpha^{(1)}(-k\tau) \quad \text{(Bunch–Davies mode)}, \quad (3.98)$$

where α was given in (3.92) and we have dropped an irrelevant phase. Using (C.23), the late-time/large-scale limit is then given by

$$v_k = \frac{1}{\sqrt{2k}}\frac{1}{(-k\tau)^{1+\epsilon-\frac{1}{2}\eta}} \quad |k\tau| \ll 1, \quad (3.99)$$

where we have approximated $\Gamma(\alpha) \approx \Gamma(\frac{3}{2}) = \frac{\sqrt{\pi}}{2}$.

In terms of the comoving curvature perturbation $\mathcal{R} = v/z$ we then find that in the late-time/large-scale limit its Fourier modes evolve to

$$\mathcal{R}_k \approx \frac{H_\star}{2\sqrt{\epsilon_\star}k^{3/2}}\left(\frac{k}{k_\star}\right)^{-\epsilon+\frac{1}{2}\eta}, \quad k_\star = a_\star H_\star. \quad (3.100)$$

This is a crucial result: The comoving curvature perturbation becomes constant at late times and in the large-scale limit (note that when k^2 is neglected, Eq. (3.92) immediately implies that one solution is $v_k \propto z$). As we hinted at when discussing the equation of motion of $\mathcal{R}$, Eq. (3.74), this constant solution will allow us to trace the curvature perturbation from its generation during inflation into the much later radiation- or matter-dominated phases. Since the solution becomes constant as the mode in question crosses the horizon, we must evaluate the relevant background quantities in (3.100) at horizon crossing $k_\star = a_\star H_\star$, which we denote by a star subscript.

We are now in a position to calculate the properties of the curvature perturbation that are needed for a comparison with observations. The vacuum expectation value is given by

$$\langle 0|\hat{\mathcal{R}}_\mathbf{k}\hat{\mathcal{R}}_{\mathbf{k}'}|0\rangle = \langle 0|\left(\mathcal{R}_\mathbf{k}^\star\hat{a}^\dagger + \mathcal{R}_\mathbf{k}\hat{a}\right)\left(\mathcal{R}_{\mathbf{k}'}\hat{a} + \mathcal{R}_{\mathbf{k}'}^\star\hat{a}^\dagger\right)|0\rangle \quad (3.101)$$

$$= \mathcal{R}_\mathbf{k}\mathcal{R}_{\mathbf{k}'}\langle 0|[\hat{a},\hat{a}^\dagger]|0\rangle \quad (3.102)$$

$$= (2\pi)^3|\mathcal{R}_k|^2\delta(\mathbf{k}+\mathbf{k}'), \quad (3.103)$$

which from Eq. (3.6) shows that the power spectrum is simply given by

$$P_\mathcal{R} = |\mathcal{R}_k|^2. \quad (3.104)$$

We can thus evaluate the variance (3.8),

$$\Delta_\mathcal{R}^2 \approx \frac{H_\star^2}{8\pi^2\epsilon_\star}\left(\frac{k}{k_\star}\right)^{3-2\alpha}, \quad k_\star = a_\star H_\star, \quad (3.105)$$

implying an amplitude

$$\mathcal{A}_\mathcal{R} \approx \frac{H_\star^2}{8\pi^2 \epsilon_\star} \qquad (3.106)$$

and a spectral index

$$n_s \approx 1 - 2\epsilon_\star + \eta_\star \approx 1 - 6\epsilon_{V\star} + 2\eta_{V\star}. \qquad (3.107)$$

We can see that the amplitude depends both on the inflationary expansion rate H and the slow-roll parameter ϵ, with the magnitude of the perturbations being enhanced by a higher expansion rate and a slower rolling of the inflaton field. As for the spectral index, we can see that for slow-roll solutions it is automatically close to scale-invariant, and tends to have a slight red tilt, since $\epsilon > 0$.

A comment: Regularly this calculation is presented in the *de Sitter limit* ($\epsilon = 0$), at least as a first "approximation," because the solution to the mode equation then reduces to the simple, explicit form

$$v_k(\tau) = \frac{1}{\sqrt{2k}} e^{-ik\tau} \left(1 - \frac{i}{k\tau}\right) \qquad \text{(Bunch–Davies, dS limit)}. \qquad (3.108)$$

However, one should note that $\epsilon \to 0$ is not a smooth limit: The action for the perturbations (3.73), which is proportional to ϵ, in fact disappears. In pure de Sitter space these perturbations do not exist – they are pure gauge modes. Turning this around, one may also see ϵ as a parameter quantifying the extent to which the de Sitter symmetry is broken. What is a correct thing to do is to consider an additional, separate scalar field evolving on a de Sitter background, in the limit where the scalar is small and does not disturb the background. In that case the mode functions are indeed given by (3.108). The scalar field is then simply a spectator, not driving the evolution of the background spacetime. We will describe such a setting in more detail in Section 6.6.

3.2.6 Quantum to classical – a preview

Before continuing, it may be useful to clarify the important fact that the (quantum) perturbations start behaving increasingly classically. We start again from the Lagrangian for each Fourier mode (we will drop the k subscript momentarily),

$$L = \frac{1}{2} v'^2 - \frac{1}{2}\left(k^2 - \frac{z''}{z}\right) v^2. \qquad (3.109)$$

The vacuum state is defined by

$$\hat{a}|0> = 0.\qquad(3.110)$$

Making use of the Wronskian normalization condition (3.88), the annihilation operator can be rewritten in terms of the field and its momentum

$$i\hat{a} = f^{*\prime}\hat{v} - f^{*}\hat{\pi}.\qquad(3.111)$$

Here we write the mode functions as f instead of v in order to avoid confusing the mode functions f and the field v itself.

The "classicality" property can perhaps best be seen in the Schrödinger picture, which we will adopt for this section. Using the definition of the vacuum above and the canonical replacement $\pi \to -i\frac{\partial}{\partial v}$, we find that the ground state Schrödinger wave function for the perturbation modes is a Gaussian

$$\Psi(v) = n\exp\left(-\frac{1}{2}Cv^2\right),\qquad(3.112)$$

where n is a normalization factor. Here C is the correlator and it is given by

$$C = -i\frac{f^{*\prime}}{f^{*}},\qquad(3.113)$$

while normalizability ($\int \Psi\Psi^{\star}dv = 1$) implies that, up to a phase,

$$n = \left(\frac{C+C^{\star}}{2\pi}\right)^{\frac{1}{4}}.\qquad(3.114)$$

The mode function was presented in (3.98). At early times, it approximately takes Minkowski form, such that $f \approx \frac{1}{\sqrt{2k}}e^{-ik\tau}$. Then we have

$$C \approx k \qquad (|k\tau| \gg 1).\qquad(3.115)$$

On the other hand, at late times, using the asymptotic forms of the Hankel functions from Appendix C, one finds that

$$C \approx -\frac{2\pi}{\Gamma(\alpha)^2\tau}\left(\frac{-k\tau}{2}\right)^{2\alpha} - i\left(\frac{1}{2} - \alpha\right)\frac{1}{\tau} \qquad (|k\tau| \ll 1)\qquad(3.116)$$

$$\approx k^3\tau^2 + \frac{i}{\tau},\qquad(3.117)$$

where we have kept the leading real and imaginary contributions and in the last line have taken the limit of very slow roll so that $\alpha \approx 3/2$. It is now useful to recall the Jeffreys–Wentzel–Kramers–Brillouin (JWKB) criterion for classicality: A wave function behaves approximately classically when it is oscillatory, i.e., when its amplitude varies slowly compared to the variation of its phase. Clearly, at early times this is not satisfied and the wave function is very quantum, while at late times the variation of its amplitude goes to zero while its phase is continually speeding up. Thus, we find our first indication that as perturbations grow larger than the horizon, they behave increasingly classically. Note also that the real part of C becomes small at late times. This means that the dispersion of the fluctuation modes becomes large – in other words, these modes get copiously produced. As Eq. (3.117) clearly shows, the transition from quantum to classical occurs precisely when $|\tau| = 1/k$, i.e., at horizon exit.

Further aspects of the quantum-to-classical transition, in particular the effects of decoherence, will be discussed in Chapter 8.

3.2.7 Tensor modes

We can now repeat the perturbation calculation for tensor fluctuations. In fact the calculation proceeds in an entirely analogous fashion, hence we can be slightly quicker here. As stated earlier, the tensor perturbations are already gauge invariant and at linear order they evolve independently of scalar and vector modes. Hence it is sufficient to consider the perturbed metric (cf. also (3.70))

$$ds^2 = -dt^2 + a^2(t)[\delta_{ij} + \gamma_{ij}(t,\mathbf{x})]dx^i dx^j, \qquad (3.118)$$

with

$$\gamma^i{}_i = \partial^i \gamma_{ij} = 0. \qquad (3.119)$$

We can transform to Fourier space and decompose the tensor perturbation into two polarization states (indicated by a plus and a cross)

$$\gamma_{ij} = \int \frac{d^3k}{(2\pi)^3} \sum_{p=+,\times} \epsilon^p_{ij} \gamma^p_{\mathbf{k}} e^{i\mathbf{k}\cdot\mathbf{x}}, \qquad (3.120)$$

with

$$\epsilon^p_{ii} = k^i \epsilon^p_{ij} = 0, \qquad \epsilon^p_{ij}\epsilon^{p'}_{ij} = 2\delta_{pp'}. \qquad (3.121)$$

The canonically normalized variable is then $v^p_{\mathbf{k}}$, defined via

$$\gamma^p_{\mathbf{k}} = \frac{2}{a} v^p_{\mathbf{k}}, \qquad (3.122)$$

which has the following Fourier space quadratic action (obtained by first expanding the action (3.65) to quadratic order in γ_{ij} and then substituting the preceding definitions)

$$S_2^t = \sum_p \int d^3k d\tau \left[(v_{\mathbf{k}}^p)'^2 - \left(k^2 - \frac{a''}{a}\right)(v_{\mathbf{k}}^p)^2 \right], \qquad (3.123)$$

and associated equation of motion

$$v_{\mathbf{k}}^{p\prime\prime} + \left(k^2 - \frac{a''}{a}\right) v_{\mathbf{k}}^p = 0. \qquad (3.124)$$

Note how similar this equation is to the scalar equivalent (3.83), except that tensor modes depend solely on the evolution of the scale factor a and not directly on ϵ as well.

The quantization now proceeds exactly as for the scalar case, following Eqs. (3.85)–(3.88). We may thus immediately proceed to analyze solutions, again when the slow-roll conditions are met. Using (3.91), to leading order in slow-roll the mode equation becomes

$$v_{\mathbf{k}}^{p\prime\prime} + \left(k^2 - \frac{\beta^2 - \frac{1}{4}}{\tau^2}\right) v_{\mathbf{k}}^p = 0, \qquad \text{where} \quad \beta = \frac{3}{2} + \epsilon. \qquad (3.125)$$

The solution satisfying the Bunch–Davies initial condition (3.95) is then given (up to an irrelevant phase) by

$$v_k^p = \frac{\sqrt{-\pi\tau}}{2} H_\beta^{(1)}(-k\tau). \qquad (3.126)$$

This leads to the late-time/large-scale limit (to be evaluated at Hubble crossing $k_\star = a_\star H_\star$)

$$\gamma_k^p = \frac{\sqrt{2}}{k^{3/2}} H_\star \left(\frac{k}{k_\star}\right)^{\frac{3}{2}-\beta}, \qquad k_\star = a_\star H_\star. \qquad (3.127)$$

In order to obtain the total variance (3.17) of the tensor perturbations we must add the contributions from both polarization states, so that we end up with the final expression that

$$\Delta_t^2 = 2 \frac{k^3}{2\pi^2} |\gamma_k^p|^2 = \frac{2}{\pi^2} H_\star^2 \left(\frac{k}{k_\star}\right)^{-2\epsilon_\star}, \qquad k_\star = a_\star H_\star, \qquad (3.128)$$

corresponding to an amplitude and spectral index given by

$$\mathcal{A}_t \approx \frac{2}{\pi^2} H_\star^2, \qquad n_t \approx -2\epsilon_\star. \qquad (3.129)$$

Note that the amplitude of the tensor perturbations depends only on the Hubble rate during inflation, but not on the slope of the potential, in contrast to the scalar fluctuations. In this way the tensor perturbations are more directly tied to the underlying geometry, and moreover an observation of their amplitude would immediately reveal the energy scale of inflation, under the assumption that inflation is driven by such simple single-field dynamics.

The tensor-to-scalar ratio (3.18) reduces to a simple expression,

$$r \equiv \frac{\mathcal{A}_t}{\mathcal{A}_\mathcal{R}} = 16\epsilon_\star . \qquad (3.130)$$

In the slow-roll approximation we thus have $r \approx -8n_t$, which is known as the *consistency relation* for (single-field) slow-roll inflation. In more complicated models this relation is not valid, or, reversing the argument, if this consistency relation could be verified observationally, that would be strong evidence for single-field slow-roll inflation.

We can also rewrite the tensor-to-scalar ratio in terms of the number of e-folds as

$$r = 16\epsilon = 8\left(\frac{d\phi}{d\mathcal{N}}\right)^2 . \qquad (3.131)$$

Assuming r to be constant and integrating, we get

$$\frac{\Delta\phi}{M_P} \approx 2 \times \sqrt{\frac{r}{0.01}} \qquad (3.132)$$

for $\mathcal{N} = 60$ e-folds of inflation. This equation shows that in order to obtain a gravitational-wave signal large enough to be observable, the scalar field must typically travel a distance of at least one Planck unit in field space – this is known as the *Lyth bound*. Models in which this is the case are often referred to as *large-field models*.

3.2.8 Non-Gaussian corrections

So far we were looking at linearized perturbations, described by Gaussian wave functions. But the nonlinearity of general relativity, and sometimes in addition explicit couplings between fields or nontrivial kinetic terms, lead to interactions that provide corrections to the Gaussian distributions calculated thus far.

It would be a vast undertaking to treat all these cases, going beyond the scope of this book. Instead, we will only provide a single example of how to calculate such corrections, based on the curvature perturbation itself – see Section 3.2.3. This should enable the reader to access the (fairly large) literature on the subject, and give an indication of how these calculations are done.[4]

---

[4] This section is more technical and condensed, and can be skipped at a first reading.

3.2 Cosmological perturbation theory

In analogy with the power spectrum, Eq. (3.6), one can define a *bispectrum*

$$\langle \mathcal{R}_{\mathbf{k_1}} \mathcal{R}_{\mathbf{k_2}} \mathcal{R}_{\mathbf{k_3}} \rangle \equiv (2\pi)^3 \delta(\mathbf{k_1} + \mathbf{k_2} + \mathbf{k_3}) \, B(k_1, k_2, k_3) \,. \tag{3.133}$$

The bispectrum quantifies 3-point functions on the sky, i.e., correlations between temperature fluctuations at three different locations on the sky. In principle this contains a wealth of information, but unfortunately it has not been observed yet (this will just be a matter of time though, since this signal is definitely expected to be there).

The bispectrum $B(k_1, k_2, k_3)$ depends only on the magnitudes of the momenta (rather than their specific directions as well), because the background space is isotropic. The relative sizes of the momenta determine the so-called *shape* of the bispectrum. If the bispectrum happens to be particularly strong for a certain configuration of momenta, this has physical meaning. The two following limits are of special significance:

- The *local shape* corresponds to the situation where one momentum vector is much smaller than the other two, say $k_1 \ll k_2 \approx k_3$. Due to momentum conservation (imposed by the delta function in (3.133)), the momentum vectors have to form a closed triangle. When one side is very small, the other two are almost parallel and equal in magnitude. This can be interpreted as the influence that a long-wavelength perturbations has on the power spectrum of two shorter-wavelength modes. This shape is especially significant when there are strong nonderivative interactions between different fields.
- The *equilateral shape*, for which all three momenta are roughly equal. This shape principally signals the presence of kinetic terms with higher derivatives.

In looking at an example, we will treat the non-Gaussian corrections as small. The wave function for a single mode of the curvature perturbation is then a product of a Gaussian part and a correction piece,

$$\Psi_\mathcal{R} = \psi_G \, \psi_{NG} \,. \tag{3.134}$$

The Gaussian part arises from free evolution. The corresponding Hamiltonian follows directly from (3.73) and reads

$$H_{\text{free}} = \frac{1}{2} \int_{\mathbf{k}} \left(\frac{\mathsf{H}^2 \tau^2}{\epsilon} \hat{\pi}_{\mathbf{k}} \hat{\pi}_{-\mathbf{k}} + \frac{\epsilon k^2}{\mathsf{H}^2 \tau^2} \mathcal{R}_{\mathbf{k}} \mathcal{R}_{-\mathbf{k}'} \right) \,. \tag{3.135}$$

The Schrödinger equation for the Gaussian part is thus

$$i \frac{\mathrm{d}}{\mathrm{d}t} \psi_G = -i \mathsf{H} \tau \frac{\mathrm{d}}{\mathrm{d}\tau} \psi_G = H_{\text{free}} \psi_G \,, \tag{3.136}$$

and it is solved by

$$\psi_\mathrm{G} = \left(\frac{A + A^\star}{\pi}\right)^{1/4} e^{-A|\mathcal{R}_\mathbf{k}|^2}, \qquad A = \frac{\epsilon k^3}{H^2} \frac{1 - \frac{i}{k\tau}}{1 + k^2\tau^2}, \qquad (3.137)$$

where A is the correlator.

The non-Gaussian part is taken to arise from a cubic interaction

$$\psi_\mathrm{NG} = e^{\int_{\mathbf{k}_1, \mathbf{k}_2, \mathbf{k}_3} \mathcal{R}_{\mathbf{k}_1} \mathcal{R}_{\mathbf{k}_2} \mathcal{R}_{\mathbf{k}_3} I(\tau)}, \qquad (3.138)$$

where we use the shorthand notation $\int_{\mathbf{k}_1,\mathbf{k}_2,\mathbf{k}_3} = \int \frac{\mathrm{d}^3\mathbf{k}_1}{(2\pi)^3} \frac{\mathrm{d}^3\mathbf{k}_2}{(2\pi)^3} \frac{\mathrm{d}^3\mathbf{k}_3}{(2\pi)^3} (2\pi)^3 \delta(\mathbf{k}_1 + \mathbf{k}_2 + \mathbf{k}_3)$. The complex, time-dependent function I characterizes the interaction, and due to the spatial isotropy of the background depends on the magnitudes k_i. The integral incorporates momentum conservation due to the delta function.

We are interested in the expectation value of three curvature perturbations, evaluated at late times, i.e., in the *interacting vacuum* when perturbations have already evolved. We can write this as

$$\langle \mathcal{R}_{\mathbf{k}_1} \mathcal{R}_{\mathbf{k}_2} \mathcal{R}_{\mathbf{k}_3} \rangle = \int \mathcal{DR}(\mathcal{R}_{\mathbf{k}_1} \mathcal{R}_{\mathbf{k}_2} \mathcal{R}_{\mathbf{k}_3}) |\Psi_\mathcal{R}|^2 \qquad (3.139)$$

$$= \int \mathcal{DR}(\mathcal{R}_{\mathbf{k}_1} \mathcal{R}_{\mathbf{k}_2} \mathcal{R}_{\mathbf{k}_3}) \psi_\mathrm{G} \psi_\mathrm{G}^\star \psi_\mathrm{NG} \psi_\mathrm{NG}^\star \qquad (3.140)$$

$$= \int \mathcal{DR}(\mathcal{R}_{\mathbf{k}_1} \mathcal{R}_{\mathbf{k}_2} \mathcal{R}_{\mathbf{k}_3}) \psi_\mathrm{G} \psi_\mathrm{G}^\star e^{\int_{\mathbf{k}_1 \mathbf{k}_2 \mathbf{k}_3} (\mathcal{R}_{\mathbf{k}_1} \mathcal{R}_{\mathbf{k}_2} \mathcal{R}_{\mathbf{k}_3}) 2 \mathrm{Re}[I]}. \qquad (3.141)$$

Because the interactions are small, one can also expand this to linear order, and the resulting expression is usually written

$$\langle \mathcal{R}(\tau, \mathbf{k}_1) \mathcal{R}(\tau, \mathbf{k}_2) \mathcal{R}(\tau, \mathbf{k}_3) \rangle$$
$$= -i \int_{-\infty}^{\tau} \mathrm{d}\tau' [\mathcal{R}(\tau, \mathbf{k}_1) \mathcal{R}(\tau, \mathbf{k}_2) \mathcal{R}(\tau, \mathbf{k}_3), H_\mathrm{int}(\tau')], \qquad (3.142)$$

where H_int is the cubic interaction Hamiltonian. The relation between I and H_int will be spelled out momentarily. The lower integration limit at $-\infty$ corresponds to the initial (Gaussian) Bunch–Davies vacuum state.

As an example we will take the following interaction term, which arises in the expansion of the Einstein–Hilbert action to cubic order,

$$H_\mathrm{int} = \frac{g}{2} \int \mathrm{d}^3 x \, a \mathcal{R}^2 \partial^2 \mathcal{R}, \qquad (3.143)$$

where $g = \mathcal{O}(\epsilon^2)$ is a coupling that is second order in slow roll. In Fourier space we have

$$H_{\text{int}}\psi = \left[\int_{\mathbf{k_1},\mathbf{k_2},\mathbf{k_3}} \mathcal{R}_{\mathbf{k_1}}\mathcal{R}_{\mathbf{k_2}}\mathcal{R}_{\mathbf{k_3}}\mathcal{H}_{\text{int}}\right]\psi \quad \text{with} \quad \mathcal{H}_{\text{int}} = \frac{g}{3\mathsf{H}\tau}(k_1^2 + k_2^2 + k_3^2), \tag{3.144}$$

where the factor $1/3$ arises from symmetrizing over the momenta and we used $a = -1/(\mathsf{H}\tau)$ to approximate de Sitter spacetime. If we now plug the wave function (3.134) into the Schrödinger equation

$$i\frac{\mathrm{d}}{\mathrm{d}t}\Psi = -i\mathsf{H}\tau\frac{\mathrm{d}}{\mathrm{d}\tau}\Psi = (H_{\text{free}} + H_{\text{int}})\Psi, \tag{3.145}$$

we obtain an equation for the interaction function I,

$$iI_{,\tau} + \alpha I + \frac{1}{\mathsf{H}\tau}\mathcal{H}_{\text{int}} = 0, \tag{3.146}$$

with

$$\alpha = -\frac{\mathsf{H}^2\tau^2}{\epsilon}\left(A(k_1,\tau) + A(k_2,\tau) + A(k_3,\tau)\right). \tag{3.147}$$

This can be solved to give

$$I(\tau) = i\int_{-\infty}^{\tau}\frac{\mathrm{d}\tau'}{\mathsf{H}\tau'}\mathcal{H}_{\text{int}}(\tau')e^{i\int_{\tau'}^{\tau}\mathrm{d}\tau''\,\alpha(\tau'')}. \tag{3.148}$$

Here the factor

$$e^{i\int_{\tau'}^{\tau}\alpha\mathrm{d}\tau} = e^{i(k_1+k_2+k_3)(\tau'-\tau)}\frac{1-ik_1\tau'}{1-ik_1\tau}\frac{1-ik_2\tau'}{1-ik_2\tau}\frac{1-ik_3\tau'}{1-ik_3\tau} \tag{3.149}$$

stems from the product of mode functions. We still need the integral

$$i\int_{-\infty}^{\tau}\frac{\mathrm{d}\tau'}{\tau'^2}e^{i(k_1+k_2+k_3)(\tau'-\tau)}(1-ik_1\tau')(1-ik_2\tau')(1-ik_3\tau')$$
$$= -\frac{i}{\tau} - \frac{k_1k_2 + k_1k_3 + k_2k_3}{k_1+k_2+k_3} - \frac{k_1k_2k_3}{(k_1+k_2+k_3)^2} + i\tau\frac{k_1k_2k_3}{k_1+k_2+k_3}, \tag{3.150}$$

where we regularize the integral by deforming the contour in the convergent part of the complex plane at early times. At late times $|k_i\tau|\to 0$, we find

$$\mathrm{Re}[I] = -\frac{g}{3\mathsf{H}^2}(k_1^2+k_2^2+k_3^2)\left(-k_{\text{tot}} + \frac{k_1k_2+k_1k_3+k_2k_3}{k_{\text{tot}}} + \frac{k_1k_2k_3}{k_{\text{tot}}^2}\right), \tag{3.151}$$

$$\mathrm{Im}[I] = -\frac{g}{3\mathsf{H}^2\tau}(k_1^2+k_2^2+k_3^2), \tag{3.152}$$

where $k_{\text{tot}} = k_1 + k_2 + k_3$. The imaginary part will be useful at a later stage, when discussing decoherence in Section 8.2.3.

To calculate the contribution to the bispectrum, we make use of the fact that the interactions are small (hence $\psi_{\text{NG}} \sim 1 + \int \zeta^3 I$) and use Wick's theorem to evaluate the resulting integral expressing the expectation value of six curvature perturbations,[5] which leads to the inclusion of a product of three power spectra, to find

$$B \supset \frac{g^2 \mathsf{H}^4}{16\epsilon^3} \frac{(k_1^2 + k_2^2 + k_3^2)}{k_1^3 k_2^3 k_3^3} \left(-k_{\text{tot}} + \frac{k_1 k_2 + k_1 k_3 + k_2 k_3}{k_{\text{tot}}} + \frac{k_1 k_2 k_3}{k_{\text{tot}}^2} \right). \tag{3.153}$$

This has a comparatively large contribution to the local shape ($\sim 1/(k_1^3 k^4)$ with $k_2 \approx k_3 = k$), but the smallness of the coupling $g = \epsilon^2$ still prevents this interaction from leading to a large bispectrum.

One can repeat similar calculations for all cubic interactions present in the theory of interest to determine the expected shape and amplitude of the 3-point function. Conversely, observations of the 3-point can in principle provide a large amount of information about the theory underlying the formation of primordial perturbations. Observational cosmologists typically define various nonlinearity parameters (typically called f_{NL}) to parametrize the size of the bispectrum for various momentum configurations, and observational results are then specified in terms of these parameters. At the time of writing, only upper bounds have been reported on all non-Gaussianity parameters. These rule out models with highly nonlinear kinetic terms and strong interactions between fields, but still allow for a wide variety of models of the early universe. For this reason we will not discuss non-Gaussianities further, but these will undoubtedly become an important subject in observational cosmology in the future.

3.3 Comparing theory and observations

Over the last three decades, observations of the CMB have made an enormous leap – see for instance Fig. 3.5. These observations have shown that the temperature fluctuations on the CMB, or equivalently the corresponding density perturbations, are of scalar nature (primordial tensor perturbations have not been detected so far), follow Gaussian statistics (up to the level of precision reached by all experiments so far), and are adiabatic (meaning that the fluctuations are effectively sourced by a single scalar field). Furthermore, the spectrum of the scalar fluctuations is close to scale invariant,

---
[5] This integral is of the form $\sim \int \mathcal{DR} \, \mathcal{R}_{\mathbf{k_1}} \mathcal{R}_{\mathbf{k_2}} \mathcal{R}_{\mathbf{k_3}} |\psi_G|^2 \int_{\mathbf{q_1}, \mathbf{q_2}, \mathbf{q_3}} \mathcal{R}^3 I$.

but measurably deviates from exact scale invariance. For convenience, let us recall from section 3.1 the current measured values for the amplitude and spectral index, as well as the upper bound on the tensor-to-scalar ratio,

$$A_\mathcal{R} = (2.09 \pm 0.04) \times 10^{-9}, \qquad (3.154)$$

$$n_s = 0.965 \pm 0.005, \qquad (3.155)$$

$$r < 0.056 \qquad \text{(at 95\% confidence level)}. \qquad (3.156)$$

We should compare these observations with the corresponding theoretical expectations stemming from single-field inflationary models. As derived in Sections 3.2.5 and 3.2.7, these are

$$\mathcal{A}_\mathcal{R} \approx \frac{H_\star^2}{8\pi^2 \epsilon_\star}, \qquad (3.157)$$

$$n_s \approx 1 - 2\epsilon_\star + \eta_\star \approx 1 - 6\epsilon_{V\star} + 2\eta_{V\star}, \qquad (3.158)$$

$$r = 16\epsilon_\star. \qquad (3.159)$$

Three obvious questions should be addressed: Are these predictions reliable, given that inflation occurred long before the CMB was emitted? If so, can single-field inflationary models account for the observed properties? And, if so, are these values reasonable?

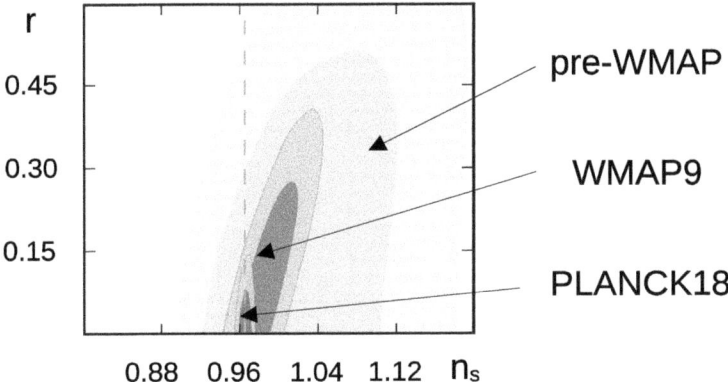

Figure 3.5 Improved observational constraints from CMB experiments, circa 2000 (pre-WMAP), 2009 (WMAP9), and 2018 (PLANCK18). Over the course of these experiments, the deviation of the spectral index n_s from exact scale invariance became established. Note that the measurement of r only constitutes an upper bound. Figure reproduced with permission from Planck Collaboration; Aghanim et al. (2020a) *Credit: ESA/Planck collaboration.*

Regarding reliability, we should look at the evolution of the causal structure of inflation – see Fig. 3.6. During inflation, the horizon $1/H$ stays roughly constant, while the universe grows exponentially. The wavelength of fluctuations (given by a/k) is initially very small, and the wave function is approximately that of vacuum fluctuations in Minkowski space. As the fluctuations get stretched, they are also amplified. Once they become larger than the horizon, they evolve into constant curvature perturbations, and they remain constant in amplitude while their wavelength becomes larger and larger. After inflation has come to an end, we enter a phase of radiation domination and the horizon starts growing faster than the wavelength gets stretched ($1/H \propto t \propto a^2$), so that progressively fluctuations start re-entering the horizon. The perturbations that were produced last re-enter first. Correspondingly, the perturbations that were produced first re-enter last. This implies that the fluctuations that we see in the CMB are those that were produced about 60 e-folds before the end of inflation. Because these perturbations were larger than the horizon, they remained constant and were oblivious to the small-scale details pertaining to the phases of reheating and early hot big bang evolution. This means that we really can trust those predictions.

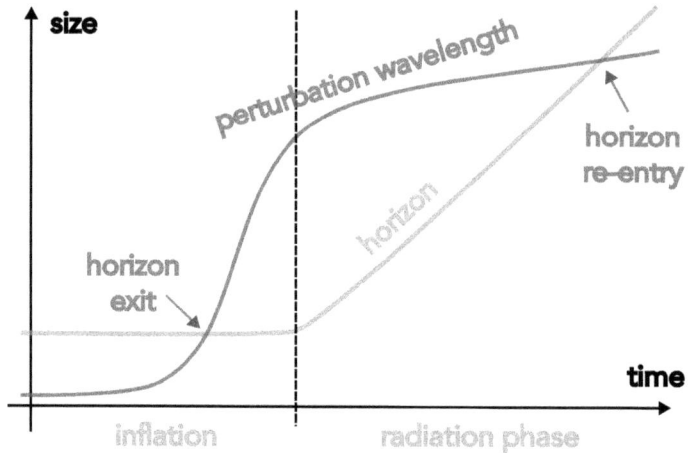

Figure 3.6 Causal evolution of the physical size of perturbations from the start of inflation, followed by horizon exit, to horizon re-entry during the radiation-dominated phase of expansion.

Now we can be more quantitative: The upper bound on the tensor-to-scalar ratio implies an upper bound on the first slow-roll parameter,

$$\epsilon_\star \lesssim 0.003\,5\,. \tag{3.160}$$

This means that the scalar field must have been in an ultra-slow-roll regime or, correspondingly, that the potential must have been very flat. If ϵ were constant (and thus $\eta = 0$), one would obtain a spectral index of $n_s = 1 - 2\epsilon_\star \gtrsim 0.993$, which is too close to scale invariance to be compatible with observations. Thus we need ϵ to have changed over time. In that case we can match the observed central value for the spectral index if

$$\eta_\star \lesssim -0.028\,, \qquad \eta_{V\star} \lesssim -0.007\,. \tag{3.161}$$

The fact that η is required to be negative has two implications. First, from the definition (2.35) we see that a negative η implies that ϵ is growing over time. Since inflation lasts until ϵ has grown to order 1, a small growth in ϵ some 50 or 60 e-folds before the end is certainly not counter-intuitive. And second, a negative η_V implies that the potential must be concave rather than convex. In other words, something like a mass term $V = \frac{1}{2}m^2\phi^2$ is ruled out, while plateau potentials such as those of Higgs or Starobinsky form can fit the data (for more details on this point, see the exercises).

From the bound on ϵ in (3.160) and the measured amplitude (3.154) we gain more information, namely an upper bound on the expansion rate, and by extension the energy scale, during inflation,

$$H_\star \lesssim 2 \cdot 10^{-5}\,. \tag{3.162}$$

Inflation thus must take place at an energy scale that is at least several orders of magnitude below the Planck scale, and hence also significantly below the conjectured grand unified scale. This means that in all likelihood, if inflation took place, then it occurred at a new energy scale, which has not played a role in physics before. One of the tasks of theoreticians will be to understand that new scale.

So, to answer the three questions we posed at the beginning of this section: Yes, inflationary predictions can be trusted, due to causality. Yes, inflation can match current observations of the CMB. And finally, the models that match observations are rather simple, being single-field slow-roll models. They can match the observations of perturbations while also lasting sufficiently long to resolve the flatness and horizon puzzles, and their energy scale is low enough so that they might not produce unwanted topological defects. However, some open issues remain, which we will focus on in the next section, and until these have been addressed as well, a definitive statement on the reasonableness of the models cannot yet be given.

Let us highlight that future observations are likely to provide us with more details. These include: A possible detection of a running of the spectral index, i.e., a change in n_s over the course of inflation; a detection of non-Gaussian corrections, i.e., evidence for self-interactions of the inflaton field,[6] or interactions with other scalar fields; and most importantly, primordial tensor perturbations might be detected. In inflation, such gravitational waves are certainly predicted to exist, but their amplitude can be low, cf. Eq. (3.129), as it is given by the Hubble rate. The advantage of such a measurement would be that it would directly point to the energy scale of inflation, and moreover it would show that primordial gravitational waves arose from the amplification of tensor fluctuations in the metric – a clear quantum gravitational effect and as such direct evidence for the quantization of spin-2 excitations.

3.4 Inflationary puzzles

We have just seen that inflation can explain the pattern of temperature fluctuations observed in the cosmic microwave background. However, in arriving at this statement, we had to make a number of assumptions, explicitly or implicitly. The open questions pertaining to inflation fall into two broad categories, having to do with either particle physics questions or cosmological ones.

Particle physics puzzles:

- Inflaton identity
  What is the nature of the inflaton field? Except for Higgs inflation, all models of inflation assume the existence of a hitherto undiscovered scalar field, and are in that sense speculative. Even for Higgs inflation, one must assume a specific nontrivial coupling to gravity and it currently remains unknown whether the Higgs indeed possesses this coupling.
- Inflaton couplings
  How does the inflaton couple to all other particles, in particular to the Standard Model particles? Again, this is only known for the Higgs field, and even in that case one has to assume an additional nontrivial coupling to gravity, as just mentioned. Once the couplings are known, the question is: Does reheating really work?
- Shape of the potential
  How does the inflationary potential arise? Can it be flat enough over the required field range? What determines its magnitude?

---

[6] The current nondetection of self-interactions in fact contributes to the self-consistency of the calculations we did above: Any loop corrections to the spectrum of perturbations (or, perhaps better said, to the propagator) are small when self-interactions are small, since loops can only occur when there is a 3-vertex. The absence of a detection of non-Gaussianities therefore implies that quantum loop corrections must also be small.

Cosmological puzzles:

- Initial conditions puzzle
  How were the initial conditions for inflation set? For inflation to start, we tacitly assumed that a roughly Hubble-sized patch of the universe exists, in which the average curvature is not too large, with subdominant anisotropies in the curvature, and in which the scalar field sits sufficiently high up the potential, with sufficiently small velocity. Only when all these conditions are met can inflation take place. Is there a theoretical framework in which such conditions come out naturally? Recall from Eq. (3.162) that observational constraints imply that the Hubble rate is significantly below the Planck scale, so that the initial patch must be comparatively large, thereby rendering a "random" occurrence of such conditions unlikely.

- The unlikeliness puzzle
  If the scalar potential is flat over various field ranges (note that the full potential could have an elaborate, high-dimensional structure), then inflation could perhaps have occurred at a higher Hubble rate, producing a bigger universe and with more galaxies (since the perturbation amplitude would also have been higher, all else remaining equal). So why was the Hubble rate comparatively low? This question is rather closely related to that of the shape of the potential: Why is the inflationary potential concave rather than convex, since the latter possibility seems much easier to achieve from a model-building perspective (as even a simple mass term would already suffice)?

- The multiverse puzzle
  Consider one Hubble patch (of size $1/H^3$) over one Hubble time $\Delta t = 1/H$ and assume slow-roll inflation. Then the classical motion of the field is $\Delta\phi_{\rm cl} = \dot\phi \Delta t = \frac{\dot\phi H}{H^2} \sim \frac{V_{,\phi}}{V}$. Meanwhile, as one can see by combining (3.55) and (3.106), the quantum evolution is of order $|\Delta\phi_{\rm qu}| \sim \frac{H}{2\pi} \sim V^{1/2}$. This can exceed the classical rolling if $V^{3/2} > |V_{,\phi}|$, or $V > \epsilon$. (Note from (3.154) that this inequality can only be satisfied higher up the potential than during the last 60 e-folds that gave rise to the observed CMB fluctuations.) Over one Hubble time the universe grows by a factor $e^3 \approx 20$. Thus, if the quantum displacement is larger than the classical rolling, after one Hubble time there will be about 10 Hubble patches where the field is kicked *up* the potential. This implies that the number of (say, Hubble-sized) regions where inflation occurs grows exponentially over time and thus, globally speaking, inflation never comes to an end (of course, at any given place inflation eventually does come to an end, but the point is that

more numerous new inflating regions are continuously created). This is known as *eternal inflation*.

As the field rolls down the potential, quantum jumps also occur, and these can change the predictions that we calculated earlier, as they can in principle change the slow-roll background that we assumed in our calculations. Because all possible quantum jumps will eventually occur, and because the number of created inflationary regions is infinite, this implies that *all* possible values for the amplitude, the spectral index, non-Gaussianities, etc., will be created. Without a measure, eternal inflation is thus not predictive, but no suitable measure has been found to describe this situation.

But is this simple picture of eternal inflation correct? For instance, is it acceptable to treat the background as classical with intermittent quantum jumps? Note, for instance, that it is highly dubious that one can use a single clock to describe the entire spacetime. Or does one need (as this author suspects) a full quantum gravitational treatment of the background for describing eternal inflation? Alternatively, given the implied infinities, might quantum gravity simply forbid eternal inflation?

We should mention that it is possible to generate the primordial temperature/density fluctuations by an analogous mechanism, i.e., quantum amplification of perturbations while the comoving horizon shrinks, with entirely different background dynamics. Two examples are an ekpyrotic contracting phase or an unstable matter-dominated contracting phase. In the ekpyrotic case, the potential is steep and thus there are naturally strong self-interactions present, which typically lead to significant non-Gaussian corrections, but which are not seen in the data. And a contracting matter phase is dynamically unstable, worsening all of the initial conditions puzzles that we were setting out to resolve. Moreover, in both cases there needs to be a bounce from the contracting to the expanding phase. The difficulties in modeling a trustworthy bounce solution – see the discussion below Eq. (2.78) – imply that such scenarios remain highly speculative to date.

For these reasons inflation, which provides a reliable history from its last 60 e-folds up to the present, is currently the most robust mechanism that can explain the primordial perturbations seen in the CMB. One may suspect that a classical treatment of the inflationary background evolution lies at the heart of the initial conditions puzzles of inflation. Put differently, one may hope that a quantum treatment of the inflationary phase, including both background and perturbations, will lead to a probability distribution of inflationary histories, and might thus elucidate the questions about initial conditions posed above. This provides the necessary motivation to tackle the quantization of gravity more directly, which brings us to the subject of the next two chapters.

Exercises

3.1 When the inflationary potential is an exponential, the first slow-roll parameter ϵ remains constant. In this case, exact solutions to the mode equations can be found. Derive these and deduce the predictions of such models for the spectral index and the tensor-to-scalar ratio. Can these models match observations?

3.2 Find out whether Starobinsky and Higgs inflation are in agreement with observations. You may want to make use of the results of Exercise 2.2.

3.3 This is an exercise to better understand the heuristic arguments regarding eternal inflation. Show that one can choose a gauge in which the spatial metric is unperturbed – this is known as *flat gauge*. Work out the constraints (3.67) and (3.68) to linear order in that gauge and substitute the slow-roll parameter ϵ where possible. Relate the comoving curvature perturbation $\mathcal{R}$ to the inflaton fluctuation $\delta\phi$ and discuss the condition for eternal inflation $|\Delta\phi_{\mathrm{qu}}| > |\Delta\phi_{\mathrm{cl}}|$ in terms of the variance of $\mathcal{R}$.

3.4 Show that a (homogeneous and isotropic) matter-dominated phase of contraction can in principle also produce scale-invariant curvature perturbations. Then discuss the dynamical stability of such a phase.

4
Quantizing gravity: canonical approach

Should gravity be quantized? Given that the other forces in nature, i.e., the nuclear and electromagnetic forces, are all described by quantum theories, it seems almost self-evident that gravity too should be described by a quantum theory at the most fundamental level. It is barely conceivable that one may obtain a consistent theory of nature in which matter is quantized but the spacetime, on which matter lives, not. For instance, how would one make sense of a many-slit experiment in which the quantum amplitude of a massive particle is a sum over all different paths the particle can take? In such an experiment, the different paths should be associated with different gravitational fields, i.e., different spacetimes, according to the quantum superposition principle. If one were instead to insist on a single background spacetime, then where exactly should that spacetime be curved? At the very least one would have to give up on the direct, causal relationship between mass–energy and gravity.

Furthermore, within general relativity itself there are already strong hints that the theory is not a final one. The best evidence is provided by the singularities inside black holes and at the big bang. These infinities point to a breakdown of classical general relativity, and one may indeed expect that a quantum theory of gravity will regularize the infinities and replace them with a predictive framework. Moreover, as remarked in the Preface, general relativity and quantum theory already display certain connections with each other, suggesting a "common ancestor."

Even though the arguments for quantum gravity are highly convincing, the precise nature of quantum gravity has remained elusive so far. Many attempts exist, some directly quantizing general relativity and some quantizing other degrees of freedom that effectively give rise to quantum gravity. The most thoroughly studied approach of the latter kind is string theory, to which we will return in Chapter 10. But whichever the ultimate theory of quantum gravity will be, one may expect that it will reduce to *semiclassical gravity* in the limit that $\hbar$ is small and that curvatures remain below the Planck scale.

Over the next two chapters, we will study the formalism of semiclassical gravity from two complementary points of view, starting in this chapter with the canonical quantization procedure, and in the next chapter adapting Feynman's path integral approach to include gravity.[1] A number of additional phenomenological aspects will be studied in later chapters. What is important for us is that semiclassical gravity may already describe quantum gravitational effects that played an important role in the history of the universe, and in this manner quantum cosmology can be fruitfully seen as a gateway to quantum gravity.

That said, recovering ordinary quantum mechanics from semiclassical gravity is already somewhat challenging. Let us briefly elaborate on this central point. First recall Schrödinger's approach to quantum theory, encapsulated in his famous equation

$$i\frac{\partial}{\partial t}\Psi = \hat{H}\,\Psi\,, \tag{4.1}$$

where $\Psi(t, x^i)$ is the wave function. Here $\hat{H}$ is the Hamiltonian (energy) operator – for the standard example of a particle of mass m in a potential $V(t, x^i)$ it is given by $\hat{H} = -\frac{1}{2m}\nabla^2 + V$. The wave function, assumed to be normalized here, can be used to calculate the probability density $\Psi^\star\Psi$ of finding the particle in the vicinity of position x^i at time t.

The specific feature to note is that the time coordinate t appears explicitly in the Schrödinger equation (4.1). But if we imagine quantizing gravity, then spacetime will be able to fluctuate, and there certainly won't be any external time coordinate. Thus we may expect to be able to recover ordinary quantum mechanics only in certain limits, when spacetime is sufficiently classical. In that sense quantum cosmology already corresponds to a generalization of ordinary quantum theory.

We have three main aims here: first, to obtain a generalization of the Schrödinger equation that includes gravity; second, to figure out how the Schrödinger equation is recovered from the new framework; and third, to start exploring the consequences for cosmology.

The strategy to achieve these goals will be to make use of the ADM formalism that we briefly encountered in Section 3.2.2. The splitting of spacetime

---

[1] The technical obstacle preventing these frameworks from being full theories of quantum gravity is the non-renormalizability of gravity. This has as a consequence that general relativity gets augmented by an infinite series of terms that contain higher powers of curvature tensors and thus, for Planckian curvatures, the formalism becomes intractable.

into time plus space allows for a Hamiltonian treatment of general relativity, which is quite close in spirit to Eq. 4.1. We will be reasonably brief here, mainly using the canonical quantization procedure as a stepping stone to the more useful (for our purposes) path integral formulation that will be the subject of the following chapter. That said, the canonical formalism is important as it illustrates many salient features of quantum gravity.

4.1 Hamiltonian formulation of general relativity

4.1.1 Action principle, variation and boundary terms

General relativity can be derived from a variational principle, with the action being given by the sum of the Einstein–Hilbert (EH) term and optionally the Gibbons–Hawking–York (GHY) term,

$$S_{\text{EH}} = \frac{1}{2} \int_M d^4x \sqrt{-g} R, \qquad (4.2)$$

$$S_{\text{GHY}} = \int_{\partial M} d^3x \sqrt{h} K = \int_{\partial M} d^3x \, \pi^{ij} h_{ij}, \qquad (4.3)$$

where the canonical momenta $\pi^{ij} = -\sqrt{h}(K^{ij} - h^{ij}K)$ were first encountered in Eq. (3.66), and where K_{ij} is the extrinsic curvature of the spatial boundary ∂M of the manifold M, defined in Eq. (3.62).[2]

The variation of these actions is a standard exercise, derived in textbooks on general relativity, with the result that

$$\delta S_{\text{EH}} = \frac{1}{2} \int_M d^4x \sqrt{-g} G_{\mu\nu} \delta g^{\mu\nu} - \int_{\partial M} d^3x \, h_{ij} \, \delta \pi^{ij}, \qquad (4.4)$$

$$\delta S_{\text{GHY}} = \int_{\partial M} d^3x \left(\pi^{ij} \, \delta h_{ij} + h_{ij} \, \delta \pi^{ij} \right). \qquad (4.5)$$

First consider the EH action alone. Its variation leads to the Einstein equations $G_{\mu\nu} = 0$ (which could be straightforwardly augmented by matter terms). But this is not all: To make the variational problem consistent, we must also set $\delta \pi^{ij} = 0$ on the initial and final hypersurfaces ∂M. In other words, we must keep the momenta, which one may think of as the expansion rates of the spacetime, fixed at the initial and final times. This is known as a *Neumann boundary condition*,

$$\delta \pi^{ij}|_{\text{boundary}} = 0 \qquad \text{(Neumann)}. \qquad (4.6)$$

---

[2] Here we consider space-like boundaries. If the boundary were time-like, one would add instead $S_{\text{GHY}}^{\text{time-like}} = -\int_{\partial M} d^3x \sqrt{-h} K$.

It implies that the EH action in fact encodes general relativity in momentum space.

Second, consider the sum $S_\text{EH} + S_\text{GHY}$. Its variation once again leads to the Einstein equations, but now the $\delta \pi^{ij}$ terms cancel, leaving us with an additional term proportional to δh_{ij} on the initial and final hypersurfaces. Thus, with the inclusion of the GHY term, we must keep the spatial metric h_{ij} fixed at the initial and final times. This is known as a *Dirichlet boundary condition*,

$$\delta h_{ij}\, |_\text{boundary} = 0 \quad \text{(Dirichlet)}, \tag{4.7}$$

and this version of the action thus corresponds to general relativity in field space.

The link between the action and boundary conditions will be important in what follows, as the action plays a central role in the quantum theory.

4.1.2 ADM framework and the Wheeler–DeWitt equation

We already encountered the usefulness of the ADM formalism for calculating cosmological perturbations in Chapter 3. Here we will extend the formalism so as to reveal a Hamiltonian treatment of general relativity. As we will see, this provides a rather direct route to quantizing gravity. When helpful, we will repeat a few equations presented earlier.

We will focus on the gravitational part of the action, with the matter content being left implicit, except that we include a cosmological constant Λ. Thus the action is taken to be ($8\pi G = 1$)

$$S = \frac{1}{2}\left[\int_M \mathrm{d}^4 x \sqrt{-g}(R - 2\Lambda) + S_\text{boundary}\right] + S_\text{matter}. \tag{4.8}$$

As we saw above in Section 4.1.1, the choice of boundary term determines which boundary conditions can be consistently imposed. In the present section we will ignore all surface terms, and re-introduce these when examining specific models later on.

Following the steps of Section 3.2.2, the action can be decomposed as

$$S = \frac{1}{2}\int \mathrm{d}^3 x\, \mathrm{d}t\; N\sqrt{h}\left[K_{ij}K^{ij} - K^2 + {}^3R - 2\Lambda\right] + S_\text{matter}. \tag{4.9}$$

This can be further rewritten in Hamiltonian form, as

$$S = \int \mathrm{d}^3 x\, \mathrm{d}t\, \left[\dot{h}_{ij}\pi^{ij} - N\mathcal{H} - N^i \mathcal{H}_i\right]. \tag{4.10}$$

Here one can see very explicitly that the lapse N and shift N^i are Lagrange multipliers, which impose constraints. There is the momentum constraint,

$$\mathcal{H}^i = -2D_j\pi^{ij} + \mathcal{H}^i_{\text{matter}} = 0, \qquad (4.11)$$

and the Hamiltonian constraint

$$\mathcal{H} = 2G_{ijkl}\pi^{ij}\pi^{kl} - \frac{1}{2}\sqrt{h}(^3R - 2\Lambda) + \mathcal{H}_{\text{matter}} = 0, \qquad (4.12)$$

where G_{ijkl} is the DeWitt metric

$$G_{ijkl} = \frac{1}{2\sqrt{h}}(h_{ik}h_{jl} + h_{il}h_{jk} - h_{ij}h_{kl}). \qquad (4.13)$$

These constraints are essentially equivalent to the time-space and time-time components, respectively, of the classical Einstein equations. Note an important feature: The constraints depend on the spatial metric and its conjugate momentum (as well as the matter fields), rather than on the full 4-dimensional metric.

Canonical quantization now amounts to imposing the constraints as operator equations, acting on a wave function $\Psi = \Psi(h_{ij}, \Phi_{\text{matter}})$ which is a function of the fields on the boundary of the spacetime. If we work in the field representation, then we may actualize the canonical commutation relations via the substitution

$$\pi^{ij} \to -i\frac{\delta}{\delta h_{ij}}, \qquad (4.14)$$

and similarly for the matter momenta. This results in four equations: the three momentum constraints

$$\hat{\mathcal{H}}^i\Psi \equiv 2iD_j\frac{\delta\Psi}{\delta h_{ij}} + \mathcal{H}^i_{\text{matter}}\Psi = 0, \qquad (4.15)$$

and the *Wheeler–DeWitt equation*

$$\hat{\mathcal{H}}\Psi \equiv \left[-G_{ijkl}\frac{\delta}{\delta h_{ij}}\frac{\delta}{\delta h_{kl}} - \sqrt{h}(^3R - 2\Lambda) + \mathcal{H}_{\text{matter}}\right]\Psi = 0. \qquad (4.16)$$

Since the constraints are so central, it is worthwhile investigating their meaning in some detail. To understand the momentum constraints (4.15) better, consider a change of coordinates on the three-surface, $x^i \to x^i - \xi^i$, focusing on the case of pure gravity. Then

$$\Psi[h_{ij} + D_{(i}\xi_{j)}] = \Psi[h_{ij}] + \int d^3\mathbf{x}\, D_{(i}\xi_{j)}\frac{\delta\Psi}{\delta h_{ij}}. \qquad (4.17)$$

4.1 Hamiltonian formulation of general relativity

Integrating by parts in the last term, and dropping the boundary term (assuming the three-manifold is compact), one finds that the change in Ψ is given by

$$\delta\Psi = -\int \mathrm{d}^3\mathbf{x}\, \xi_j D_i \left(\frac{\delta\Psi}{\delta h_{ij}}\right) = -\frac{1}{2i}\int \mathrm{d}^3\mathbf{x}\, \xi_i \mathcal{H}^i \Psi\,. \qquad (4.18)$$

This will be zero when the momentum constraint is imposed. Hence it expresses spatial diffeomorphism invariance. This argument can be straightforwardly extended to the case where matter is included.

Let us now discuss some of the properties of the Wheeler–DeWitt (WDW) equation (4.16):

1. The WDW equation, due to its association with the lapse function, is similarly related to time reparametrization invariance ($\delta S/\delta N = 0 \to \mathcal{H} = 0$).
2. It combines geometry and matter into a single quantum equation, as expected in quantum gravity.
3. The arguments of the wave function $\Psi(h_{ij}, \Phi)$ are the 3-geometry h_{ij} and matter field values Φ, on what may be considered to be an initial or final hypersurface – see also Fig. 4.1. In the momentum representation, the arguments would similarly be the initial or final momenta. Let us emphasize this point by noting that the 4-geometry (spacetime) is *not* an appropriate variable. This is a clear distinction from classical general relativity. A 4-geometry only arises with a certain probability from a succession of 3-geometries. We will say more about this important feature in Section 4.2.
4. Let us elaborate a little further: The configuration space is the set of all inequivalent 3-geometries – geometries that are merely related by a change of coordinates are identified. This configuration space is called *superspace*. (It is not to be confused with the superspace of supersymmetric theories, which is an extension of Riemannian geometry to include fermionic/Grassmannian coordinates.) The wave function is a function on superspace. A classical 4-geometry may then be seen as a 1-dimensional curve through superspace. But due to the uncertainty principle, the wave function necessarily has a certain spread. An approximately classical 4-geometry then arises when the wave function has support along a fairly narrow path through superspace. Note that a 4-geometry plays a similar role here to that of a particle trajectory in quantum mechanics; both are approximate concepts, only meaningful in a semiclassical approximation.
5. One variable stands out: the scale factor of the universe, given in 3 spatial

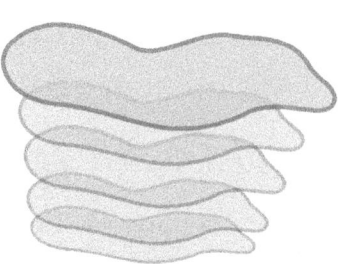 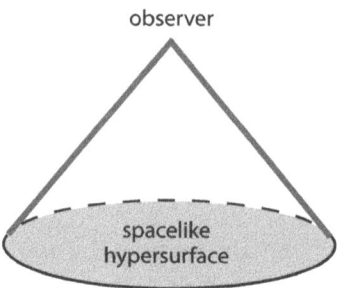

Figure 4.1 *Left panel:* A 4-geometry can be sliced (in many inequivalent ways) into a series of 3-geometries, like the slicing of a potato into chips. These 3-geometries, as well as matter configurations on them, are the arguments of the wave function. A 4-geometry then only emerges as an approximate concept, in a semiclassical approximation of the dynamics. *Right panel:* An observer only sees their past light cone. (In this sketch time is vertical and space horizontal.) However, if we know the physics on a space-like hypersurface we may infer what happened on the light cone, which lies in its causal future, so that it is justified to concentrate on space-like hypersurfaces as arguments of the wave function. That said, it would be of interest to construct a theory that directly deals with light cones only.

dimensions by $(\det(h_{ij}))^{1/6}$ (or one could just as well use the volume). It enters with a negative sign in the DeWitt metric. It is straightforward to work this property out directly, and we will also see it in explicit examples – see, e.g., (4.27). Other metric deformations, as well as matter fields, enter with a positive sign. The negative sign of the scale factor term has its root in the inherent instability of gravity: A gravitating system, left to its own device and in the absence of any other forces, will collapse irretrievably.

6. The WDW equation does not contain any explicit dependence on coordinates, in particular on time. This is different from ordinary quantum mechanics, where time plays a privileged role. It makes sense here, since we do not measure time directly, but rather correlations between field configurations (e.g., the small arrow on the watch points towards the 12 and the sun is high in the sky). This illustrates the fact that quantum gravity is an operational theory, in principle dealing only with measurable quantities, and not allowing questions that do not have an operational/experimental meaning.

7. The WDW equation is linear, hence the superposition principle continues to hold when gravity is included. That is, if Ψ_1 and Ψ_2 both satisfy the WDW equation, then so does $\alpha_1\Psi_1 + \alpha_2\Psi_2$, with $\alpha_{1,2} \in \mathbb{C}$. Thus

canonical quantum gravity includes superpositions of geometries with corresponding matter configurations.

8. There is an ambiguity, usually called the problem of *factor ordering*, as to the precise placement of the derivatives when implementing (4.14). For the example of the scale factor, should one put $\frac{\delta}{\delta(a^2)}\frac{\delta}{\delta(a^2)}$ or perhaps $a^{p-2}\frac{\delta}{\delta a}a^{-p}\frac{\delta}{\delta a}$, for some p? In explicit examples, sensible choices can often be found, in particular by requiring invariance under field redefinitions.

9. We have not said yet how probabilities are supposed to emerge from the wave function. This is tricky, as there is no external time, and probabilities in ordinary quantum mechanics are directly tied to the evolution in external time t. The best we will be able to do is to identify an effective time in the semiclassical approximation, when it will make sense to talk about an effective 4-geometry. In such a situation, relative probabilities can be defined. This will be discussed in Section 4.2 below.

4.1.3 Restriction to minisuperspace

A general 3-surface contains an infinite number of possible deformations. This means that solving the WDW equation (4.16) for $\Psi(h_{ij}, \Phi)$ in full generality is technically impossible. But in a cosmological context, we are not necessarily interested in this most general solution. Observations of homogeneity on large scales, for instance, motivate us to restrict the analysis at first to homogeneous metrics, containing only a finite (typically small) number of specified anisotropic deformations. This restriction of superspace is known as *minisuperspace*, and it can render the WDW equation much more manageable. One potential issue when one restricts to just a subset of possible metric deformations is that one has to set all other deformations, as well as their associated momenta, to zero, which is in conflict with the uncertainty principle. In practice one has to check in every case that (at least) small, linear perturbations around the minisuperspace geometries are suppressed. If this is the case, the minisuperspace restriction will be retrospectively justified (although one should remain aware that there might still be nonperturbative instabilities).

In a minisuperspace context, of which we will see explicit examples shortly, the action has the following general structure

$$S = \int dt N \left(\frac{1}{2} G_{AB} \frac{1}{N} \frac{dq^A}{dt} \frac{1}{N} \frac{dq^B}{dt} - U(q^A) \right), \qquad (4.19)$$

where $q^A(t)$ are fields that depend solely on time, G_{AB} is the field space metric, N the lapse function, and $U(q^A)$ a potential. The label A runs over all fields, encompassing both metric and matter degrees of freedom. When one of the fields is the scale factor, then the field space metric G_{AB} has indefinite signature $(-++\cdots+)$, as remarked in the previous section.

Pretty much the simplest example is the case where one retains only the scale factor of the universe plus a (purely time-dependent) scalar field, so that $q^A = (a, \phi)$. Then with the closed FLRW metric

$$ds^2 = -N^2 dt^2 + a(t)^2 d\Omega_3^2, \tag{4.20}$$

the action for general relativity coupled to the minimally coupled scalar field is given by

$$S = \int dt \mathcal{L} = 2\pi^2 \int N dt \left[-\frac{3}{N^2} a \dot{a}^2 + \frac{1}{2N^2} a^2 \dot{\phi}^2 + 3a - a^3 V(\phi) \right], \tag{4.21}$$

where $V(\phi)$ is the scalar potential. The physical time is given by the integral $\int dt\, N$. From the action we can read off the field space metric as

$$G_{aa} = -12\pi^2 a, \qquad G_{\phi\phi} = 2\pi^2 a^3. \tag{4.22}$$

The canonical momenta are given by

$$p_a = \frac{\partial \mathcal{L}}{\partial \dot{a}} = -12\pi^2 a \frac{\dot{a}}{N}, \qquad p_\phi = \frac{\partial \mathcal{L}}{\partial \dot{\phi}} = 2\pi^2 a^3 \frac{\dot{\phi}}{N}. \tag{4.23}$$

In terms of these, the Hamiltonian may be written as

$$\mathcal{H} = \frac{1}{2} G^{AB} p_A p_B + U, \tag{4.24}$$

where the effective potential takes the form

$$U = 2\pi^2 \left(-3a + a^3 V \right). \tag{4.25}$$

Classically, the vanishing of the Hamiltonian corresponds to the Friedmann equation. Canonical quantization (in the field basis) corresponds to replacing momenta with derivatives, $p_A \to -i\hbar \frac{\partial}{\partial q^A} \equiv -i\hbar \partial_A$, and this leads to the WDW equation

$$\hat{\mathcal{H}} \Psi = \left(-\frac{\hbar^2}{2} \Box + U \right) \Psi = 0, \tag{4.26}$$

where $\Psi = \Psi(a, \phi)$ is the wave function of the universe. We have fixed the factor ordering ambiguity by writing the d'Alembertian in covariant form in field space, $\Box = G^{AB} \nabla_A \nabla_B$, since then its form is unchanged under field redefinitions. Explicitly, the WDW equation reads

$$\left[\frac{\hbar^2}{48\pi^4} \left(\frac{1}{a} \frac{\partial^2}{\partial a^2} + \frac{1}{a^2} \frac{\partial}{\partial a} - \frac{6}{a^3} \frac{\partial^2}{\partial \phi^2} \right) - 3a + a^3 V(\phi) \right] \Psi(a, \phi) = 0. \tag{4.27}$$

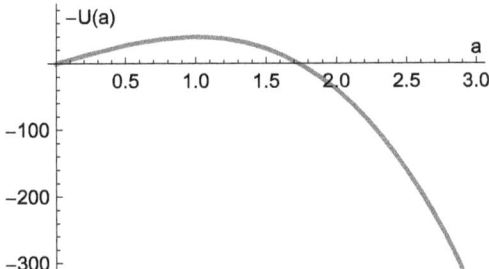

Figure 4.2 The effective potential $-U(a) = 2\pi^2(3a - a^3 V)$, shown here with constant potential $V = \Lambda = 1$, has the shape of a tunneling potential at small scale factor a and falls off asymptotically at large a.

At first, we will be most interested in the qualitative nature of solutions, especially in the context of inflation. A first approximation to inflation is to consider a constant potential $V(\phi) = \Lambda$. Then the scalar field is simply constant and one can neglect ∂_ϕ in (4.27). The effective potential looks as shown in Fig. 4.2. It is reminiscent of a tunneling problem.

We will temporarily reinstate $\hbar$, since it is instructive to do so. Then, to leading order in $\hbar$, the solution to (4.27) is very simple,

$$\Psi \approx e^{\pm \frac{12\pi^2}{\hbar \Lambda}\left(1 - \frac{a^2 \Lambda}{3}\right)^{3/2}}. \tag{4.28}$$

Any linear combination of the two solutions above will also be a solution. Before analyzing the solution further, let us point out that the single-derivative term proportional to ∂_a in (4.27) only contributes at sub-leading order in $\hbar$. Note that this term arose precisely from the choice of factor ordering – its coefficient could easily have been changed (and, e.g., made to vanish) by choosing a different ordering of derivatives. This illustrates the general point that the ambiguity in factor order is a subleading effect, and thus typically unimportant in a semiclassical approximation.

Now, for large values of the scale factor, it is clearer to extract a factor of -1 from the root, and write the approximate solutions as

$$\Psi \approx e^{\pm i \frac{4\pi^2}{\hbar}\sqrt{\frac{\Lambda}{3}}a^3} \quad \left(a \gg a_{\rm dS} \equiv \sqrt{\frac{3}{\Lambda}}\right). \tag{4.29}$$

The main property is that these solutions oscillate, and they do so ever faster with the expansion of the universe. They essentially describe the classical expansion of the universe, as we will justify shortly.

Meanwhile, at small a the solution (4.28) is exponentially growing or damped. This is the regime where the scale factor takes nonclassical values, smaller than the classical minimum radius $a_{\rm dS} = \sqrt{3/\Lambda}$ of a de Sitter universe. Recalling the analogy with a particle under a potential barrier (for which the wave function is damped under the barrier, and oscillating outside of it), it is tempting to interpret such a regime as describing the quantum tunneling of the universe from zero size to $a = a_{\min}$, thereby overcoming the potential barrier seen in Fig. 4.2. In Chapter 7, we will see that the no-boundary proposal provides a framework to make this intuitive picture much more precise.

Before proceeding to a more careful consideration of the interpretation of the wave function, let us point out that a difficulty with solving the WDW equation is that one typically does not know how to fix the integration constants that arise in the solutions. This issue, as we will see in Chapter 5, is much clearer in the path integral approach.

4.2 How to reconstruct the universe from the wave function

If, in principle, the wave function contains all the information we can possibly obtain about the universe, then an important question is how we can extract this information from the wave function. In particular, how does a 4-dimensional spacetime with matter arise from the wave function, which only deals with 3-geometries and associated 3-dimensional matter configurations? Also, how are general relativity and quantum mechanics recovered in appropriate limits? What, indeed, is the appropriate limit?

Let us start with a mathematical observation: In its minisuperspace form (4.26), the WDW equation looks very much like the Klein–Gordon equation for a scalar field. As such, it admits a conserved current, defined via

$$J^A = -\frac{i\hbar}{2}\left(\Psi^\star \nabla^A \Psi - \Psi \nabla^A \Psi^\star\right). \tag{4.30}$$

In this section, the reduced Planck constant $\hbar$ will play an important organizing role, which is why we will write it out explicitly. It is straightforward to check that $\nabla_A J^A = 0$ subject to using the WDW equation (4.26). This current is the quantity that is conserved, while $\Psi^\star \Psi$ (which one might have naïvely expected to use as a probability density) is not. Note that the current is conserved in field space, not in time.

To proceed, it is useful to put the wave function in a form such that its amplitude and phase are written out separately,

$$\Psi = e^{\frac{1}{\hbar}(\mathcal{W}+i\mathcal{S})}. \tag{4.31}$$

Here $\mathcal{W}, \mathcal{S}$ are real functions of the fields q^A, $\mathcal{W}$ is called the *weighting*, and

$\mathcal{S}$ the *phase*. We can then expand the WDW equation (4.26) as a series in $\hbar$, treating $\mathcal{W}$ and $\mathcal{S}$ to be $\mathcal{O}(\hbar^0)$, to find at leading and subleading orders, respectively,

$$-\frac{1}{2}(\nabla\mathcal{W})^2 + \frac{1}{2}(\nabla\mathcal{S})^2 + U = 0, \quad \nabla\mathcal{W}\cdot\nabla\mathcal{S} = 0, \qquad (4.32)$$

$$\Box\mathcal{W} = 0, \quad \Box\mathcal{S} = 0. \qquad (4.33)$$

The contractions above are effected with the field space metric, e.g., $\nabla\mathcal{W}\cdot\nabla\mathcal{S} \equiv G^{AB}\nabla_A\mathcal{W}\nabla_B\mathcal{S}$. There are two equations at each order, one being the real part and the other the imaginary part of the equations. With these definitions, the conserved current (4.30) becomes

$$J^A = e^{\frac{2}{\hbar}\mathcal{W}}\nabla^A\mathcal{S}. \qquad (4.34)$$

Now recall some intuition from the Jeffreys–Wentzel–Kramers–Brillouin (JWKB) semiclassical approximation often used in quantum mechanics: A wave function behaves semiclassically if its amplitude varies slowly compared to the variation of the phase. If we assume the same here in Eq. (4.32), namely that $\mathcal{W}$ varies slowly compared to $\mathcal{S}$, then we actually recover the classical Hamilton–Jacobi equation,

$$(\nabla\mathcal{W})^2 \ll (\nabla\mathcal{S})^2 \quad \rightarrow \quad \frac{1}{2}(\nabla\mathcal{S})^2 + U \approx 0, \qquad (4.35)$$

subject to identifying $\mathcal{S}$ with the classical action (hence the suggestive notation). The canonical momenta then also satisfy the standard relation

$$p_A = \frac{\partial\mathcal{S}}{\partial q^A}. \qquad (4.36)$$

The remaining equations, i.e., the right-hand side of (4.32) and all of (4.33), then express the conservation of the current (4.34) to leading and subleading order.

To define a probability measure, one has to choose a foliation of superspace – see Fig. 4.3. One may picture classical trajectories as curves in superspace. The foliation should be such that it intersects each classical trajectory only once. There are various ways of doing this. One possibility is to foliate along surfaces of constant classical action $\mathcal{S}$ (since $\mathcal{S}$ describes a congruence of classical trajectories). One then picks a normal vector n^A to these surfaces, such that $\nabla_n\mathcal{S} > 0$. Note that this will not be possible everywhere in superspace, as for instance the locus $\nabla\mathcal{S} = 0$ describes the breakdown of the semiclassical approximation. One can then define a relative probability measure, valid only in the semiclassical realm, as

$$\mathcal{P} = e^{\frac{2}{\hbar}\mathcal{W}}\nabla_n\mathcal{S}. \qquad (4.37)$$

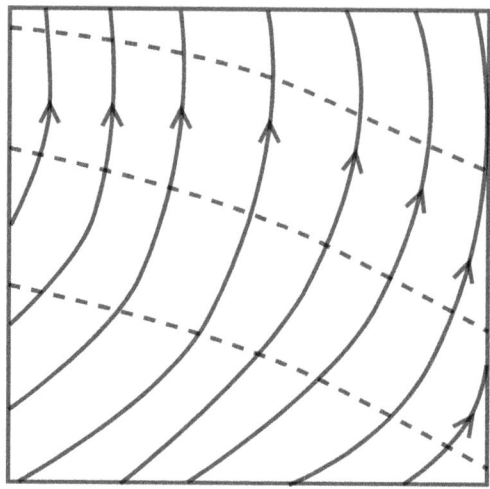

Figure 4.3 Solid lines depict classical trajectories in superspace. The dashed lines provide a possible foliation, intersecting each classical trajectory only once. Such a foliation is useful in defining semiclassical probabilities.

It is not currently known how an overall normalization can be implemented in a mathematically precise way, hence the probabilities should be thought of as being relative to each other. But, importantly, these relative probabilities are positive and conserved along classical trajectories. In the semiclassical approximation, i.e., for small $\hbar$, they are proportional to the standard probabilities in quantum mechanics,

$$\mathcal{P} \propto e^{\frac{2}{\hbar}\mathcal{W}} = \Psi^*\Psi. \tag{4.38}$$

Different foliations will give rise to the same relative probabilities as long as they are chosen so as to intersect each classical trajectory only once.

Further note that, when the JWKB approximation holds, the wave function (4.31) satisfies

$$p_A \Psi = -i\hbar\, \partial_A \Psi \approx \partial_A \mathcal{S}\, \Psi, \tag{4.39}$$

to a good approximation, in agreement with the assignment (4.36). This relation shows that the semiclassical wave function is peaked on solutions described by the first-order relation $p_A = \partial_A \mathcal{S}$. In other words, the universes contained in the semiclassical wave function are first integrals of the classical momentum–action relationship.

The ability to assign relative probabilities is clear progress, but it is insufficient to make predictions yet, since the probabilities for different classical trajectories will depend on the particular solutions to the WDW equation

4.2 How to reconstruct the universe from the wave function

that one chooses, in particular the choice of integration constants. Thus one also needs a theory of initial/boundary conditions – we will defer this topic to Chapter 7.

For now, we can make further progress in understanding subsystems, regardless of the overall probability of that particular universe. Let us denote the part of the Hamiltonian that acts on the subsystem by H_2. We will assume that the subsystem exerts negligible backreaction on the universe as a whole – it is appropriate to think of H_2 as describing small perturbations in the universe, for example curvature perturbations during inflation. Then the WDW equation (4.26) simply gets extended by a term,

$$\left(-\frac{\hbar^2}{2}\Box + U + H_2\right)\Psi = 0. \tag{4.40}$$

If we expand the WDW equation once again in powers of $\hbar$, then at subleading order we obtain

$$i\hbar \nabla \mathcal{S} \cdot \nabla \psi_2 = H_2 \psi_2, \tag{4.41}$$

where we factorized $\Psi = \psi_0 \psi_2$ into a product of ψ_0, depending only on the background variables (i.e., this is what we called Ψ so far), and ψ_2, depending on both the background fields and the perturbations. The crucial realization is that the directional derivative $\nabla \mathcal{S} \cdot \nabla$ may be identified with a time derivative

$$\nabla \mathcal{S} \cdot \nabla \equiv \frac{\partial}{\partial t}. \tag{4.42}$$

If one does that, the subsystem obeys the standard Schrödinger equation

$$i\hbar \frac{\partial}{\partial t} \psi_2 = H_2 \psi_2. \tag{4.43}$$

This is remarkable. As long as the JWKB approximation holds, we can see that an effective classical spacetime background emerges, which provides a time coordinate and in turn allows ordinary quantum mechanics to be valid on this curved spacetime background. In this way both general relativity and quantum theory emerge in the semiclassical approximation.

With the explicit example (4.29) we had above, we may check that the identification of time is sensible. With $\mathcal{S} \approx -4\pi^2 \sqrt{\frac{\Lambda}{3}} a^3$ and $G_{aa} = -12\pi^2 a$, one gets

$$\nabla \mathcal{S} \cdot \nabla = G^{aa} \partial_a \mathcal{S} \, \partial_a = \sqrt{\frac{\Lambda}{3}} a \, \partial_a \stackrel{!}{=} \frac{\partial}{\partial t}, \tag{4.44}$$

which indeed agrees with the corresponding classical de Sitter solution expressed in physical time t, $a = \sqrt{\frac{3}{\Lambda}} e^{\sqrt{\frac{\Lambda}{3}} t}$.

The recovery of the Schrödinger equation for subsystems, as well as the assignment of relative probabilities for the background, hinges crucially on the wave function being approximated by a single JWKB term. However, in general the wave function will be a superposition of a great many terms. And even if all these terms are such that their amplitudes vary slowly compared to their phases, the assignments above nevertheless do not work. So a crucial question is whether we are justified in treating such terms individually.

The answer is that once one includes matter and perturbations, then decoherence will occur and the different terms will effectively evolve as separate universes, each with its own effective spacetime. We will discuss decoherence in detail in Chapter 8.

Another question is under what conditions we may expect the wave function to evolve into JWKB terms in the first place. This is of central importance if we want to understand how an effectively classical spacetime could have arisen at all. Put differently, what does it take for a classical spacetime to emerge from a fully quantum state? This requires something more than decoherence – it requires a dynamical mechanism that causes the wave function to become oscillating, with a slowly evolving amplitude. For this, let us go back to the WDW equation in the presence of a scalar field, Eq. (4.27) (ignoring the "factor-ordering" term in ∂_a). Then there are two known possibilities for obtaining an oscillating solution, and these can be understood heuristically:

- *Inflation:* $V(\phi) > 0$ and $\frac{|V_{,\phi}|}{V} < \sqrt{2}$.
  During inflation, the scalar field rolls slowly down the potential compared to the rapid expansion of space. Thus to a first approximation one may neglect derivatives with respect to the scalar field, $\frac{1}{a^2}\frac{\partial^2}{\partial \phi^2} \ll \frac{\partial^2}{\partial a^2}$. For scale factor values larger than the de Sitter radius, we are then left with the approximate WDW equation

$$\left(\frac{\hbar^2}{48\pi^4 a} \frac{\partial^2}{\partial a^2} + a^3 V \right) \Psi = 0 \qquad (4.45)$$

  which is a wave equation for the scale factor, and admits oscillating solutions $\Psi \sim e^{\pm i a^3 / \hbar}$.

- *Ekpyrosis:* $V(\phi) < 0$ and $\frac{|V_{,\phi}|}{|V|} > \sqrt{6}$.
  During ekpyrosis, we have the reverse situation in that the scalar field rolls fast and the scale factor contracts very slowly. Thus we can ignore the scale factor derivatives this time, leaving us with the approximate

WDW equation
$$\left(\frac{\hbar^2}{8\pi^4 a^3}\frac{\partial^2}{\partial \phi^2} - a^3 V\right)\Psi = 0. \tag{4.46}$$

Since the potential $V \sim -e^{-c\phi}$ is negative (and steep) during an ekpyrotic phase, we once again obtain oscillating solutions $\Psi \sim e^{\pm i a^3 \sqrt{-V}/\hbar}$.

This reveals a perhaps unexpected property: The two dynamical mechanisms known to be able to robustly smooth out a crumpled universe, namely inflation and ekpyrosis, are also the two known mechanisms that drive the quantum gravitational wave function to JWKB form, and that can thus dynamically produce an effective classical spacetime out of a fully quantum state. This fact provides compelling additional evidence for taking these mechanisms seriously as possible phases of early cosmological evolution.

Exercises

4.1 Consider the commutation between the scale factor and the associated momentum. If you translate this into an uncertainty relation, and assume the uncertainty is equally distributed between the two variables, then at what scale factor value does this happen? Discuss.

4.2 With the metric (4.20), write the action for gravity plus a cosmological constant in the form $S = \int \mathrm{d}t \left(\frac{K}{N} - NP\right)$. Identify the kinetic term K and the potential P. The Hamiltonian constraint is obtained by varying the lapse N, and it can be solved straightforwardly for N. Plug this expression back into the metric, which now contains the scale factor as a time variable. By changing coordinates, express this metric in terms of physical time and show that the constraint is solved by the de Sitter metric.

4.3 We saw that inflation and ekpyrosis are dynamical mechanisms that drive the wave function to JWKB semiclassicality. In this exercise we will calculate the rate at which this happens (in a slightly abridged manner), for inflationary models with constant slow-roll parameter ϵ.

A constant slow-roll parameter ϵ is obtained with an exponential potential $V = e^{-\sqrt{2\epsilon}\phi}$, where we set $V_0 = 1$ by shifting the origin of ϕ. This model admits the scaling solution (2.26), which is an attractor, as shown in Section 2.1.3. Start by rewriting the constant a_0 in terms of a and $V(\phi)$. This constant can be used to label different asymptotic solutions. Then calculate the on-shell action at late times (large a), and use the expression for a_0 to rewrite this in terms of the final $a_f \equiv b, \phi_f \equiv \chi$ values. This provides us with the phase $\mathcal{S}$ of the semiclassical wave function. For the weighting, use the ansatz $\mathcal{W} = b^{n_1} V^{n_2}$ and solve the second equation in (4.32), $\nabla \mathcal{S} \cdot \nabla \mathcal{W} = 0$, to find a relation between n_1 and n_2. What does this imply for the asymptotic time dependence of weighting? A more refined argument, derived in Chapter 7 and making use of the symmetries of the action, shows that in fact $n_1 = 2\epsilon/(\epsilon - 1)$. Then derive the scaling of the JWKB classicality condition $|(\nabla \mathcal{W})^2/(\nabla \mathcal{S})^2|^{1/2}$ and compare it to the efficiency of the attractor.

5
Quantizing gravity: path integral approach

Feynman discovered an extremely fruitful reformulation of quantum mechanics in terms of sums over particle paths. This can be motivated heuristically by starting with the double-slit experiment, which shows interference fringes building up when single particles successively pass through a wall with two slits – see Fig. 5.1. Famously, the observed interference patterns can be explained by summing contributions from paths through both slits, even though only single particles are sent at any time. If one extends this setup to include many walls, into which one cuts more and more slits until the walls have disappeared, then one arrives at the picture that the wave function should be calculated by summing over all possible particle paths. This reformulation may be shown to incorporate the Schrödinger equation and has the advantage of providing an intuitive picture in which particle paths can be used to perform calculations.

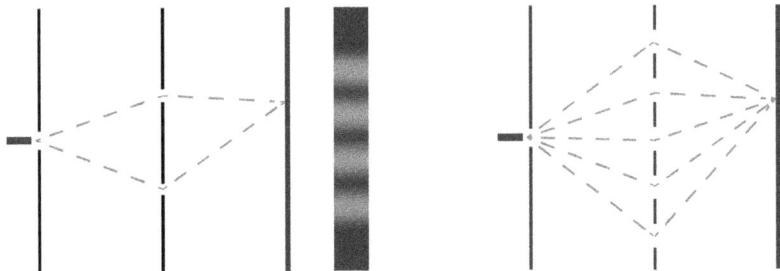

Figure 5.1 *Left panel:* In the double-slit experiment a particle is emitted on the left and hits a screen on the right. On the way to the screen one places a wall with two open slits. The interference fringes (shown on the right) can be explained by considering the wave function to be a sum of contributions from two paths, one for each slit the particle can pass through. *Right panel:* the extension to multiple slits.

The question that interests us here is how this formulation can be extended to include gravity. As already mentioned in the previous chapter, the analogue of particle paths are now 4-geometries and their matter configurations – see Fig. 5.2. Hence we are led to construct a sum over such geometries and matter configurations. More precisely, a transition amplitude between an initial configuration $|in\rangle$ consisting of a spatial hypersurface with matter living on it, and an analogous final configuration $|out\rangle$, should then be regarded as a sum over interpolating 4-geometries and matter configurations. Loosely speaking we are interested in a transition amplitude (propagator) of the form

$$\langle out| \cdots |in\rangle = \int_{in}^{out} \mathcal{D}[g_{\mu\nu}]\mathcal{D}[\Phi]\, e^{iS[g_{\mu\nu},\Phi]}, \qquad (5.1)$$

which we will try to make increasingly precise over the coming pages. We will also be interested in the question of what the most likely state of the universe is – this "wave function of the universe" can be seen as a transition amplitude where one only has a final state, and the question of what/where one should be summing *from* needs to be addressed. Amazingly enough, we will see that there is a direct link between such sums over geometries and the Wheeler–DeWitt equation that we obtained using canonical quantization.

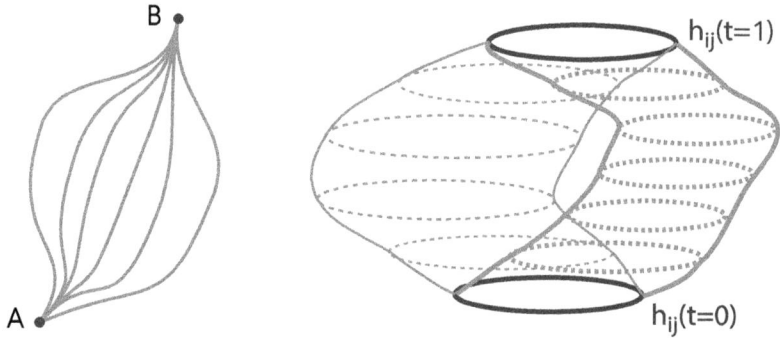

Figure 5.2 *Left panel:* Feynman's path integral approach to quantum mechanics involves calculating transition amplitudes from A to B by summing over particle paths linking the two events. *Right panel:* When gravity is included, the role of particle paths is played by entire 4-geometries (and their matter configurations) interpolating between two hypersurfaces, here labeled by the coordinate locations $t = 0$ and $t = 1$.

The advantage of the "path integral" formulation is not only that it is more intuitive. It is also potentially more precise, in at least two ways:

It allows for a clear specification of boundary conditions, and the choice of geometries that are summed over strongly impacts the results of calculations.

When trying to define integrals over geometries, one faces the challenge that naïvely the integral over spacetimes is oscillatory and thus only conditionally convergent. A proposed remedy, popular in the 1980s, was to use Euclidean time and define the integral over Euclidean (rather than Lorentzian) geometries. As we will see, such Euclidean path integrals are however divergent when gravity is included.

A proper definition requires absolutely convergent integrals, otherwise the meaning of the integrals remains ambiguous. A mathematically precise and powerful way of dealing with such oscillatory integrals is called Picard–Lefschetz theory, and we will make use of it in what follows. For convenience a brief overview of Picard–Lefschetz (PL) theory is given in Appendix B, which you are encouraged to read now. The way PL theory works is that it extends the integrals to be over complex paths/geometries and then shifts the integration contour to a sum of manifestly convergent contours. This elegant approach raises the pertinent question of which complex geometries might be sensible to include. In the present chapter we will not overly concern ourselves with this issue, but we will return to it in Chapter 7, where we will see that it may have crucial physical consequences. Here our plan is to first show how integrals over geometries can be constructed in a simple setting, and then we will look at examples of transition amplitudes when the only matter present is a cosmological constant. Finally, we will explore the relationship between the path integral formulation and the WDW equation.

5.1 Gauge fixing the sum over spacetimes

When defining a sum over geometries one has to make sure that each geometry is counted precisely once – in particular, two metrics related by a coordinate transformation are to be counted as a single geometry. We will outline how this can be done, using the simplest example where we restrict to a metric of FLRW form. Even for such a simple form, the question is how to parametrize it correctly.[1]

We will start with a metric of the form $\mathrm{d}s^2 = -N(t)^2\mathrm{d}t^2 + a(t)^2\mathrm{d}\Omega_3^2$, where we allow for time dependence not only in the scale factor $a(t)$, but also in the lapse $N(t)$. Denoting the momentum conjugate to a as p_a and recalling

---

[1] This section is more technical and condensed, and can be omitted at a first reading.

the discussion in Section 4.1.3, we know that the action can be written as

$$S = \int_0^1 dt\, (p\dot{a} - NH)\,, \tag{5.2}$$

with the Hamiltonian $H = \frac{1}{2}p^2 + U(a)$. The time coordinate is scaled such that t runs over the interval $[0, 1]$. We must be careful not to overcount geometries due to time reparametrization invariance. In other words, we must fix a gauge. The gauge fixing condition must be such that any time evolution can be brought into a form that satisfies the condition, and moreover this condition must fix the gauge completely, leaving no residual gauge freedom. A convenient and appropriate choice is

$$\dot{N} = f(a, p_a, N)\,, \tag{5.3}$$

where f is an arbitrary function of the fields, but not of their time derivatives. Imposing such a gauge condition can be done by introducing a Lagrange multiplier $\Pi(t)$ and multiplying it with the gauge condition,

$$S_{gf} = \int_0^1 dt\, \Pi\left(\dot{N} - f\right)\,. \tag{5.4}$$

Varying with respect to Π then imposes the gauge fixing relation. In order to see whether this procedure is sensible, we also have to check whether the path integral is in fact independent of the choice of gauge function $f(a, p_a, N)$. To do so, we can make use of the general formalism developed by Batalin, Fradkin, and Vilkovisky, extending a method first introduced by Fadeev and Popov. This is standard textbook material in quantum field theory, which the interested reader is encouraged to study independently. Since we will not make use of this formalism in the rest of the book, we will simply outline the steps of the calculation here.

The main idea is to add (fictitious) ghost fields that extend the theory so that the desired symmetry is made manifest and that, once integrated out, fix the gauge. A subtlety is that these new ghost fields $C, \bar{C}$, and their conjugate momenta $P, \bar{P}$, are taken to be anticommuting fields (their values are Grassmann numbers, which one may think of as fermionic numbers). The appropriate choice for the action of these new fields is

$$S_{gh} = \int_0^1 dt\, \left(\bar{P}\dot{C} + \bar{C}\dot{P} - \bar{P}P + C\{f, H\}\bar{C} + P\frac{\partial f}{\partial N}\bar{C}\right)\,, \tag{5.5}$$

where $\{f, H\} = \frac{\partial f}{\partial a}\frac{\partial H}{\partial p} - \frac{\partial f}{\partial p}\frac{\partial H}{\partial a}$ denotes a Poisson bracket. This action is in-

5.1 Gauge fixing the sum over spacetimes

variant under a so-called BRS symmetry, with anticommuting parameter λ,

$$\delta a = \lambda C \frac{\partial H}{\partial p}\,,\ \delta p = -\lambda C \frac{\partial H}{\partial a}\,,\ \delta N = \lambda P\,,\ \delta \Pi = 0\,,$$
$$\delta C = 0\,,\ \delta P = 0\,,\ \delta \bar{C} = -\lambda \Pi\,,\ \delta \bar{P} = -\lambda H\,, \quad (5.6)$$

where one imposes that Π, C, and $\bar{C}$ vanish at the end points $t = 0, 1$.

Putting everything together, the full phase-space path integral is obtained by integrating with the Liouville measure (noting that Π has become the momentum conjugate to the lapse),

$$\Psi = \int \mathcal{D}a\mathcal{D}p\mathcal{D}N\mathcal{D}\Pi\mathcal{D}C\mathcal{D}P\mathcal{D}\bar{C}\mathcal{D}\bar{P}\, e^{\frac{i}{\hbar}(S+S_{gf}+S_{gh})}\,. \quad (5.7)$$

We are now in a position to verify whether the gauge fixing function f can be replaced by a different choice, say $\tilde{f}$. This can be done explicitly by choosing the specific gauge parameter

$$\lambda = \frac{i}{\hbar} \int_0^1 dt(f - \tilde{f})\,. \quad (5.8)$$

The action is clearly invariant under this transformation, since the transformation was constructed precisely such that it is an invariance of the action. This is not quite enough though, since we must also make sure that the measure transforms properly. The Jacobian factor of this transformation can be evaluated to be

$$J = e^{\frac{i}{\hbar}\int dt\left(C\{\tilde{f}-f,H\}\bar{C}+P\frac{\partial(\tilde{f}-f)}{\partial N}\bar{C}\right)}, \quad (5.9)$$

which effectively replaces f by $\tilde{f}$ in the ghost action (5.5). This confirms that the path integral is independent of the choice of gauge f, which we can therefore fix for convenience. Different choices will turn out to be useful in different situations.

For now, we will proceed with the simplest possible choice, namely $f = 0$ or $\dot{N} = 0$, but in the next section we will already encounter a second, very useful choice. With the choice $f = 0$, the integrations of ghosts factorize out since they do not depend on any other fields. In fact, this implies that they will give rise to a purely numerical factor. Using the properties of anticommuting numbers, these integrals are quickly performed to find

$$\int \mathcal{D}C\mathcal{D}P\mathcal{D}\bar{C}\mathcal{D}\bar{P}e^{\frac{i}{\hbar}\int_0^1 dt(\bar{P}\dot{C}+\bar{C}\dot{P}-\bar{P}P)} = e^{[(\bar{C}(1)-\bar{C}(0))(C(1)-C(0))]} = 1\,, \quad (5.10)$$

where the last equality is obtained by making use of the boundary conditions $C(0) = C(1) = \bar{C}(0) = \bar{C}(1) = 0$.

As a consequence, the integral over the lapse also simplifies considerably,

$$\int \mathcal{D}N\mathcal{D}\Pi \, e^{\frac{i}{\hbar}\int_0^1 dt \dot{N}\Pi} = \int \mathcal{D}N \delta(N) = \int dN \,. \qquad (5.11)$$

Thus the functional integral over the lapse has effectively collapsed to an ordinary integral. This last property is a major reason for the success of minisuperspace models, since the lapse integral is most closely associated with the emergence of the 4th dimension linking the initial and final hypersurfaces. The fact that this integral has become an ordinary one renders many minisuperspace models tractable.

When the dust settles, we are left with

$$\Psi = \int dN \mathcal{D}a \mathcal{D}p \, e^{\frac{i}{\hbar}\int_0^1 dt(p\dot{a}-NH)} \,. \qquad (5.12)$$

And sometimes it will be even simpler to switch from this phase-space integral to a field-space one, as we will see in concrete examples.

5.2 Transitions in a universe with positive cosmological constant

To become a little more familiar with a quantum view of cosmology, we will study the simple example of transitions in a universe dominated by gravity and a positive cosmological constant. It should be stressed that this is not the main application of quantum cosmology, as a realistic treatment of such transitions would have to include other matter and a more general metric, and would be strongly affected by the effects of decoherence. Nevertheless, the examples below serve both to check the formalism built up so far, and to build intuition in dealing with a quantum treatment of spacetime.

The restriction to gravity and a positive cosmological constant $\Lambda > 0$ goes somewhat beyond being a simple toy model, as it provides a good first approximation to the present-day universe, and to a possible early inflationary phase. The action is given by (setting $8\pi G = 1$)

$$S = \int_{\mathcal{M}} d^4x \sqrt{-g} \left(\frac{R}{2} - \Lambda\right) + \gamma \int_{\partial \mathcal{M}} d^3y \sqrt{h} K \,, \qquad (5.13)$$

where we included a GHY surface term, which is required if one wants to hold the metric fixed on the initial ($\gamma = -1$) or final ($\gamma = +1$) hypersurface, as discussed in Section 4.1.1. By contrast, if one would like to fix the momentum instead, one should dispense with the GHY term and set $\gamma = 0$.

As we saw in the previous section, when we restrict to the FLRW metric

$$ds^2 = -N^2 dt^2 + a^2(t) d\Omega_3^2 \,, \qquad (5.14)$$

5.2 Transitions in a universe with positive cosmological constant

we can set the lapse N to a constant in order to fix the gauge. Here we will focus on spatial metrics that are 3-spheres,[2] since they have finite volume. One can equally well work with flat or open spatial metrics, but in that case one must assume a nontrivial topology (for a spatially flat universe, one can take it to be a 3-torus, for example) such that the spatial volume is finite – otherwise the action will be infinite, and no sensible result can be obtained.

The action then becomes

$$\frac{S}{2\pi^2} = \int_0^1 dt\, N \left(3a^2 \frac{\ddot{a}}{N^2} + 3a \frac{\dot{a}^2}{N^2} + 3a - a^3 \Lambda \right) - \sum_{t=0,1} \gamma \frac{3a^2 \dot{a}}{N}, \qquad (5.15)$$

where the surface term can be nontrivial on both boundaries $t = 0, 1$. Since we take the 3-geometry to be spherical, the only degree of freedom left is the scale factor (or its conjugate momentum), which can be fixed on the two boundaries. The object that we are interested in is thus the transition amplitude from, say, scale factor value a_0 to scale factor value a_1, and it is given by the path integral

$$G[a_1; a_0] = \int_\mathcal{C} dN \int_{a=a_0}^{a=a_1} \mathcal{D}a\, e^{\frac{i}{\hbar} S(N, a)}. \qquad (5.16)$$

A few comments are in order. First, we have called the path integral $G[a_1; a_0]$ rather than Ψ. This is because it is in fact a propagator between two spatial configurations (the G is used for Green's function). At least in the semiclassical approximation, its modulus squared would be interpreted as specifying the relative probability for such a transition. Second, one can decompose the path integral into two parts. The functional integral over the scale factor $\int_{a=a_0}^{a=a_1} \mathcal{D}a\, e^{\frac{i}{\hbar} S(N,a)}$ represents the amplitude to evolve from a_0 to a_1 in a fixed proper time N, while the lapse integral then sums over all possible proper times that this transition could take. And third, the integration contours for both lapse and scale factor have not been specified yet. We will turn to this question shortly, as it turns out to be tied to the specific choice of boundary conditions.

But before doing that, there is one more simplification that we can perform, and that is a clever change of variables. We rescale the lapse function N by an inverse factor of the scale factor, $\tilde{N}/a$, and call the new lapse $\tilde{N}$. This corresponds to the gauge fixing condition $\dot{N} = \frac{p_a}{12\pi^2 a^2} N^2$. Moreover, we will work with the scale factor squared as the basic variable, $a(t)^2 \equiv q(t)$, but we will retain t as the label for the time coordinate, as this should not lead

---

[2] Recall the explicit form $d\Omega_3^2 = d\chi^2 + \sin^2 \chi \left(d\theta^2 + \sin^2 \theta\, d\phi^2 \right)$, with $0 \leq (\chi, \theta) \leq \pi$, $0 \leq \phi \leq 2\pi$. The unit 3-sphere has volume $2\pi^2$.

to confusion. All in all, we end up with the metric

$$ds^2 = -\frac{\tilde{N}^2}{q(t)}dt^2 + q(t)d\Omega_3^2.\tag{5.17}$$

This rescaling of the lapse does not carry any particular physical meaning. Rather, its advantage is entirely computational, as it renders the action quadratic in q:

$$\frac{S}{2\pi^2} = \int_0^1 dt\left(\frac{3}{2\tilde{N}}q\ddot{q} + \frac{3}{4\tilde{N}}\dot{q}^2 + \tilde{N}(3-\Lambda q)\right) - \sum_{t=0,1}\gamma\frac{3}{2\tilde{N}}q\dot{q}.\tag{5.18}$$

This will allow for a great simplification of the path integral over the scale factor, as we will see in the specific examples we will analyze next.

5.2.1 Dirichlet boundary conditions

We will first analyze the case where we hold the scale factor fixed both on the initial and on the final hypersurfaces – that is to say, we fix $q(0) = q_0$ and $q(1) = q_1$. We should point out that the model we are looking at is so simple that no direction of time is implied by the formalism, so calling a particular hypersurface "initial" or "final" is arbitrary. We only do this because it is convenient to refer to the two hypersurfaces by different names, and because we have in mind applications to an expanding universe. Thus we will assume $q_1 > q_0$.

Fixing q at $t = 0, 1$ requires us to include the GHY surface term. Hence we will set $\gamma = -1$ on the initial and $\gamma = +1$ on the final hypersurface. Varying the action with respect to q and $\tilde{N}$ results in

$$\frac{\delta S}{2\pi^2} = \int dt\left[\delta q\left(\frac{3}{2\tilde{N}}\ddot{q} - \tilde{N}\Lambda\right) + \delta\tilde{N}\left(\frac{3}{4\tilde{N}^2}\dot{q}^2 + 3 - \Lambda q\right)\right] - \frac{3}{2\tilde{N}}\dot{q}\delta q|_{t=0}^{t=1}.\tag{5.19}$$

This confirms that we can indeed fix q on the boundaries ($\delta q = 0$). The equation of motion and constraint following from the variation of the fields are then, respectively,

$$\ddot{q} = \frac{2\Lambda}{3}\tilde{N}^2,\tag{5.20}$$

$$\frac{3}{4\tilde{N}^2}\dot{q}^2 = \Lambda q - 3.\tag{5.21}$$

With the imposed boundary conditions, the equation of motion can be solved

5.2 Transitions in a universe with positive cosmological constant

straightforwardly,

$$\bar{q} = \frac{\Lambda}{3}\tilde{N}^2 t^2 + \left(-\frac{\Lambda}{3}\tilde{N}^2 + q_1 - q_0\right)t + q_0. \tag{5.22}$$

At this point, this is just a solution to the classical equation of motion; this solution does not necessarily also satisfy the constraint (5.21). However, we can use the solution to the equation of motion to perform the path integral over q. The idea is to shift the integration variable such that the new variable Q describes (completely arbitrary) variations away from the solution (5.22).

$$q(t) \equiv \bar{q}(t) + Q(t). \tag{5.23}$$

Since $\bar{q}$ already satisfies the imposed boundary conditions, we must impose $Q(0) = Q(1) = 0$ on the new, shifted variable. The path integral over q now conveniently splits up into two pieces,

$$G[q_1; q_0] = \int_{\mathcal{C}} d\tilde{N} e^{2\pi^2 i S_0/\hbar} \int_{Q(0)=0}^{Q(1)=0} \mathcal{D}Q e^{2\pi^2 i S_2/\hbar}, \tag{5.24}$$

where

$$S_0 = \int_0^1 dt\left(-\frac{3}{4\tilde{N}}\dot{\bar{q}}^2 + 3\tilde{N} - \tilde{N}\Lambda\bar{q}\right), \quad S_2 = -\frac{3}{4\tilde{N}}\int_0^1 dt\, \dot{Q}^2. \tag{5.25}$$

Since $\bar{q}$ solves the equation of motion ($\delta S/\delta q = 0$), no terms linear in Q appear in the action. This now implies that the path integral over q is given by a piece already fully specified (the integral involving S_0) plus a quadratic, Gaussian integral involving S_2.

The path integral over S_2 can be evaluated in various ways. It has a unique equivalence class of convergent integration contours, in direct analogy with the Fresnel (e^{ix^2}) integral, with the thimble rotated by an angle $\pi/4$ from the real line. To evaluate it, one possibility is to subdivide it into small time intervals and relate it to ordinary Gaussian integrals. There is also a neat way using the zeta function. We will review this latter method now: Start by working with the rescaled coordinate $\tilde{t} = t\tilde{N}$, with range $0 \leq \tilde{t} \leq \tilde{N}$. The integral we are interested in reads (up to a trivial numerical rescaling of $\tilde{N}$)

$$\begin{aligned}F_{\mathrm{DD}}(\tilde{N}) &= \int_{Q(0)=0}^{Q(1)=0} D[Q] e^{i\int_0^{\tilde{N}} d\tilde{t}\, Q_{,\tilde{t}}^2} \\ &= \int_{Q(0)=0}^{Q(1)=0} D[Q] e^{-i\int_0^{\tilde{N}} Q\frac{d^2}{d\tilde{t}^2}Q} \\ &= \sqrt{\frac{2}{\pi i}}\left[\det\left(-\frac{d^2}{d\tilde{t}^2}\right)\right]^{-1/2}. \end{aligned} \tag{5.26}$$

The operator $-\frac{d^2}{d\tilde{t}^2}$ satisfies the eigenvalue equation $-\frac{d^2}{d\tilde{t}^2}x_n = \lambda_n x_n$ with eigenfunctions x_n and eigenvalues λ_n. With our boundary conditions these are

$$x_n = a_n \sin\left(\frac{n\pi}{\tilde{N}}\tilde{t}\right), \quad \lambda_n = \left(\frac{n\pi}{\tilde{N}}\right)^2, \quad n \in \mathbb{N}. \tag{5.27}$$

One can then define an analog of the zeta function $\zeta(s) = \sum_{n \in \mathbb{N}_*} n^{-s}$ in which n is replaced by the eigenvalues,

$$\zeta_\lambda(s) \equiv \sum_{n \in \mathbb{N}_*} \lambda_n^{-s} = \left(\frac{\tilde{N}}{\pi}\right)^{2s} \sum_{n \in \mathbb{N}_*} \frac{1}{n^{2s}} = \left(\frac{\tilde{N}}{\pi}\right)^{2s} \zeta(2s). \tag{5.28}$$

We now exploit the fact that the determinant equals the product of all eigenvalues. From the definition just given, we can infer that

$$\left[\det\left(-\frac{d^2}{d\tilde{t}^2}\right)\right] = e^{-\zeta'_\lambda(0)}. \tag{5.29}$$

The zeta function can be analytically continued to $s = 0$, where $\zeta(0) = -\frac{1}{2}$ and $\zeta'(0) = -\frac{1}{2}\ln(2\pi)$. Using these values and the explicit expression for the generalized zeta function (5.28) we obtain

$$e^{-\zeta'_\lambda(0)} = 2\tilde{N}, \tag{5.30}$$

so that in the end

$$F_{\rm DD}(\tilde{N}) = \sqrt{\frac{1}{i\pi\tilde{N}}}. \tag{5.31}$$

Reinstating the numerical factor we scaled out above, we find the fluctuation integral

$$\int_{Q[0]=0}^{Q[1]=0} DQ e^{2\pi^2 i S_2} = \sqrt{\frac{3\pi i}{2\tilde{N}}}. \tag{5.32}$$

It has the effect of multiplying the measure by $\tilde{N}^{-1/2}$, so that $\tilde{N} = 0$ becomes a branch point. We will discuss this feature further below.

We can now continue with our evaluation of the path integral (5.24). The time integral in (5.25) can be evaluated explicitly, given the solution (5.22). We are then left solely with the lapse integral

$$G[q_1; q_0] = \sqrt{\frac{3\pi i}{2\hbar}} \int_C \frac{d\tilde{N}}{\tilde{N}^{1/2}} e^{2\pi^2 i S_0}, \tag{5.33}$$

with

$$S_0 = \tilde{N}^3 \frac{\Lambda^2}{36} + \tilde{N}\left(-\frac{\Lambda}{2}(q_0 + q_1) + 3\right) + \frac{1}{\tilde{N}}\left(-\frac{3}{4}(q_1 - q_0)^2\right). \tag{5.34}$$

5.2 Transitions in a universe with positive cosmological constant

This integral is rather nontrivial, not least because it is only conditionally convergent when the integral is taken over real $\tilde{N}$ values (since the modulus of the integrand is then 1 everywhere) and thus over Lorentzian geometries. As discussed in Appendix B on Picard–Lefschetz (PL) theory, conditionally convergent integrals are problematic because their value may in general depend on the order of integration. The proper way to handle them is to use PL theory and deform the integration to absolutely convergent contours known as *Lefschetz thimbles*.

In the present case, it just so happens that the lapse integral can be evaluated exactly, the solution being given by a bilinear expression in Airy functions. The actual solution, with the appropriate deformation of the integration contour to run over thimbles, and satisfying the imposed boundary conditions, turns out to be

$$G[q_1; q_0] = \frac{2^{1/3}(-3)^{2/3}\pi^{5/3}}{\Lambda^{1/3}} \left(Ai[z_0] - iBi[z_0]\right) Ai[z_1], \quad (5.35)$$

$$\text{where} \quad z = \frac{e^{i\frac{\pi}{3}}(12\pi^4)^{1/3}(\Lambda q - 3)}{\Lambda^{2/3}}.$$

However, we will not make use of it here. There are two reasons for this: the first is that it is a great coincidence that the lapse integral can be solved exactly in this case – in more general models no exact solutions exist. And secondly, it is actually more useful to evaluate the integral approximately, in the saddle-point approximation. This is because the saddle points provide us with physical intuition about the solution, as we shall see shortly.

To implement the PL procedure, we must first determine the saddle points of the integrand. This is done by solving $\partial S_0/\partial \tilde{N} = 0$, leading to the four saddle points $\sigma = 1, 2, 3, 4$,

$$\tilde{N}_\sigma = c_1 \frac{3}{\Lambda} \left[\left(\frac{\Lambda}{3}q_0 - 1\right)^{1/2} + c_2 \left(\frac{\Lambda}{3}q_1 - 1\right)^{1/2} \right], \quad (5.36)$$

with $c_1, c_2 \in \{-1, 1\}$. The action at the saddle points is easily determined,

$$S_{0,\sigma}(\tilde{N}_\sigma) = -c_1 \frac{6}{\Lambda} \left[\left(\frac{\Lambda}{3}q_0 - 1\right)^{3/2} + c_2 \left(\frac{\Lambda}{3}q_1 - 1\right)^{3/2} \right]. \quad (5.37)$$

The important question is which of these saddle points contribute to the path integral. Here we must distinguish different cases, depending on whether $\frac{\Lambda}{3}q - 1$ is positive, zero, or negative. We will refer to these as classical, borderline, or nonclassical values, respectively, recalling that the classical de Sitter solution in the closed slicing has a minimum radius at $q = a_{\min}^2 = \frac{3}{\Lambda}$.

We will analyze two cases in detail: that of a transition between two classical boundary conditions $\frac{3}{\Lambda} < q_0 < q_1$, and the transition from a nonclassical value to a classical one $q_0 < \frac{3}{\Lambda} < q_1$.

Classical boundary conditions ($\frac{3}{\Lambda} < q_0 < q_1$)

We will first look at classical boundary conditions, where both the initial and final scale factors are larger than the minimal classically allowed value. In this case the saddle points (5.36) are located at real lapse values, and the actions (5.37) at the saddle points are also real. The saddle points come in pairs with opposite values of the lapse. This is simply because the metric (5.17) depends on $\tilde{N}^2$. In principle one could try to fix the sign of $\tilde{N}$ to break this degeneracy, but since we will analytically continue all variables anyway, we will not put in such a restriction here.

The geometries at the saddle points are shown in Fig. 5.3 for a representative example. These geometries are identical for the saddle points with opposite lapse values. The solutions are real – see also (5.22) – and therefore correspond to Lorentzian spacetimes. Since we are at the saddle points, they satisfy the full Einstein equations, i.e., both the equation of motion (5.20) and the constraint (5.21). In fact, because our metric ansatz is spatially homogeneous and isotropic, the saddle-point geometries are portions of de Sitter spacetime.

There are two qualitatively distinct evolutions: The saddle point with smaller lapse value corresponds to a purely expanding geometry, while the one with larger lapse (and thus longer time evolution) corresponds to a geometry that contracts first, smoothly bounces at the minimal radius $q_{\min} = 3/\Lambda$, and then expands to the final value q_1. When visualized on the hyperboloidal embedding of de Sitter spacetime, the two saddle points correspond to evolution on one side of the waist, or across the waist, respectively – this is shown in the right panel of Fig. 5.3.

Which of these saddles contribute to the path integral is determined by the choice of integration contour for the lapse integral. In order to determine the possible contours, we must first find the thimbles. The steepest ascent ($\mathcal{K}$) and descent ($\mathcal{J}$) contours represent the loci where the phase of the integrand remains constant – i.e., they are the curves on which the phase remains identical to that at the respective saddle points (see again Appendix B).[3] A numerical example is shown in Fig. 5.4. They all start and end either at infinity or at the essential singularity at the origin of the complex lapse

---

[3] When determining such flows numerically, one has to pay attention to eliminate "spurious" flows that are not connected to any saddle point, but simply happen to have the same phase as one of the saddle points.

5.2 Transitions in a universe with positive cosmological constant 131

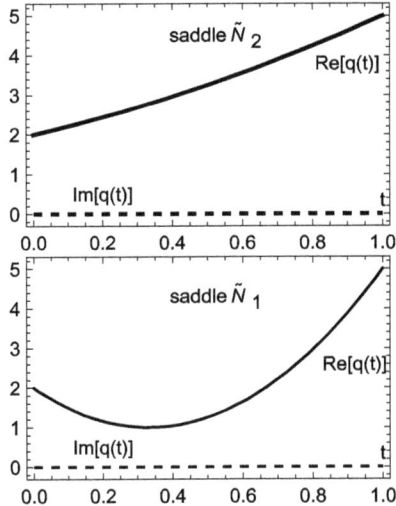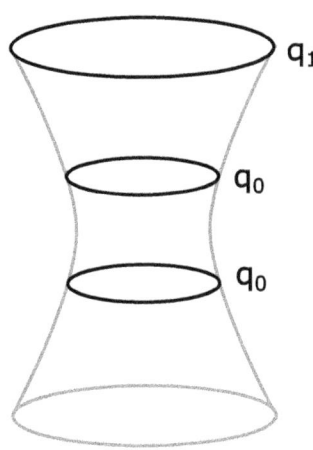

Figure 5.3 In this example of classical Dirichlet boundary conditions we used the values $\Lambda = 3, q_0 = 2, q_1 = 5$, and the saddle points are located at $\tilde{N}_\sigma = \pm 1, \pm 3$. *Left panels:* the geometries at the saddle points $\tilde{N}_2 = 1$ (corresponding to pure expansion) and $\tilde{N}_1 = 3$ (corresponding to contraction followed by expansion). *Right panel:* These geometries may also be represented on the de Sitter hyperboloid, where they can be seen as either remaining on one side of the waist of the hyperboloid, or transitioning across it.

plane. The singularity at $\tilde{N} = 0$ stems from the term proportional to $1/\tilde{N}$ in the exponent (5.34). Physically, this singularity may be interpreted as being due to the fact that, for unequal initial and final scale factors, it is impossible to find a geometry that links them in zero time evolution.

Mathematically speaking, the integration contour must run between regions where the weighting diverges to minus infinity in order for the lapse integral to be properly defined. At large $\tilde{N}$, Eq. (5.34) shows that the integral is dominated by a term of the form $e^{i\tilde{N}^3}$, and thus the regions of asymptotic convergence are the three wedges with angles $0 < \text{Arg}(\tilde{N}) < \frac{\pi}{3}$, $\frac{2\pi}{3} < \text{Arg}(\tilde{N}) < \pi$, and $-\frac{2\pi}{3} < \text{Arg}(\tilde{N}) < -\frac{\pi}{3}$. Near the essential singularity at the origin, the integral is dominated by a term of the form $e^{-i/\tilde{N}}$, and correspondingly the region of convergence lies in the upper half plane, $0 < \text{Arg}(\tilde{N}) < \pi$. These regions of asymptotic convergence are indicated by a shading in Fig. 5.4. All thimbles $\mathcal{J}_\sigma$ interpolate between two of them. But mathematics alone does not tell us which thimbles to pick.

Physically, it would be ideal if the integration contour could be fixed by fundamental principles. It is probably fair to say that opinions diverge on

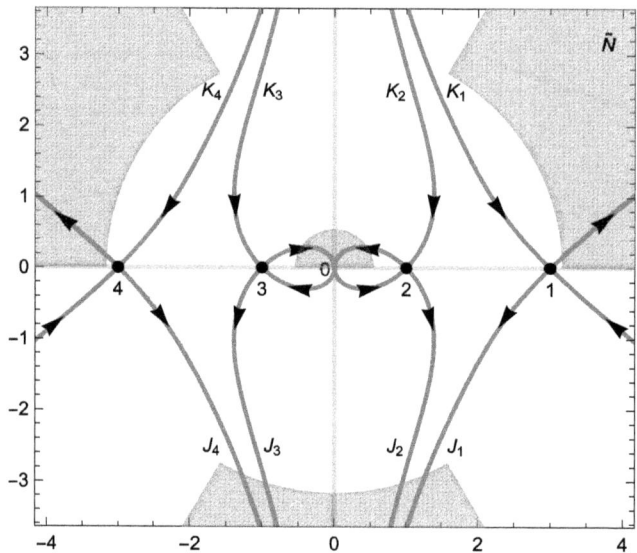

Figure 5.4 This graph shows the location of the steepest descent ($\mathcal{J}$) and ascent ($\mathcal{K}$) contours associated with the four saddle points of the integral (5.33), in the complex plane of the lapse function $\tilde{N}$, for classical Dirichlet boundary conditions. Arrows indicate downward flow. The asymptotic regions where the integral converges are shaded, and they include the upper half-plane region near the singularity at $\tilde{N} = 0$. Sensible integration contours must run between two regions of asymptotic convergence. They can then be deformed into sums of thimbles $\mathcal{J}_\sigma$. The geometries at the saddles are shown in Fig. 5.3.

what exactly those principles should be. There are several general attitudes towards this issue. Based on currently well-established physics, it would make most sense to integrate over Lorentzian metrics – i.e., one would try to integrate over real (and perhaps only positive) $\tilde{N}$. A second suggestion, motivated by approaches to define quantum field theories rigorously, is to integrate over Euclidean metrics, i.e., to integrate over the imaginary $\tilde{N}$ axis (or half-axis). And a third viewpoint is that, since we are led to analytically continue to complex metrics in order to find the most convergent integration contours, we should not be prejudiced and a priori allow any convergent contour of integration. Let us discuss these approaches in turn, to see what merits and/or problems they have.

For the integration over Lorentzian metrics, we first have to decide whether to integrate over the entire real line or only the half line starting at the origin (we will discuss the mathematical difference between full and half-line contours in more detail in Section 5.3). The half-line contour (excluding $\tilde{N} = 0$)

5.2 Transitions in a universe with positive cosmological constant 133

is the most conservative choice. The saddle points 2 and 1 lie directly on the real $\tilde{N}$ line and are automatically picked up by this contour of integration. From Fig. 5.4 we can immediately see that the positive real line can be deformed into the sum of the associated thimbles without any additional contributions (since $\mathcal{K}_3$ and $\mathcal{K}_4$ do not intersect the positive real line),

$$\mathcal{C}_+ = (0, +\infty) = \mathcal{J}_2 + \mathcal{J}_1, \tag{5.38}$$

with the appropriate choice of orientation of the thimbles. Note also that the arc at infinity connecting the real line with the thimble (which shoots off to infinity at an angle of $\frac{\pi}{6}$) does not contribute to the integral, precisely because it resides in a region of convergence where the integrand tends to zero asymptotically, and this region of convergence is adjacent to the real line. An analogous situation arises for the arc at the origin, connecting the real line to the $\mathcal{J}_2$ thimble there. Once again, no additional contribution arises due to the rapid fall-off of the weighting in this region.

The minisuperspace path integral is defined as a sum over a large class of geometries, within which the saddle points provide the dominant contributions. Following the discussion in Appendix B, the propagator (5.33) can then be approximated by (reinstating $\hbar$ momentarily)

$$
\begin{aligned}
G_+[q_1; q_0] &= \sum_{\sigma=1,2} \sqrt{\frac{3\pi i}{2\hbar}} \int_{\mathcal{J}_\sigma} \frac{d\tilde{N}}{\tilde{N}^{1/2}} e^{2\pi^2 i S_0/\hbar} \\
&\approx \sum_{\sigma=1,2} \sqrt{\frac{3\pi i}{2\hbar}} \frac{e^{2\pi^2 i S_{0,\sigma}(\tilde{N}_\sigma)/\hbar}}{\tilde{N}_\sigma^{1/2}} \int_{\mathcal{J}_\sigma} d\tilde{N} e^{\frac{i\pi^2}{\hbar} S_{0,\tilde{N}\tilde{N}}(\tilde{N}-\tilde{N}_\sigma)^2} \\
&\approx \sum_{\sigma=1,2} n_\sigma \sqrt{\frac{3\pi i}{2\hbar}} \frac{e^{2\pi^2 i S_{0,\sigma}(\tilde{N}_\sigma)/\hbar}}{\tilde{N}_\sigma^{1/2}} e^{i\theta_\sigma} \int_{\mathcal{J}_\sigma} dn e^{-\frac{\pi^2}{\hbar}|S_{0,\tilde{N}\tilde{N}}|n^2} \\
&\approx \sum_{\sigma=1,2} \sqrt{\frac{3i}{2\tilde{N}_\sigma |S_{0,\tilde{N}\tilde{N}}|}} e^{i\theta_\sigma} e^{2\pi^2 i S_{0,\sigma}(\tilde{N}_\sigma)/\hbar} \\
&= \frac{e^{i\pi/4} 3^{1/2}}{[(\Lambda q_0 - 3)(\Lambda q_1 - 3)]^{1/4}} \cos\left(\frac{12\pi^2}{\hbar \Lambda}\left(\frac{\Lambda}{3}q_0 - 1\right)^{3/2} + \frac{\pi}{4}\right) \\
&\quad \times e^{-i\frac{12\pi^2}{\hbar \Lambda}\left(\frac{\Lambda}{3}q_1 - 1\right)^{3/2}}, \tag{5.39}
\end{aligned}
$$

where we defined $\tilde{N} - \tilde{N}_\sigma \equiv n e^{i\theta_\sigma}$ with n real and θ_σ being the angle of the Lefschetz thimble with respect to the positive real $\tilde{N}$ axis (i.e., we aligned the Gaussian integral over n with the thimbles). This calculation includes the contributions from small fluctuations in $\tilde{N}$ around the saddle points. In one or two instances, we will make use of this more precise expression for the propagator, but most of the time it is enough to consider such amplitudes to leading order in $\hbar$,

$$G_+[q_1;q_0] \approx \cos\left(\frac{12\pi^2}{\hbar\Lambda}\left(\frac{\Lambda}{3}q_0 - 1\right)^{3/2}\right) e^{-i\frac{12\pi^2}{\hbar\Lambda}\left(\frac{\Lambda}{3}q_1 - 1\right)^{3/2}}. \quad (5.40)$$

It consists of two parts: The factor involving q_1 oscillates rapidly as the universe expands. And the factor q_0 stems from the interference between the two saddle-point contributions. The fact that two saddle points contribute to the propagator is perfectly natural given our boundary conditions: All we said was that the universe starts at scale factor q_0, but due to the uncertainty principle we cannot simultaneously fix the expansion rate – thus we cannot specify whether the universe is initially expanding or contracting. Since both behaviors correspond to legitimate solutions of the equations of motion they can, and do, both contribute as saddle points. Note however that this interference is likely unobservable in practice, since in a universe filled with matter decoherence would take place and the two saddle-point contributions would evolve as independent universes (we will study this effect in Chapter 8).

Let us now consider the integral over the entire real lapse line. Here we have to make an additional choice, namely whether to pass above or below the singularity at $\tilde{N} = 0$ (since the measure $d\tilde{N}/\tilde{N}^{1/2}$ has a branch point at $\tilde{N} = 0$, one would put the associated branch cut such that it points away from the contour of integration – then we do not need to consider this cut any further). This is a significant difference, since the region just above $\tilde{N}$ is one where the integrand is very small, while below $\tilde{N} = 0$ it is very large. For the contour passing above the singularity, all four saddle points are picked up,

$$\mathcal{C}_\cap = (-\infty, +\infty)_{\text{above} 0} = \mathcal{J}_4 + \mathcal{J}_3 + \mathcal{J}_2 + \mathcal{J}_1. \quad (5.41)$$

As we will discuss in Section 5.3, for an infinite contour the transition amplitude is conventionally called a wave function rather than a propagator, and thus we will denote it by Ψ rather than G. It can be approximated by

$$\Psi_\cap[q_1;q_0] \approx \cos\left(\frac{12\pi^2}{\hbar\Lambda}\left(\frac{\Lambda}{3}q_0 - 1\right)^{3/2}\right) \cos\left(\frac{12\pi^2}{\hbar\Lambda}\left(\frac{\Lambda}{3}q_1 - 1\right)^{3/2}\right), \quad (5.42)$$

which is in fact just the half-line expression (5.40) plus its complex conjugate from the negative-lapse half line (up to a negligible numerical factor). In other words, there is no contribution from the arc passing above the singular point $\tilde{N} = 0$, which is consistent with this being the region of convergence discussed earlier. At face value, the expression (5.42) describes a massive interference, between all four saddle-point geometries. As such, one would certainly not expect the wave function (which in addition is real valued) to predict any JWKB classical behavior at all. However, as for the half-line expression (5.40), we will have to add matter and/or fluctuations, and look at the effects of decoherence, in order to have a more realistic setup. For now, let us note that with the simple model at hand the full real-line integral is overcounting geometries since the metric depends only on $\tilde{N}^2$. That said, one might expect such a procedure to be more sensible when adding fermionic fields, as these couple to the local frame, which is sensitive to the sign of the lapse.

The last "Lorentzian" integral we have left to consider is the one where we pass below the $\tilde{N} = 0$ singularity. In fact, as one can see from Fig. 5.4, such a contour in fact avoids the steepest ascent contours $\mathcal{K}_3$ and $\mathcal{K}_2$,

$$\mathcal{C}_\cup = (-\infty, +\infty)_{\text{below } 0} = \mathcal{J}_4 + \mathcal{J}_1. \tag{5.43}$$

Thus it constitutes are rather large departure from the preceding two integrals, and that is the reason for putting the word Lorentzian in quotation marks above. The contribution from the region below the singularity has a decisive influence on this integral, and the metrics in that region are certainly not Lorentzian (they are fully complex). From the rewriting of the contour in terms of the thimbles $\mathcal{J}_4$ and $\mathcal{J}_1$ we can in fact infer that this contour does not make much sense, as it contains solely the bouncing solutions. Imagine a situation where q_0 is very close to q_1 – then, instead of a brief expansion, we would predict that the universe first contracts all the way to the minimum de Sitter radius before expanding back to a slightly larger size. This is clearly in disagreement with observations of the universe from one day to the next. What this goes to show is that the choice of contour is far from innocuous.

What about other prescriptions for integration contours? Perhaps the most often cited prescription when gravity is involved is to integrate over Euclidean geometries. This is motivated by several attractive properties of Euclidean cosmological and black hole solutions, as we will discuss in Chapters 6, 7, and 9. However, glancing once again at Fig. 5.4, we can see that the integral along the imaginary $\tilde{N}$ axis contains regions of divergence both at $+i\infty$ and below the $\tilde{N} = 0$ singularity. This means that no Euclidean integral, either over the full imaginary axis or over half of it, is convergent. Put blandly, the Euclidean path integral does not exist. One can trace the origin of these divergences back to the conformal factor problem, namely the problem that the kinetic term for the scale factor has the wrong sign – see, e.g., Eq. (4.21) – and causes the action to be unbounded below. Note the important point that the nonexistence of the Euclidean path integral does not necessarily imply that there is anything wrong with Euclidean saddle points. As we will see, Euclidean or complex saddle points can full well contribute to Lorentzian (or other) path integrals – there is no need for saddle points to lie directly on the defining contour of integration.

Finally, let us mention the possibility of defining the integral directly by integrating over complex paths. This provides us with the interesting possibility of defining the integral such that only the expanding saddle(s) contribute(s). When using a half-line contour, one could let it run from $-i\infty$ to $\tilde{N} = 0$, approaching the singularity from the upper half plane and passing to the right of the singularity. This would single out the thimble $\mathcal{J}_2$ associated with the saddle point shown in the upper-left panel of Fig. 5.3,

$$\mathcal{C}_1 = (-i\infty, 0)_{\text{right of } 0} = \mathcal{J}_2, \qquad (5.44)$$

and would give us the propagator

$$G_1[q_1; q_0] \approx e^{\frac{i}{\hbar} 2\pi^2 S_{0,2}(\tilde{N}_2)} \approx e^{i \frac{12\pi^2}{\hbar \Lambda} \left(\frac{\Lambda}{3} q_0 - 1\right)^{3/2} - i \frac{12\pi^2}{\hbar \Lambda} \left(\frac{\Lambda}{3} q_1 - 1\right)^{3/2}}. \qquad (5.45)$$

The oscillations in q_0 and q_1 simply reflect the classical expansion that this amplitude predicts. By analogy, one would pick up both expanding saddle points 3 and 2 either by considering a contour that starts and ends at $-i\infty$ and loops around the singularity at $\tilde{N} = 0$, or by considering the sum of two contours, both starting at $-i\infty$ and ending up at $\tilde{N} = 0$, but with one passing left and the other right of the singularity. We leave it as an exercise to show that these two choices would give amplitudes that are proportional to either the real or to the imaginary part of (5.45).

Nonclassical to classical transition ($q_0 < \frac{3}{\Lambda} < q_1$)

Up to now we discussed the quantum treatment of spacetime amplitudes between initial and final values of the scale factor that are classically allowed. In other words, we investigated the quantum treatment of classical general relativistic solutions. But now that we have this formalism set up, we can go beyond this and ask what happens when we set one of the scale factor values to a small value $q_0 < \frac{3}{\Lambda}$ that classically does not exist (in the closed slicing of de Sitter spacetime).

The steps of the calculation are entirely analogous to the previous case, but now we can see that the saddle points (5.36) as well as the actions at

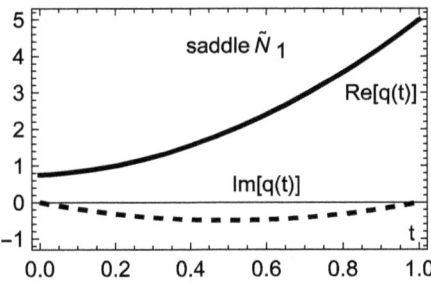 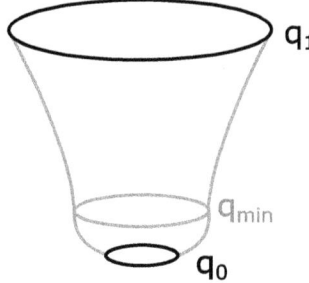

Figure 5.5 *Left panel:* saddle-point geometry at the saddle point $\tilde{N}_1 = 2 + \frac{1}{2}i$. The scale factor starts and ends with real values (as required by the boundary conditions), but is complex in between. In this example we used the values $\Lambda = 3, q_0 = \frac{3}{4}, q_1 = 5$, and the saddle points are located at $\tilde{N}_\sigma = \pm 2 \pm \frac{1}{2}i$. *Right panel:* The same geometry may also be described as a Euclidean extension of the de Sitter hyperboloid, in which part of a 4-sphere is glued onto the waist of the hyperboloid.

the saddle points (5.37) become complex,

$$N_\sigma = c_1 \frac{3}{\Lambda} \left[i \left(1 - \frac{\Lambda}{3} q_0\right)^{1/2} + c_2 \left(\frac{\Lambda}{3} q_1 - 1\right)^{1/2} \right], \tag{5.46}$$

$$S_{\sigma,0} = c_1 \frac{6}{\Lambda} \left[i \left(1 - \frac{\Lambda}{3} q_0\right)^{3/2} - c_2 \left(\frac{\Lambda}{3} q_1 - 1\right)^{3/2} \right], \tag{5.47}$$

again with $c_1, c_2 \in \{-1, 1\}$. Consequently, the geometry at the saddle points is also complex – see Fig. 5.5 for a numerical example. The figure shows that, as required, at the boundaries the scale factor values are real, but the interpolating geometry is complex. One may also visualize it (this can be achieved by an appropriate complex deformation of the time coordinate, in agreement with Cauchy's theorem) as a portion of the de Sitter hyperboloid glued at the waist to a portion of the 4-sphere ending at radius q_0, shown in the right panel of the figure. One could say that the geometry is the closest one can get to an expanding solution, given that a purely real solution is not available.

The steepest ascent and descent lines associated with the saddle points (5.46) are shown in Fig. 5.6, and we can use it to once again assess the implications of various choices of integration contour for the lapse. The asymptotic regions of convergence have not changed, and hence the possibilities for designing integration contours remain identical to the case analyzed above. This means for instance that a Euclidean integral remains undefined. Let

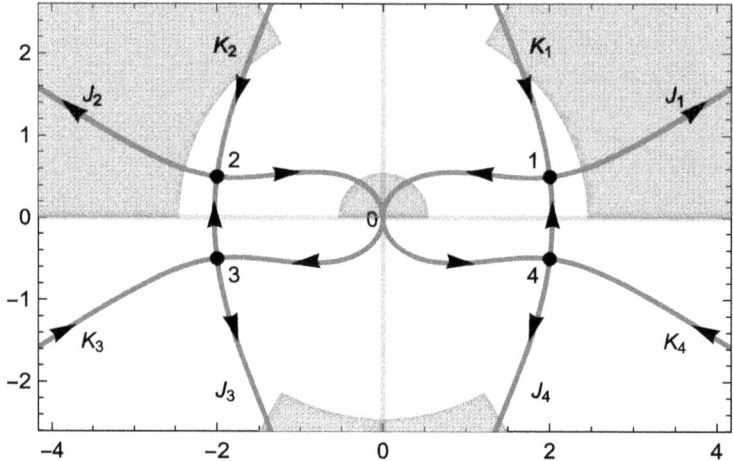

Figure 5.6 In analogy with Fig. 5.4, this graph shows the location of the steepest descent ($\mathcal{J}$) and ascent ($\mathcal{K}$) contours associated with the four saddle points of the integral (5.33), in the complex plane of the lapse function $\tilde{N}$, for transitions between a nonclassical scale factor value and a classical one. The geometry at saddle point 1 is shown in Fig. 5.5.

us then analyze the consequences of the Lorentzian and complex choices considered previously.

Before doing so, we should note one new feature: There are Stokes rays present, which link saddle points 2 and 3, and also 1 and 4. These lines are steepest ascent contours from the point of view of saddles 1 and 2, yet they are steepest descent contours for saddles 3 and 4. They arise due to the symmetry of the saddle points under reflection across the real $\tilde{N}$ axis, which leads to the saddle-point actions having equal real parts. Their presence presents no difficulty here, though if one feels uncomfortable with such a degeneracy one may remove it by adding a symmetry-breaking term (e.g., a term of the form $\epsilon i \tilde{N}$), which disentangles the flow lines, and then one lets the regulator vanish ($\epsilon \to 0$) at the end of the calculation. Here the Stokes lines imply that any integration contour that picks up saddle 4 will also pick up saddle 1 (and likewise for saddles 3 and 2).

The Lorentzian positive half-line contour is only intersected by the steepest ascent flow originating from saddle point 1. Thus this is the only saddle point contributing to the corresponding integral,

$$\mathcal{C}_+ = (0, +\infty) = \mathcal{J}_1 . \tag{5.48}$$

The same arguments as those given above then imply that there are no additional contributions from the arcs at infinity and near zero. The propagator may thus be approximated as

$$G_+[q_1; q_0] \approx e^{\frac{i}{\hbar}2\pi^2 S_{0,1}(\tilde{N}_1)} \tag{5.49}$$

$$\approx e^{-\frac{12\pi^2}{\hbar\Lambda}\left(1-\frac{\Lambda}{3}q_0\right)^{3/2}} e^{-i\frac{12\pi^2}{\hbar\Lambda}\left(\frac{\Lambda}{3}q_1-1\right)^{3/2}}. \tag{5.50}$$

The dominant geometry was shown in Fig. 5.5. The fact that it is complex reflects the classical impossibility of this transition. Rather, one may interpret this transition as a quantum tunneling amplitude to go from q_0 to q_1. The suppression factor $e^{-\frac{12\pi^2}{\hbar\Lambda}\left(1-\frac{\Lambda}{3}q_0\right)^{3/2}}$ lends support to this interpretation, as it indicates that the transition is less likely to occur than classical evolution. Moreover, for smaller q_0 values the suppression is stronger. By contrast, the amplitude is oscillatory in the final scale factor q_1, which is an indication of classical evolution that fits with the classical nature of the boundary condition on the final hypersurface.

Let us highlight what may be a surprising feature: Even though the saddle points are solutions to the classical equations of motion, they can describe quantum effects. This is because in this case the boundary conditions are classically impossible, forcing upon us a *complex* solution to the classical field equations.

If we now consider the full real-lapse line integral, passing above the singularity, then just as in the case of classical boundary conditions, this just adds the complex conjugate of (5.50) (in the form of the thimble $\mathcal{J}_2$ with saddle-point number 2), yielding a wave function that is proportional to the real part of (5.50). And we would once again expect (and show in Chapter 8 that this is indeed the case) that decoherence would disentangle the two spacetimes associated with the two saddle points upon inclusion of additional matter and/or fluctuations.

As for the integration contour following the real lapse line except passing below the singularity at the origin, we can see from Fig. 5.6 that this can be deformed into the sum over all four thimbles. However, we know from the analysis above that this contour becomes nonsensical when q_0 surpasses the de Sitter radius (squared) – hence we will consider this contour to be unphysical and discard it.

It is useful to note an important feature here: If the integration contour is over purely Lorentzian metrics, then it can only pick up saddle points with either zero or negative weighting. This is because in order to contribute to the integral, one must be able to flow down from the original integration contour to a relevant thimble. Thus, for Lorentzian integrals, it is impossible to pick up saddles 3 or 4, since the latter ones have a positive weighting $+\frac{12\pi^2}{\hbar\Lambda}\left(1-\frac{\Lambda}{3}q_0\right)^{3/2}$. The real line integral passing below the singularity can pick up these saddle points precisely because the region below the singularity is a region of divergence in which the weighting is large and positive. Put differently, this integral is certainly not Lorentzian as its dominant contribution comes precisely from the region where the contour departs from the real line to circumnavigate the singularity.

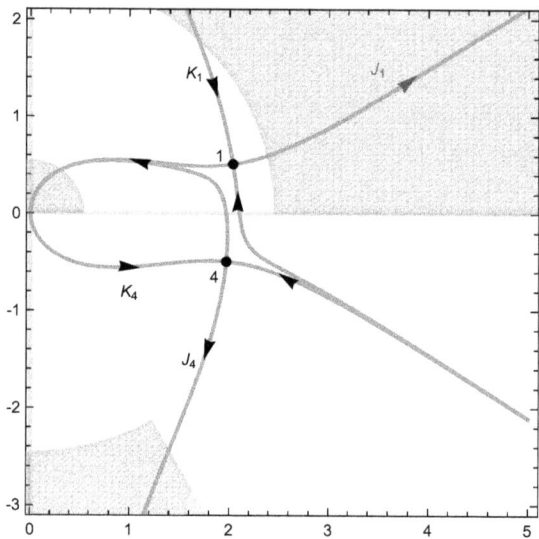

Figure 5.7 This graph shows an example of how the flow lines are deformed in the complex $\tilde{N}$ plane when adding a small degeneracy-breaking term $\epsilon i \tilde{N}$ (here with $\epsilon = -0.1$) to the action. The thimble $\mathcal{J}_4$ then passes next to saddle 1 and continues toward the origin. Other deformations could break the degeneracy in different ways, but the end result is always that when saddle 4 is picked up it contributes dominantly to the path integral.

Following on from these considerations, let us note that a complex contour certainly can pick up the enhanced saddles. As an example, consider again the semi-infinite contour starting at $-i\infty$ and ending up at the origin, approaching the origin from the upper half plane and passing the singularity to the right. This time this contour can be deformed to follow the thimble $\mathcal{J}_4$. Due to the Stokes line, it looks as if this thimble, after reaching saddle point 1, follows half of $\mathcal{J}_1$ en route to the origin. However, if we add a deformation to break the degeneracy, then we can see that thimble $\mathcal{J}_4$ in fact goes all the way to the origin by itself (see Fig. 5.7),

$$\mathcal{C}_| = (-i\infty, 0)_{\text{right of } 0} = \mathcal{J}_4. \tag{5.51}$$

This gives us the propagator

$$G_|[q_1; q_0] \approx e^{\frac{i}{\hbar} 2\pi^2 S_{0,4}(\tilde{N}_4)} \approx e^{+\frac{12\pi^2}{\hbar\Lambda}\left(1-\frac{\Lambda}{3}q_0\right)^{3/2} - i\frac{12\pi^2}{\hbar\Lambda}\left(\frac{\Lambda}{3}q_1-1\right)^{3/2}}, \tag{5.52}$$

which has an enhanced weighting, the more so the smaller q_0 happens to be. We will see in Chapter 7 that a very similar kind of saddle point may be a useful toy model for describing the nucleation of the universe.

For now, let us add the remark that, had we not used a deformation to break the degeneracy of the flow lines, it would have appeared that saddle 1 also contributed to the transition amplitude. However, the weighting of saddle 1 is exponentially suppressed, and as such it gives an entirely negligible contribution compared to that of saddle 4. It does not make sense to include this "correction" here, because it is in fact smaller than corrections (which we have already neglected) to (5.52) stemming from fluctuations around that saddle point (and which would have contributed a prefactor to (5.52), compared to which the contribution of saddle 1 is suppressed by a further exponential factor). Thus we may conclude that the contour $\mathcal{C}_1$ gives an integral best approximated by the saddle-point contribution 4 alone.

Just as for classical boundary conditions, one could now add or subtract the mirrored contour starting at $-i\infty$, passing the singularity on the left before approaching it from the upper half plane. This would give a wave function proportional to the real or imaginary parts of (5.52), respectively.

5.2.2 Neumann boundary conditions

In the previous section, we looked at transitions between fixed scale factor values. For consistency, this required adding a GHY boundary term to the action. Meanwhile, the pure, unadorned Einstein–Hilbert action leads to a variational problem in which one may fix the momentum conjugate to the scale factor. Because this setting does not require any additional surface term, the Neumann boundary condition is arguably the most natural boundary condition for gravity. Let us study a particular example.

For definiteness, we will impose a Neumann condition on the $t = 0$ boundary – i.e., in the action (5.18) we will set $\gamma = 0$ on the initial hypersurface. It turns out that in the highly simplified model we are looking at it would be overconstraining to also impose a Neumann condition on the final hypersurface. Hence we will keep the Dirichlet condition there – i.e., we will keep $\gamma = 1$ on the $t = 1$ hypersurface. The minisuperspace action then reads

$$S = 2\pi^2 \int_0^1 dt \left[\frac{3}{2\tilde{N}} q\ddot{q} + \frac{3}{4\tilde{N}} \dot{q}^2 + \tilde{N}(3 - \Lambda q) \right] - \frac{3\pi^2}{\tilde{N}} q\dot{q}|_{t_q = 1}. \quad (5.53)$$

Varying with respect to the scale factor q, we obtain the same equation of

142   *Quantizing gravity: path integral approach*

motion as before, Eq. (5.20), together with the boundary conditions

$$-\frac{3\pi^2}{\tilde{N}} q\delta(\dot{q}) = 0|_{t=0}, \quad -\frac{3\pi^2}{\tilde{N}} \dot{q}\delta q|_{t=1}, \quad (5.54)$$

confirming that we can specify the momentum $p_0 \equiv -\frac{3\pi^2}{\tilde{N}}\dot{q}(t=0)$ and the scale factor $q_1 \equiv q(t=1)$, as desired. With these boundary conditions, the solution to the equation of motion (though not necessarily the constraint) is

$$\bar{q} = \frac{\Lambda}{3}\tilde{N}^2 t^2 - \frac{p_0 \tilde{N}}{3\pi^2} t + q_1 - \frac{\Lambda}{3}\tilde{N}^2 + \frac{p_0 \tilde{N}}{3\pi^2}. \quad (5.55)$$

In order to evaluate the integral over q, we can now use the same trick as in Section 5.2.1, and shift the integration variable to be a generic perturbation away from the background solution, $q = \bar{q} + Q(t)$. This time, the fluctuation must satisfy the boundary conditions $\dot{Q}(t=0) = 0$ and $Q(t=1) = 0$. The integral then splits up once again into a background piece and a fluctuation integral, in direct analogy with (5.24). The fluctuation integral we are interested in this time is of the form

$$F_{\mathrm{ND}}(\tilde{N}) = \int_{\dot{Q}(0)=0}^{Q(1)=0} D[Q] e^{i\int_0^{\tilde{N}} d\tilde{t} \, \dot{Q}_{,\tilde{t}}^2} = \sqrt{\frac{2}{\pi i}} \left[\det\left(-\frac{d^2}{d\tilde{t}^2}\right)\right]^{-1/2}. \quad (5.56)$$

This time the eigenfunctions and eigenvalues of the operator $-\frac{d^2}{dt^2}$ are given by

$$x_n = a_n \cos\left[\frac{(2n+1)\pi}{2N}\tilde{r}\right], \quad \lambda_n = \left[\frac{(2n+1)\pi}{2N}\right]^2, \quad n \in \mathbb{N}. \quad (5.57)$$

The determinant, which is the product of all eigenvalues, can again be evaluated by defining a generalized zeta function

$$\zeta_\lambda(s) \equiv \sum_{n \in \mathbb{N}} \lambda_n^{-s} = \left(\frac{2N}{\pi}\right)^{2s} \sum_{n \in \mathbb{N}} \frac{1}{(2n+1)^{2s}}. \quad (5.58)$$

The last term is analogous to the ordinary zeta function, except that here one sums only over odd terms. We can rewrite it in terms of the zeta function by combining terms as follows,

$$1 + \frac{1}{3^{2s}} + \frac{1}{5^{2s}} + \cdots = 1 + \frac{1}{2^{2s}} + \frac{1}{3^{2s}} + \cdots - [\frac{1}{2^{2s}} + \frac{1}{4^{2s}} + \cdots]$$
$$= 1 + \frac{1}{2^{2s}} + \frac{1}{3^{2s}} + \cdots - \frac{1}{2^{2s}}[1 + \frac{1}{2^{2s}} + \frac{1}{3^{2s}} + \cdots], \quad (5.59)$$

thereby obtaining

$$\zeta_\lambda(s) = \left(\frac{2N}{\pi}\right)^{2s} \left(1 - 2^{-2s}\right) \zeta(2s). \tag{5.60}$$

This allows us to deduce the value of the determinant,

$$\left[\det\left(-\frac{d^2}{d\tilde{r}^2}\right)\right] = e^{-\zeta'_\lambda(0)} = 2, \tag{5.61}$$

where we have made use of $\zeta(0) = -\frac{1}{2}$. The end result for the fluctuation factor is surprisingly simple,

$$F_{\text{ND}}(\tilde{N}) = \frac{1}{\sqrt{\pi i}}. \tag{5.62}$$

And with the appropriate normalization of the lapse, we can use this result to write out the Neumann–Dirichlet fluctuation integral

$$\int_{\dot{Q}(0)=0}^{Q(1)=0} \mathcal{D}[Q] e^{-\frac{i}{\hbar} \int_0^1 dt \frac{3\pi^2 \dot{Q}^2}{2\tilde{N}}} = \sqrt{\frac{3\pi i}{2\hbar}}. \tag{5.63}$$

Compared to the Dirichlet–Dirichlet case, we note that it does not contain any dependence on the lapse, and thus does not introduce a branch point at the origin into the path integral measure.

Integrating the background action over the solution (5.55), we end up with the lapse integral

$$G[p_0; q_1] = \sqrt{\frac{3\pi i}{2\hbar}} \int_\mathcal{C} d\tilde{N}\, e^{\frac{i}{\hbar} 2\pi^2 \left[\frac{\Lambda^2}{9} \tilde{N}^3 - \frac{p_0 \Lambda}{6\pi^2} \tilde{N}^2 + (3 + \frac{p_0^2}{12\pi^4} - q_1 \Lambda)\tilde{N} + \frac{p_0 q_1}{2\pi^2}\right]}. \tag{5.64}$$

We will evaluate this integral in the saddle-point approximation, using PL theory. This requires us to investigate the saddle points and associated flow lines. Before doing so, note that the lapse integral manifests a clear distinction to the Dirichlet case, which is that it is regular at $\tilde{N} = 0$. This makes sense, as we fix the initial momentum and not the initial size. The path integral then effectively sums over different initial sizes, and this sum includes the desired final size, which can thus be reached with zero lapse evolution (see (5.55) with $\tilde{N} = 0$). Also note that the asymptotic regions of convergence at large $\tilde{N}$ are unchanged compared to the Dirichlet case – they are once again given by the three wedges with angles $0 < \text{Arg}(\tilde{N}) < \frac{\pi}{3}$, $\frac{2\pi}{3} < \text{Arg}(\tilde{N}) < \pi$, and $-\frac{2\pi}{3} < \text{Arg}(\tilde{N}) < -\frac{\pi}{3}$. All acceptable integration contours must join two such regions.

This time, the integrand in (5.64) only admits two saddle points, which

144   *Quantizing gravity: path integral approach*

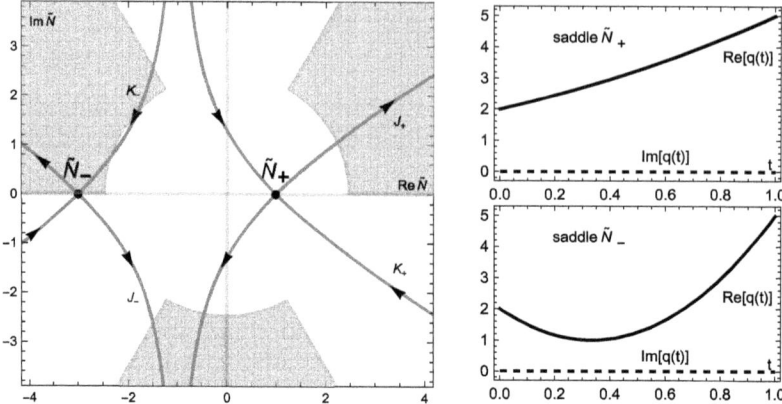

Figure 5.8 *Left panel:* saddle points and their steepest descent/ascent lines in the complex plane of the lapse function, for the situation where we have a transition with mixed Neumann–Dirichlet boundary conditions. Here $\Lambda = 3$, $p_0 = -6\pi^2$, $q_1 = 5$, so that $\tilde{N}_{\mp} = (-3, +1)$. *Right panel:* the corresponding saddle-point geometries.

we will denote with $\pm$ subscripts,

$$\tilde{N}_\pm = \frac{p_0}{2\pi^2 \Lambda} \pm \frac{3}{\Lambda}\sqrt{\frac{\Lambda}{3}q_1 - 1}, \qquad (5.65)$$

and with saddle point action

$$S(\tilde{N}_\pm) = \frac{3p_0}{\Lambda} + \frac{p_0^3}{36\pi^4 \Lambda} \mp \frac{12\pi^2 \left(\frac{\Lambda}{3}q_1 - 1\right)^{3/2}}{\Lambda}. \qquad (5.66)$$

In order to proceed, we can look at an example. To be able to compare with the Dirichlet case, we will choose a momentum value that corresponds to one of the saddle points of the classical example we looked at in Section 5.2.1, namely saddle 2 (with a purely expanding geometry) from Fig. 5.3, for which $p_0 = -3\pi^2 \dot{\tilde{q}}/\tilde{N}_2(t=0) = -6\pi^2$. We will also keep $q_1 = 5$. With these boundary conditions, we obtain the saddle-point geometries and flow lines shown in Fig. 5.8. From the figure, we can see that saddle point $\tilde{N}_+$ actually reproduces exactly the same expanding geometry that we expected to be present. What is more surprising is saddle $\tilde{N}_-$, because this is a bouncing geometry. How can a bouncing geometry arise when we imposed a boundary condition evaluated from a purely expanding geometry where $\dot{\tilde{q}}/\tilde{N} > 0$? The reason is that for this saddle point the lapse is negative. Hence the combination of contraction $\dot{q} < 0$ and negative lapse $\tilde{N} < 0$ can indeed satisfy the boundary condition.

This time there are fewer choices of integration contour for the lapse. Since there is no singularity at finite $\tilde{N}$, we can only have infinite contours, and thus we will refer to the amplitudes as wave functions. The three regions of asymptotic convergence were already described above. They imply that all integration contours must be built from just two distinct ones, which we can take to be the deformation of the real line into the upper half plane, and a contour joining $-i\infty$ to the region of convergence just above positive real infinity.

The real line contour is easily seen to intersect both steepest ascent lines, and can thus be deformed into a sum of the two thimbles,

$$\mathcal{C} = (-\infty, \infty) = \mathcal{J}_- + \mathcal{J}_+ . \tag{5.67}$$

This yields an approximate wave function that is a sum over two phases,

$$\Psi_{\mathrm{ND},\mathcal{C}}(p_0, q_1) \approx e^{\frac{i3p_0}{\hbar\Lambda} + \frac{ip_0^3}{36\pi^4 \hbar \Lambda}} \cos\left(\frac{12\pi^2}{\hbar\Lambda}\left(\frac{\Lambda}{3}q_1 - 1\right)^{3/2}\right). \tag{5.68}$$

This amplitude simply represents the classical evolution associated with the two saddle-point geometries.

The contour starting at $-i\infty$ and ending in the region of convergence above $+\infty$ only intersects $\mathcal{K}_+$, and thus the corresponding wave function simply corresponds to the phase factor (5.66) stemming from $\tilde{N}_+$.

The simple examples provided in this section and the previous one serve to illustrate the general framework of quantum cosmology. We will fill in some gaps later on (in Chapter 7), especially a consideration of cosmological fluctuations and of nonclassical momentum conditions. But first we will relate the path integral formalism derived in this chapter with the canonical formalism that we explored in the preceding one.

5.3 Relation to the Wheeler–DeWitt equation and composition law

The canonical and path integral quantization schemes look rather different, but are in fact closely related. We will derive the connection here, by showing that the path integral transition amplitudes either satisfy the WDW equation or a close cousin thereof, the inhomogeneous WDW equation.

With our current choice of metric (5.17), one can follow the same steps as those taken in Section 4.1.3 to arrive at the WDW equation,

$$\hat{H}\Psi = 0 \to \frac{\partial^2 \Psi}{\partial q^2} + 12\pi^4(\Lambda q - 3)\Psi = 0, \tag{5.69}$$

with $\Psi(q)$ being the wave function of the universe, and where we have chosen the simplest factor ordering. The inhomogeneous WDW equation admits a delta function on the right-hand side, and is given by

$$\hat{H}G = -i6\pi^2 \delta(q_0 - q_1),\quad (5.70)$$

where the Hamiltonian operator acts either on q_0 or on q_1 (the factor $6\pi^2$ is also sometimes absorbed into the definition of $\hat{H}$). We will discuss its meaning shortly.

Let us first see how one can explicitly derive the WDW equation from the path integral. We will do this for the Dirichlet–Dirichlet case, where the lapse integral was given in Eqs. (5.33) and (5.34). For definiteness, we will focus on the case where the defining integration contour for the lapse is the positive real line $\mathcal{C}_+$. Then we can take successive derivatives of the propagator as follows,

$$\frac{\partial G_+}{\partial q_1} = \sqrt{\frac{3\pi i}{2}} \int \frac{d\tilde{N}}{\tilde{N}^{1/2}} 2\pi^2 i S_{0,q_1} e^{2\pi^2 i S_0}$$

$$= \sqrt{\frac{3\pi i}{2}} \int \frac{d\tilde{N}}{\tilde{N}^{1/2}} 2\pi^2 i \left[-\frac{\tilde{N}}{2}\Lambda - \frac{3}{2\tilde{N}}(q_1 - q_0)\right] e^{2\pi^2 i S_0}, \quad (5.71)$$

and analogously

$$\frac{\partial^2 G_+}{\partial q_1^2} = \sqrt{\frac{3\pi i}{2}} \int \frac{d\tilde{N}}{\tilde{N}^{1/2}} \left[2\pi^2 i S_{0,q_1 q_1} - 4\pi^4 S_{0,q_1}^2\right] e^{2\pi^2 i S_0}$$

$$= \sqrt{\frac{3\pi i}{2}} \int \frac{d\tilde{N}}{\tilde{N}^{1/2}} \left[-4\pi^4 \left(\frac{\tilde{N}}{2}\Lambda + \frac{3}{2\tilde{N}}(q_1 - q_0)\right)^2 - \frac{3\pi^2 i}{\tilde{N}}\right] e^{2\pi^2 i S_0}. \quad (5.72)$$

Now note the following relation, obtained by using integration by parts,

$$\left[\tilde{N}^{-\frac{1}{2}} e^{2\pi^2 i S_0}\right]_0^\infty = \int d\tilde{N} \frac{d}{d\tilde{N}} \left[\tilde{N}^{-\frac{1}{2}} e^{2\pi^2 i S_0}\right]$$

$$= -\frac{1}{2} \int \frac{d\tilde{N}}{\tilde{N}^{\frac{3}{2}}} e^{2\pi^2 i S_0} + 2\pi^2 i \int \frac{d\tilde{N}}{\tilde{N}^{\frac{1}{2}}} S_{0,\tilde{N}} e^{2\pi^2 i S_0}. \quad (5.73)$$

This relation can be substituted back into (5.72), with the result

$$\frac{\partial^2 G_+}{\partial q_1^2} = \sqrt{\frac{3\pi i}{2}} \left[\int \frac{d\tilde{N}}{\tilde{N}^{1/2}} \left[-12\pi^4 (\Lambda q_1 - 3k)\right] e^{2\pi^2 i S_0}\right.$$

$$\left. + 6\pi^2 i \left[\tilde{N}^{-\frac{1}{2}} e^{2\pi^2 i S_0}\right]_0^\infty\right]$$

$$= -12\pi^4 (\Lambda q_1 - 3) G_+ + 6\pi^2 i \sqrt{\frac{3\pi i}{2}} \left[\tilde{N}^{-\frac{1}{2}} e^{2\pi^2 i S_0}\right]_0^\infty. \quad (5.74)$$

5.3 Relation to the Wheeler–DeWitt equation and composition law

This is already very close to the WDW equation. Since we are integrating over a Lefschetz thimble, the contribution at large lapse vanishes. Near the origin, there is a subtlety, which is most easily obtained by rewriting $\tilde{N} = in$, so that one can take a standard limit,

$$\lim_{\tilde{N}\to 0} \frac{e^{2\pi^2 i S_0}}{\sqrt{\tilde{N}}} = \sqrt{\frac{2\pi}{i}} \lim_{n\to 0} \frac{e^{-3\pi^2 \frac{(q_1-q_0)^2}{2n}}}{\sqrt{2\pi n}} = \sqrt{\frac{2}{3\pi i}} \delta(q_0 - q_1). \quad (5.75)$$

Hence we indeed obtain the inhomogeneous WDW equation

$$\frac{\partial^2 G_+}{\partial q_1^2} + 12\pi^4 (\Lambda q_1 - 3) G_+ = -6\pi^2 i \delta(q_0 - q_1). \quad (5.76)$$

If we had integrated over an infinite lapse contour from $\tilde{N} = -\infty$ to $\tilde{N} = +\infty$ (passing around the singularity at $\tilde{N} = 0$), the right-hand side would have vanished and we would have recovered the standard WDW equation (5.69).

There exists a more formal argument showing that this relation between the path integral and the WDW equation must hold. The argument proceeds by explicitly splitting up the propagator or wave function (denoted generically by G here) into the lapse integral and an amplitude with fixed lapse,

$$G[q_1; q_0] = \int_0^\infty d\tilde{N}\, G[q_1; q_0; \tilde{N}]. \quad (5.77)$$

The integrand,

$$G[q_1; q_0; \tilde{N}] = \int_{q=q_0}^{q=q_1} \mathcal{D}q\, e^{\frac{i}{\hbar} S(\tilde{N}, q)}, \quad (5.78)$$

represents the quantum amplitude to evolve from q_0 to q_1 in a span of "time" $\tilde{N}$. In this case the lapse plays the role of an external time, and this propagator thus satisfies a standard Schrödinger-type equation

$$i\frac{\partial G[q_1; q_0; \tilde{N}]}{\partial \tilde{N}} = \hat{H} G[q_1; q_0; \tilde{N}], \quad (5.79)$$

supplemented with the usual initial (coincidence) condition

$$\lim_{\tilde{N}\to 0} G[q_1; q_0; \tilde{N}] = 6\pi^2 \delta(q_0 - q_1). \quad (5.80)$$

The factor of $6\pi^2$ has no fundamental meaning, it simply depends on conventions. From the "Schrödinger" equation, it is then a quick step to arrive at the inhomogeneous WDW equation,

$$\hat{H} G[q_1; q_0] = \int_0^\infty d\tilde{N}\, \hat{H} G[q_1; q_0; \tilde{N}]$$

148   *Quantizing gravity: path integral approach*

$$= i \int_0^\infty d\tilde{N} \, \frac{\partial G[q_1; q_0; \tilde{N}]}{\partial \tilde{N}}$$
$$= i \, G[q_1; q_0; \tilde{N}]\Big|_{\tilde{N}=0}^{\tilde{N}=\infty}$$
$$= -i \, 6\pi^2 \, \delta(q_0 - q_1) \,. \tag{5.81}$$

Again, if the integration domain for the lapse had been the infinite lapse line, the right-hand side would have vanished and we would have recovered the homogeneous WDW equation.

This allows us now to better understand the distinction between a propagator and a wave function. Indeed, if one integrates over positive lapse only, then one restricts the possible diffeomorphisms that one has access to. In particular, consider the case where the initial and final hypersurfaces are close together. Using time reparametrization invariance, one can move the initial hypersurface forward and bring it closer to the final one. However, one cannot move it past the final one, since this would require a negative lapse. This explains the delta function at zero, where the Hamiltonian constraint (which, as we saw in Chapter 4 precisely expresses time reparametrization invariance) is not satisfied. Moreover, restricting to positive lapse imposes a primitive notion of causality, whereby one can say that the initial hypersurface indeed occurred before the final one. Thus one must in some sense choose between causality and gauge invariance. The propagator stands on the side of causality.

With an infinite integration contour, over all lapse values, one loses this notion of causality but recovers full time reparametrization invariance everywhere. In that case, as we saw, typically saddle points with opposite time evolutions appear together. The wave function stands on the side of gauge invariance.

The appropriate choice of contour, half-infinite or infinite, depends on the physical situation one is interested in: If one is looking at transitions in a pre-existing universe, then such transitions to a later configuration are meaningfully described by using a propagator. Meanwhile, if one is interested in describing the creation/emergence/nucleation of the universe, it may make sense to use an infinite contour, since no notion of pre-existence then exists, and full gauge invariance may be more essential.

Composition Law

If the universe has grown from size q_0 to q_1, and then to q_2, how do the corresponding propagators, $G[q_1, q_0]$ and $G[q_2, q_1]$, combine? We can figure this out by looking at the WDW equations they satisfy at the intermediate

5.3 Relation to the Wheeler–DeWitt equation and composition law

value q_1,

$$\frac{\partial^2 G[q_2, q_1]}{\partial q_1^2} + 12\pi^4(\Lambda q_1 - 3)G[q_2, q_1] = -6\pi^2 i \delta(q_1 - q_2), \qquad (5.82)$$

$$\frac{\partial^2 G[q_1, q_0]}{\partial q_1^2} + 12\pi^4(\Lambda q_1 - 3)G[q_1, q_0] = -6\pi^2 i \delta(q_0 - q_1). \qquad (5.83)$$

Multiplying (5.83) by $G[q_2, q_1]$, and subtracting off (5.82) multiplied by $G[q_1, q_0]$, leads to

$$\frac{\partial}{\partial q_1}\left[G[q_2, q_1] \overleftrightarrow{\partial_{q_1}} G[q_1, q_0] \right]$$
$$= -6\pi^2 i \left[\delta(q_0 - q_1) G[q_2, q_1] - \delta(q_1 - q_2) G[q_1, q_0] \right]. \qquad (5.84)$$

We can integrate this relation, but need to figure out the value of the resulting integration constant. When q_1 is either smaller than the initial size q_0 or larger than the final size q_2, the dependence of the two propagators on q_1 is identical, and thus $\left[G[q_2, q_1] \overleftrightarrow{\partial_{q_1}} G[q_1, q_0] \right]$ will in fact be zero. This fixes the integration constant, and we find the composition law

$$\left[G[q_2, q_1] \overleftrightarrow{\frac{\partial}{\partial q_1}} G[q_1, q_0] \right] = \begin{cases} -3i V_3 G[q_2, q_0] & \text{if } q_1 \in [q_2, q_0], \\ 0 & \text{otherwise}. \end{cases} \qquad (5.85)$$

In the expression above, we kept the spatial volume V_3 explicitly. In the closed slicing, it is $V_3 = 2\pi^2$, but the composition law is valid more generally (and similarly, in the inhomogeneous WDW equation, the right-hand side would contain $3V_3$ instead of the factor $6\pi^2$).

One can explicitly verify that the composition law holds for the propagator (5.39). In fact, we can gain a little more information about the physics of this system by looking at that calculation in detail. The propagator (5.39) was derived by summing the contributions from two thimbles, i.e., there were two distinct saddle-point geometries that contributed significantly, one a purely expanding one (due to saddle 2 at a smaller lapse value) and one a bouncing geometry (due to the saddle 1 at a larger lapse value). It is useful for our present purposes to write out their contributions separately, $G_+[q_2, q_1] = G_{\exp}[q_2, q_1] + G_{\text{bounce}}[q_2, q_1]$, with

$$G_{\exp}[q_2, q_1] = \frac{i 3^{1/2}}{2[(\Lambda q_1 - 3)(\Lambda q_2 - 3)]^{1/4}} e^{+i \frac{12\pi^2}{\Lambda}(\frac{\Lambda}{3} q_1 - 1)^{\frac{3}{2}}} e^{-i \frac{12\pi^2}{\Lambda}(\frac{\Lambda}{3} q_2 - 1)^{\frac{3}{2}}}, \qquad (5.86)$$

$$G_{\text{bounce}}[q_2, q_1] = \frac{3^{1/2}}{2[(\Lambda q_1 - 3)(\Lambda q_2 - 3)]^{1/4}} e^{-i\frac{12\pi^2}{\Lambda}(\frac{\Lambda}{3}q_1 - 1)^{\frac{3}{2}}} e^{-i\frac{12\pi^2}{\Lambda}(\frac{\Lambda}{3}q_2 - 1)^{\frac{3}{2}}}.$$

(5.87)

The main observation now is that the dependence of $G_{\text{bounce}}[q_2, q_1]$ on q_1 is identical to the q_1 dependence of the incoming propagator $G[q_1, q_0]$. Thus the composition of these two vanishes

$$G_{\text{bounce}}[q_2, q_1] G_+[q_1, q_0]_{,q_1} - G_{\text{bounce}}[q_2, q_1]_{,q_1} G_+[q_1, q_0] = 0. \quad (5.88)$$

Meanwhile, the q_1 dependence of $G_{\exp}[q_2, q_1]$ is precisely such that the composition with $G_+[q_1, q_0]$ eliminates the dependence on q_1 and yields the desired result

$$G_{\exp}[q_2, q_1] G_+[q_1, q_0]_{,q_1} - G_{\exp}[q_2, q_1]_{,q_1} G_+[q_1, q_0] = -6\pi^2 i G_+[q_2, q_0]. \quad (5.89)$$

Physically speaking, this is very reasonable: If one does not know what the universe was doing beforehand, then it is natural that the quantum amplitude should contain both an expanding and a bouncing saddle point, since the universe could have been expanding or contracting beforehand. However, in combining with a second period of expansion, we see that only the expanding saddle contributes. This is as it should be, since at that point one knows that the universe was already in the process of expansion. It is a good check of the formalism that this physical intuition is reflected so clearly in the mathematics.

5.4 Discussion and some open questions

In the last two chapters, introducing the canonical and path integral approaches to quantum gravity, we have started exploring the consequences of bringing general relativity and quantum theory together. As we saw, the two formulations look quite different, but are in fact equivalent when treated carefully. That said, it should have become clear that certain physical questions can be addressed more easily and transparently in one or the other approach.

The Wheeler–DeWitt equation, for instance, clearly shows that quantum gravity has to operate in the absence of an external time parameter. Only the fields themselves, and in particular correlations between them, have meaning. Moreover, it shows (perhaps surprisingly) that spacetime can play no fundamental role in quantum gravity, rather it must make its appearance as an emergent phenomenon. By contrast, transition amplitudes can be calculated as path integrals summing over both spacetimes and matter configurations. In that sense, they are closer to our intuition, and this is their strength. Nevertheless, as we have shown, path integrals satisfy the WDW equation and are thus equivalent.

The fact that the WDW equation simply expresses time reparametrization invariance implies that just about any approach to quantum gravity must recover this equation in some limit. At the same time, the non-renormalizability of gravity informs us that this equation cannot be the final word, as it is built on pure general relativity, which is bound to become extended/replaced at ultrahigh curvatures (and perhaps in other regimes too). Thus it is clear that the approach we have followed so far pertains to the domain of semiclassical gravity. This is precisely the regime in which an effective spacetime description emerges, where (relative) probabilities can be defined, and where quantum gravity connects with established physics. To put it succinctly, the plan for the rest of this book will be to explore the physical consequences of semiclassical gravity.

Before proceeding, we should point out that even semiclassical gravity contains many open questions, which provide ample opportunities for future research. Let us mention a few questions that become most pertinent in the path integral quantization scheme. The most obvious and important question is which spacetimes and which matter configurations should actually be summed over? Already in ordinary quantum mechanics, the question is nontrivial, as it is known that one must include nondifferentiable particle paths in order to retain the commutation relations of quantum variables. But the quantum gravity equivalent of nondifferentiable paths are spacetimes for which the curvature blows up everywhere. It is not currently known how to handle such spaces mathematically. What we resorted to here is the minisuperspace restriction, in which one purposefully sets most variables and their conjugate momenta to zero, in order to obtain a tractable system. This is certainly useful, if only in order to gain an understanding of the saddle points, which are smooth and differentiable. One must then check that when extending the system to include perturbations it remains stable – we will do so in later chapters.

But even within the minisuperspace simplification, the question of which spacetimes one should sum over remains. As we saw, the old suggestion that it should be Euclidean spaces does not work due to the conformal factor problem (which, as we will see, does not preclude Euclidean saddle points from potentially playing important roles in the history of the universe – a Euclidean saddle point can contribute to a non-Euclidean path integral). A sum over Lorentzian spacetimes appears more promising, although to define it properly one must in fact deform the integral into the complex plane.

This not only brings the question of integration contours back to the fore, but it also engenders the question of which complex "spacetimes" should be included and which should be discarded. This is an active area of research, which we will return to at various instances in this book.

A further question is whether one should also sum over different topologies. As we will see, black hole physics as well as studies of the no-boundary proposal suggest that the answer is affirmative. But when different topologies are included, new puzzles (having to do with wormholes) appear. This is certainly one of the most interesting and ill-understood questions of quantum gravity.

The examples of transition amplitudes we studied in this chapter can be seen as our first foray into describing inflation quantum mechanically, including the background spacetime. We will delve deeper into this topic both in the exercises (where we will explore Robin boundary conditions in some detail) and in the coming chapters, first by noting the similarity between fluctuations around black holes and fluctuations in inflation, and later by trying to set up a theory of initial conditions for inflation. This will bring us a little closer to the holy grail of cosmology, namely a quantum gravitational understanding of the big bang.

Exercises

5.1 Consider Einstein–Hilbert gravity with the addition of the following (Robin) surface term on the final boundary,

$$S_R = -\xi \int_{\partial \mathcal{M}} \mathrm{d}^3 y \sqrt{h} \,. \tag{5.90}$$

Show that in an FLRW universe of arbitrary spatial curvature, this terms allows one to fix the Hubble rate on the boundary, an example of a Robin boundary condition.

5.2 An alternative to imposing boundary conditions is to assume an initial state ψ_0, which thus provides the initial condition for subsequent evolution.

$$G[q_1; \psi_0] = \int G[q_1; q_0]\psi_0(q_0)\mathrm{d}q_0 = \int_{0^+}^{\infty} \mathrm{d}\tilde{N} \int \mathrm{d}q_0 \, G[q_1; q_0; \tilde{N}]\psi_0(q_0) \,. \tag{5.91}$$

In the following three exercises, we will investigate the impact of assuming the universe to be in an initial coherent state at the onset of inflation. This turns out to be equivalent to imposing a Robin condition. So, assume an initial coherent state

$$\psi_0(q_0) = \frac{1}{\sqrt[4]{2\pi\Delta^2}} e^{\frac{i}{\hbar}p_i(q_i)q_0 - \frac{(q_0-q_i)^2}{4\Delta^2}}, \tag{5.92}$$

expressing the idea that the universe had size q_i (with uncertainty Δ) and momentum p_i. Work out the value of p_i that we should use in an expanding spatially flat universe containing vacuum energy Λ, working with a metric of the form (5.17). Then convolve the initial state with the propagator,

$$G[q_1, \bar{q}_0; \tilde{N}] = \int G[q_1; q_0; \tilde{N}]\psi_0(q_0)\mathrm{d}q_0, \tag{5.93}$$

and find the effective initial size $\bar{q}_0$. Check what happens to it in the limit $\Delta \to 0$.

5.3 For the lapse integral that was just derived, find the saddle points (for general Δ), and also derive the saddle-point geometries when $\Delta = 0$. Check (numerically) which saddle points contribute to the integral, starting with small position uncertainty Δ (large momentum uncertainty) and gradually increasing Δ (i.e., decreasing the uncertainty in the momentum). It is sufficient to focus on the saddles with $\mathrm{Re}[\tilde{N}] > 0$. Interpret the results.

5.4 As just shown, there is a critical value at which the flow lines undergo a topological change (Stokes phenomenon) and beyond which the bouncing saddle point does not contribute anymore. Find the minimum size of the universe for which this happens.

6
Linking gravity, quantum theory, and thermodynamics

Horizons are special in that they shield one region of the universe from the rest. In other words, they represent a limit to what one can observe, and therefore lead to a fundamental level of ignorance about parts of the universe. Statistical physics is also built on ignorance, and the resulting coarse-grained quantities, like heat or entropy, are useful in describing many-particle systems. This loose analogy turns out to be much more than that. In fact, astonishingly, general relativity seems to "know" about such concepts as temperature and entropy, even though it is a classical field theory. This is one of the most remarkable features of general relativity, and it shows that general relativity is in no way in conflict with quantum physics (as is sometimes claimed), but rather suggests that the two theories fit together nicely. In this chapter we will explore three occurrences of horizons: those of accelerating observers (leading to the Unruh effect), those of black holes (leading to Hawking radiation), and those of observers in de Sitter spacetime (leading to the Gibbons–Hawking temperature). They are closely related, and in fact we will start with a heuristic calculation illustrating the chain of thought that leads rather generically from horizons to temperatures.

6.1 Heuristics: horizons, imaginary time, and temperature

When gravity becomes strong, it becomes increasingly difficult for rocks, space ships, or particles to leave the influence of the gravitating object. Increasing the mass of the object further eventually leads to a situation in which no other object, no matter how light, can leave a certain region of spacetime. This region is bounded by a surface on which only light can still propagate, but from which even light cannot escape outward anymore.

To model the geometry near a horizon, we will use coordinates such that a horizon exists at a certain fixed radius, in a spherically symmetric setup (although the observer could be either outside of the horizon, or surrounded by it). For simplicity we will assume that the geometry is static (time independent) outside of the horizon. Thus we are led to a metric of the form

$$ds^2 = -h(r)dt^2 + f(r)dr^2 + (r+c)^2 d\Omega_2^2, \tag{6.1}$$

where $d\Omega_2^2$ is the metric on the 2-sphere (one could trivially generalize this calculation to a different 2-dimensional space here), and where c is a constant that allows us to shift the origin of r such that $r = 0$ corresponds to the horizon. The horizon is the surface on which only light can still propagate, though without being able to cross it. The horizon is thus a null surface. On a null surface, the normal is also null (if this result is unfamiliar, see the spacetime diagram in Fig. 6.2 for intuition). Since the horizon is at fixed r, the normal is given by $\partial_\mu r$, and it is null when

$$g^{\mu\nu} \partial_\mu r \partial_\nu r = g^{rr} = \frac{1}{f(r)} = 0 \quad \text{at } r = 0. \tag{6.2}$$

However, we do not want the spacetime to degenerate at $r = 0$ and be singular there. The 4-volume element is proportional to $\sqrt{-g} \propto \sqrt{g_{tt} g_{rr}}$, and we can avoid its degeneration if $g_{tt} \propto 1/g_{rr}$ at $r = 0$, or equivalently $h(r) = 4a^2/f(r)$ near $r = 0$, where we set the proportionality constant to $4a^2$ for later convenience (here a is a number, and not the scale factor). Furthermore, we would like the Einstein equations to be nonsingular at the horizon. A quick calculation shows that both the R_{tt} and R_{rr} components of the Ricci tensor contain second-derivative terms proportional to $h_{,rr}$. These terms should provide no contribution at the horizon (otherwise one would have to assume a specific stress tensor source on the horizon, which goes against our assumptions), and together with the requirement that $h(r = 0) = 0$, we can see that we can model a near-horizon geometry by taking $h(r) = r$, leading to

$$ds^2 = -4a^2 r\, dt^2 + \frac{dr^2}{r} + (r+c)^2 d\Omega_2^2. \tag{6.3}$$

Note a curious feature: Space and time switch roles as one crosses the horizon. Indeed, when $r < 0$, t becomes a space-like direction and r becomes time-like. Thus inside of the horizon the metric is (in general) not static at all. Rather, it corresponds to an anisotropic evolving universe, shielded from the static outside region by the horizon.

When $r \geq 0$ we can perform a coordinate change to simplify the metric, by setting $\frac{dr}{r^{1/2}} \equiv dR$. This implies that $2r^{1/2} = R$, where we fixed the integration constant such that $R = 0$ corresponds to the location of the horizon.

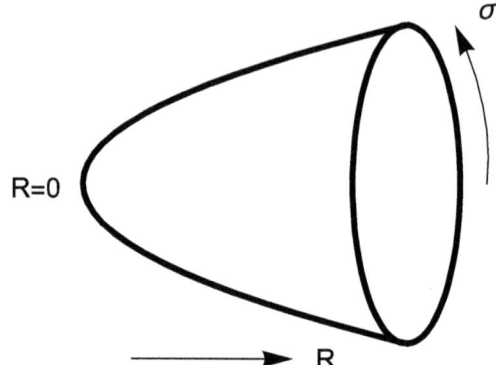

Figure 6.1 The Euclidean version of a near-horizon geometry caps off smoothly at the horizon (located at $R = 0$), provided the Euclidean time coordinate σ has the appropriate periodicity. The periodicity turns out to be the inverse of the temperature of the radiation associated with the horizon.

At the same time, we will analytically continue the metric to Euclidean time $d\sigma = idt$, for reasons that will become apparent momentarily. The metric then reads

$$ds^2 = a^2 R^2 \, d\sigma^2 + dR^2 + \left(\frac{1}{4}R^2 + c\right)^2 d\Omega_2^2. \tag{6.4}$$

The σ-R part of the metric looks suggestive. Recall the metric of flat space in polar coordinates, $ds^2 = dR^2 + R^2 d\theta^2$, which is nonsingular at $R = 0$ as long as θ is an angular variable with periodicity $\Delta\theta = 2\pi$ (for any other periodicity, there would be a conical defect, i.e., a singularity, at the tip of the cone $R = 0$). This tells us that if we assume that σ is periodic with period $\Delta\sigma = \frac{2\pi}{a}$, then the Euclidean near-horizon metric is in fact entirely nonsingular. It caps off smoothly at $R = 0$ – see Fig. 6.1.

The question now is what the meaning of this smooth Euclidean solution might be. Let us first give a hand-wavy argument, and then try to make it more precise. Time evolution in quantum mechanics occurs via the Hamiltonian, whose eigenvalues are the energy states E_n, via $e^{iHt} = \sum e^{iE_n t}$. If we let the time become imaginary, $t = i\sigma$, then the time evolution factor becomes a suppression factor,

$$e^{iE_n t} = e^{-E_n \sigma} \sim e^{-\frac{E_n}{\mathcal{T}}}, \tag{6.5}$$

which is reminiscent of a Boltzmann factor $e^{-E_n/\mathcal{T}}$, where $\mathcal{T}$ is the temperature (the Boltzmann factor gives the relative probability of a state of energy E_n in an equilibrium system at temperature $\mathcal{T}$). In our case the Euclidean time coordinate is periodic with period $\Delta\sigma$, leading to the tentative identification $\Delta\sigma = 1/\mathcal{T}$, or

$$\mathcal{T} = \frac{1}{\Delta\sigma} = \frac{a}{2\pi}. \tag{6.6}$$

Note that the periodicity in Euclidean time is consistent with the statement that this corresponds (at least approximately) to an equilibrium configuration.

The relationship between quantum mechanics and statistical mechanics is thus formally simply a continuation to (periodic) imaginary time. We can make it more precise by looking at density matrices. When it is diagonal, a density matrix ρ can be thought of as an ensemble of pure states, each occurring with probability p_n,

$$\rho = \sum_n p_n |n\rangle\langle n|, \tag{6.7}$$

where the probabilities are nonnegative and sum to unity $\sum_n p_n = 1$. The number p_n then represents the probability of being in the state $|n\rangle$.

If this density matrix has rank 1, i.e., when it can be written as $\rho = |\psi\rangle\langle\psi|$ for some state $|\psi\rangle$, then the state it describes is *pure* (a pure state satisfies $\rho = \rho^2$ and $\text{Tr}\rho^2 = 1$). Otherwise it is a *mixed* state. The expectation value of an observable A is calculated as

$$\langle A \rangle = \text{Tr}(\rho A) = \sum_n p_n \langle n|A|n\rangle, \tag{6.8}$$

i.e., it is the weighted average of the values of the operator in the constituent $|n\rangle$ states.

Now, for us the most important state to consider is the *thermal state*

$$\rho_\beta = \frac{e^{-\beta H}}{Z(\beta)} = \frac{1}{Z(\beta)} \sum_n e^{-\beta E_n} |n\rangle\langle n|, \tag{6.9}$$

in which the relative probability of being in the energy eigenstate $|n\rangle$ is given by the Boltzmann factor $e^{-E_n \beta}$ at temperature $\mathcal{T} = 1/\beta$, and where E_n is the energy eigenvalue of $|n\rangle$. The normalization factor is known as the *partition function*

$$Z(\beta) = \text{Tr}\, e^{-\beta H} = \sum_n e^{-\beta E_n}. \tag{6.10}$$

A system in equilibrium at temperature $\mathcal{T} = 1/\beta$ is also sometimes described as the canonical ensemble. The partition function is an extremely useful quantity in thermodynamics, as it can be used to find other quantities of interest, in particular the energy E and entropy S of the system, via the relations

$$E = -\partial_\beta \ln Z, \qquad S = (1 - \beta \partial_\beta) \ln Z. \tag{6.11}$$

If we consider a simple harmonic oscillator with frequency ω (which one should also think of as a single frequency mode of a free quantum field), then we may explicitly calculate the occupation number of each state. A standard quantum mechanics result is that the energy eigenvalues of the nth level are $E_n = (n + \tfrac{1}{2}\omega)$, and thus the expectation value of the number operator N is

$$\langle N \rangle = \mathrm{Tr}(\rho_\beta N) = \frac{\sum_n n e^{-n\beta\omega}}{\sum_n e^{-n\beta\omega}} = \frac{1}{e^{\beta\omega} - 1}. \tag{6.12}$$

This is the well-known Bose–Einstein distribution.

The link between thermal states and imaginary time periodicity is then expressed by the Kubo–Martin–Schwinger (KMS) condition

$$\langle A_t B \rangle = \langle B A_{t+i\beta} \rangle, \tag{6.13}$$

where A_t is a time-dependent operator evolving as $A_t = e^{iHt} A_0 e^{-iHt}$, and B is a time-independent one. We can verify that for a thermal state this condition is satisfied:

$$\begin{aligned}
\langle A_t B \rangle_\beta &= \frac{1}{Z(\beta)} \mathrm{Tr}(e^{-\beta H} A_t B) \\
&= \frac{1}{Z(\beta)} \mathrm{Tr}(e^{-\beta H + iHt} A_0 e^{-iHt} B) \\
&= \frac{1}{Z(\beta)} \mathrm{Tr}(A_{t+i\beta} e^{-\beta H} B) \\
&= \frac{1}{Z(\beta)} \mathrm{Tr}(e^{-\beta H} B A_{t+i\beta}) \\
&= \langle B A_{t+i\beta} \rangle_\beta,
\end{aligned} \tag{6.14}$$

where the second to last line is obtained by using the property that the trace is invariant under cyclic permutations. This condition thus establishes a precise link between thermal states and evolution in imaginary time, in quantum mechanics and quantum field theory. A fruitful idea has been to assume that such a link remains true when gravity is included. This then provides a very direct connection between gravity, quantum mechanics and thermodynamics.

Let us sketch how it is implemented. As we saw in Chapter 5, a transition amplitude between a field configuration ϕ_0 at time t_0 and a configuration

6.1 Heuristics: horizons, imaginary time, and temperature

ϕ_1 at time t_1 is obtained by the path integral

$$\langle \phi_1, t_1 | \phi_0, t_0 \rangle = \int \mathcal{D}[\phi]\, e^{iS[\phi]}, \qquad (6.15)$$

summing over all paths that join the initial and final configurations. The transition amplitude can also be written in terms of Hamiltonian time evolution,

$$\langle \phi_1, t_1 | \phi_0, t_0 \rangle = \langle \phi_1 | e^{-iH(t_1-t_0)} | \phi_0 \rangle. \qquad (6.16)$$

When the time interval is replaced (and analytically continued) to a finite periodic Euclidean time, $t_1 - t_0 = -i\Delta\sigma = -i\beta$, and if we identify initial and final states $\phi_1 = \phi_0$, then we are in effect calculating the partition function

$$Z(\beta) = \text{Tr}\, e^{-\beta H} = \int_{\phi,0}^{\phi,\beta} \mathcal{D}[\phi]\, e^{-S_E[\phi]}, \qquad (6.17)$$

where $S_E = -iS$ is the Euclidean action. The path integral is now over field configurations that are periodic in imaginary time with period β.

When gravity is included, we retain the same definition (at least tentatively – though much evidence has accumulated that this gives reasonable results), except that now the spacetime background is not fixed. Rather, as we know from Chapters 4 and 5, we must fix a 3-dimensional spatial boundary configuration, and then sum over metrics with Euclidean time periodicity β, and fill in the rest of the 4-space (we will return to the question of how the integral is defined more precisely later – and see also the discussions in Chapter 5). Just as we saw when dealing with path integrals earlier, the main contribution (in the semiclassical limit) will come from saddle points, i.e., from solutions to the classical equations of motion, having finite action and satisfying the specified boundary conditions. In the presence of a horizon, the Euclidean metric having the form (6.4) near the horizon, and having finite total action, will provide just such a contribution, confirming that a horizon indeed leads to a thermal state with temperature $\mathcal{T} = 1/\beta$, as given in (6.6). In the following sections, we will analyze several explicit examples of this phenomenon, both from the point of view of Euclidean saddle points, and in a more standard quantum field theory setting (for which the necessary techniques will be explained in the next section).

In closing this section, let us highlight that conceptually the gravitational solution fits well with the intuition coming from quantum statistical physics: The Euclidean version of the solution ends at the horizon. This means that it does not contain any information about what lies behind the horizon, and this in turn encapsulates the idea that in this prescription we are tracing over all degrees of freedom behind the horizon. Out of this ignorance we obtain a thermal distribution.

6.2 Bogolyubov transformations

The fact that horizons lead to thermal radiation can be seen as part of the larger framework of quantum effects in curved spacetimes. This section will go over the tools required for our purposes.

In quantum theories, the dynamics is specified with respect to a time coordinate, cf. the Schrödinger equation $i\partial_t \psi = H\psi$. But relativity has taught us that the times perceived by different observers need not coincide. For example, accelerated motion or the curvature of spacetime can influence what an observer's clock registers. Different observers may therefore experience different proper times. They may thus describe quantum evolution differently – though of course the accounts of different observers must be consistent with each other.

In this section, we will review the framework of quantum field theory in curved spacetime, which allows us to relate the points of view of two different observers in a (in general) curved, but fixed spacetime. This will be useful for understanding the link between horizons and temperature in significantly more detail. In Section 6.5 we will then go one step further (albeit in a simple minisuperspace context) by presenting a calculation in which the background metric is also quantized.

We will assume a fixed background metric $g_{\mu\nu}$ that is explicitly split between time t and spatial coordinates x^i (essentially the ADM decomposition – see Section 3.2.2),

$$\mathrm{d}s^2 = g_{00}\,\mathrm{d}t^2 + 2\,g_{0i}\,\mathrm{d}t\mathrm{d}x^i + h_{ij}\,\mathrm{d}x^i\mathrm{d}x^j\,, \tag{6.18}$$

so that the spacetime is foliated by hypersurfaces Σ_t of constant time coordinate. The normal to these hypersurfaces is denoted n^μ and satisfies $n^\mu n_\mu = -1$.

We will consider a scalar field ϕ with a potential $V(\phi)$ in this fixed spacetime, with action

$$S = \int \mathrm{d}^4x\,\mathcal{L} = \int \mathrm{d}^4x\,\sqrt{-g}\left[-\frac{1}{2}g^{\mu\nu}\partial_\mu\phi\partial_\nu\phi - V(\phi)\right]\,, \tag{6.19}$$

and thus with the Klein–Gordon equation of motion

$$\Box\phi - V_{,\phi} = 0\,. \tag{6.20}$$

For (in general complex) solutions of the equation of motion one can define an inner product on a spatial hypersurface Σ_t,

$$(\phi_1, \phi_2) = -i \int_\Sigma d^3x \sqrt{h} n^\mu \left(\phi_1 \nabla_\mu \phi_2^* - (\nabla_\mu \phi_1) \phi_2^*\right). \tag{6.21}$$

This (Klein–Gordon) inner product, which can be shown to be independent of the choice of hypersurface, is covariantly conserved by virtue of the equation of motion.

The momentum conjugate to ϕ is given by

$$\pi = \frac{\delta \mathcal{L}}{\delta(\partial_0 \phi)} = -\sqrt{-g}\, g^{0\mu} \partial_\mu \phi = \sqrt{h} n^\mu \partial_\mu \phi. \tag{6.22}$$

In order to quantize, both the field and the momentum are elevated to operators, satisfying the canonical, equal-time, commutation relations on the hypersurfaces Σ_t,

$$[\hat{\phi}(t, \mathbf{x}), \hat{\phi}(t, \mathbf{x}')] = 0, \tag{6.23}$$

$$[\hat{\pi}(t, \mathbf{x}), \hat{\pi}(t, \mathbf{x}')] = 0, \tag{6.24}$$

$$[\hat{\phi}(t, \mathbf{x}), \hat{\pi}(t, \mathbf{x}')] = i\delta(\mathbf{x} - \mathbf{x}'), \tag{6.25}$$

where the spatial delta function is normalized with respect to the spatial volume only,

$$\int_\Sigma d^3x \sqrt{h}\, f(\mathbf{x}') \delta(\mathbf{x} - \mathbf{x}') = f(\mathbf{x}). \tag{6.26}$$

We would now like to expand the field in terms of a complete set of modes,

$$\hat{\phi} = \sum_i \left[\hat{a}_i^- f_i + \hat{a}_i^+ f_i^*\right], \tag{6.27}$$

where i is taken to be a discrete label here – i.e., we are assuming that the spatial hypersurfaces are compact (the continuous case is a direct generalization, replacing sums with appropriate integrals). For this it is useful to choose modes (by which we mean solutions to the equation of motion) $f_i(x^\mu)$ that are orthonormal, and that thus satisfy

$$(f_i, f_j) = -i \left(f_i \dot{f}_j^* - \dot{f}_i f_j^*\right) = \delta_{ij}, \quad (f_i, f_j^*) = 0. \tag{6.28}$$

Such modes are said to have positive norm (in Chapter 3, we called this the Wronskian – see (3.88)). Their complex conjugates will then have negative norm since $(f_i^*, f_j^*) = -\delta_{ij}$, which is a feature of the Klein–Gordon inner product.

Our definitions imply the simple relation

$$(\phi, f_i) = a_i^-, \tag{6.29}$$

which, together with (6.23)–(6.25), can be used to derive the commutation relations for the annihilation/creation operators $\hat{a}^-/\hat{a}^+$,

$$[\hat{a}_i^-, \hat{a}_j^-] = 0, \quad [\hat{a}_i^+, \hat{a}_j^+] = 0, \quad [\hat{a}_i^-, \hat{a}_j^+] = \delta_{ij}. \quad (6.30)$$

One may then proceed to define a vacuum state, annihilated by all $\hat{a}_i^-$,

$$\hat{a}_i^- |0\rangle_f = 0, \quad (6.31)$$

while states containing what we will refer to as f-particles are obtained by using the creation operators,

$$|n_i, n_j, \cdots\rangle_f = \frac{1}{\sqrt{n_i! n_j! \cdots}} \left(\hat{a}_i^+\right)^{n_i} \left(\hat{a}_j^+\right)^{n_j} \cdots |0\rangle_f. \quad (6.32)$$

The state above then contains n_i f-particles with momentum i, n_j particles with momentum j, and so on. The number of particles with momentum i can be counted by making use of the number operator

$$\hat{n}_{f,i} = \hat{a}_i^+ \hat{a}_i^-. \quad (6.33)$$

Now we turn our attention to a second observer, who finds it natural (due to their own, generally different, proper time) to use a different basis of modes,

$$\hat{\phi} = \sum_i \left[\hat{b}_i^- g_i + \hat{b}_i^+ g_i^*\right]. \quad (6.34)$$

These new modes g are also taken to be orthonormal, and they have their own annihilation and creation operators $\hat{b}^{-,+}$ associated with them, satisfying

$$[\hat{b}_i^-, \hat{b}_j^-] = 0, \quad [\hat{b}_i^+, \hat{b}_j^+] = 0, \quad [\hat{b}_i^-, \hat{b}_j^+] = \delta_{ij}. \quad (6.35)$$

Following analogous arguments to those presented above, the g-vacuum may be defined as

$$\hat{b}_i^- |0\rangle_g = 0, \quad (6.36)$$

while states with g-particles are constructed with the creation operators,

$$|n_i, n_j, \cdots\rangle_g = \frac{1}{\sqrt{n_i! n_j! \cdots}} \left(\hat{b}_i^+\right)^{n_i} \left(\hat{b}_j^+\right)^{n_j} \cdots |0\rangle_g. \quad (6.37)$$

Just as above, the number of g-particles with momentum i may then be extracted using the number operator

$$\hat{n}_{g,i} = \hat{b}_i^+ \hat{b}_i^-. \quad (6.38)$$

6.2 Bogolyubov transformations

The important realization is that since the f-modes provide a complete basis, the new modes may be re-expressed as a linear combination of f and f^*, and vice versa,

$$g_i = \sum_j \left(\alpha_{ij} f_j + \beta_{ij} f_j^* \right), \tag{6.39}$$

$$f_i = \sum_j \left(\alpha_{ji}^* g_j - \beta_{ji} g_j^* \right), \tag{6.40}$$

where the first line is a definition of the Bogolyubov coefficients α_{ij}, β_{ij}, which are subject to the normalization conditions

$$\sum_k (\alpha_{ik} \beta_{jk} - \beta_{ik} \beta_{jk}) = 0, \quad \sum_k (\alpha_{ik} \alpha_{jk}^* - \beta_{ik} \beta_{jk}^*) = \delta_{ij}. \tag{6.41}$$

Note that we can also extract the Bogolyubov coefficients using the inner product,

$$\alpha_{ij} = (g_i, f_j), \quad \beta_{ij} = -(g_i, f_j^*). \tag{6.42}$$

The Bogolyubov coefficients then allow us to relate the two sets of annihilation and creation operators to each other,

$$\hat{a}_i^- = \sum_j \left(\alpha_{ji} \hat{b}_j^- + \beta_{ji}^* \hat{b}_j^+ \right), \tag{6.43}$$

$$\hat{b}_i^- = \sum_j \left(\alpha_{ji}^* \hat{a}_j^- - \beta_{ij}^* \hat{a}_j^+ \right). \tag{6.44}$$

This has the somewhat astonishing consequence that what constitutes the vacuum state for the first observer may in fact contain particles from the point of view of the second observer. More concretely, we can use the number operator for g-particles but evaluate its expectation value on the f-vacuum,

$$_f\langle 0| \hat{n}_{g,i} |0\rangle_f = {}_f\langle 0| \hat{b}_i^+ \hat{b}_i^- |0\rangle_f \tag{6.45}$$

$$= {}_f\langle 0| \sum_j \sum_k \left(\alpha_{ji} \hat{a}_j^+ - \beta_{ij} \hat{a}_j^- \right) \left(\alpha_{ki}^* \hat{a}_k^- - \beta_{ik}^* \hat{a}_k^+ \right) |0\rangle_f \tag{6.46}$$

$$= \sum_j \beta_{ij} \beta_{ij}^*, \tag{6.47}$$

where we used $_f\langle 0| \hat{a}_i^- \hat{a}_j^+ |0\rangle_f = \delta_{ij}$. This means that the f-vacuum, as seen from the point of view of the second observer, contains $\sum_j |\beta_{ij}|^2$ particles of g-type. Thus the purported emptiness of a vacuum state is a subjective concept. What is the vacuum for one observer may be filled with particles for another observer, who has a different notion of time. If we go back to the relation between f- and g-modes, we see that we can trace the appearance of g-particles back to the fact that the positive-frequency f-modes split up into a combination of both positive- and negative-frequency g-modes, and it is the negative-frequency modes, as seen from the point of view of the second observer, that signal the presence of particle creation. In the next sections, we will illustrate this rather formal description with concrete examples. More specifically, we will present what are arguably the three most important instances where a connection between gravity (or acceleration) and thermodynamics has been (theoretically) established: the motion of an accelerated observer, the radiation emitted by black holes, and the production of particles in a de Sitter spacetime.

6.3 The Unruh effect

A simple, but important setting in which the relative nature of the vacuum state has striking physical consequences is that of a constantly accelerated observer in a flat spacetime. Here it is not the curvature of the spacetime that leads to particle creation, but rather the fact that an accelerated observer sees a horizon and thus effectively lives in a subset of Minkowski spacetime (although, of course, the equivalence principle implies that acceleration could also be described as gravity). Recalling the discussion of the previous section, the crucial point is that an accelerated explorer experiences a different proper time than an inertial, nonaccelerated companion.

Below we will consider constant acceleration in a particular coordinate direction, say the x-direction. Then all of the interesting physics takes place in the 2-dimensional $t - x$ subspace, and for simplicity as well as clarity we will simply suppress the remaining spatial coordinates momentarily. Thus we will take the $(1+1)$-dimensional Minkowski spacetime to be given by the metric

$$\mathrm{d}s^2 = -\mathrm{d}t^2 + \mathrm{d}x^2. \tag{6.48}$$

In this background spacetime we consider the motion of an observer, whose velocity and acceleration are given by derivatives with respect to proper time t_p,

$$v^\mu = \frac{\mathrm{d}x^\mu}{\mathrm{d}t_p}, \quad a^\mu = \frac{\mathrm{d}v^\mu}{\mathrm{d}t_p} = \frac{\mathrm{d}^2 x^\mu}{\mathrm{d}t_p^2}. \tag{6.49}$$

Since the spacetime is flat, the covariant derivatives reduce to ordinary ones. For a time-like trajectory we have $v_\mu v^\mu = -1$, and for constant acceleration

a we must impose $a_\mu a^\mu = a^2$. These relations, as you are encouraged to verify, can be solved by

$$t(t_p) = \frac{1}{a}\sinh(a\,t_p)\,, \qquad x(t_p) = \frac{1}{a}\cosh(a\,t_p)\,, \qquad (6.50)$$

implying that a constantly accelerated rocket follows (part of) a hyperbolic trajectory $x^2 - t^2 = \frac{1}{a^2}$. Such a trajectory is drawn as the thick dashed curve in region I in Fig. 6.2. Also note the relations $x \pm t = \frac{1}{a} e^{\pm a\,t_p}$.

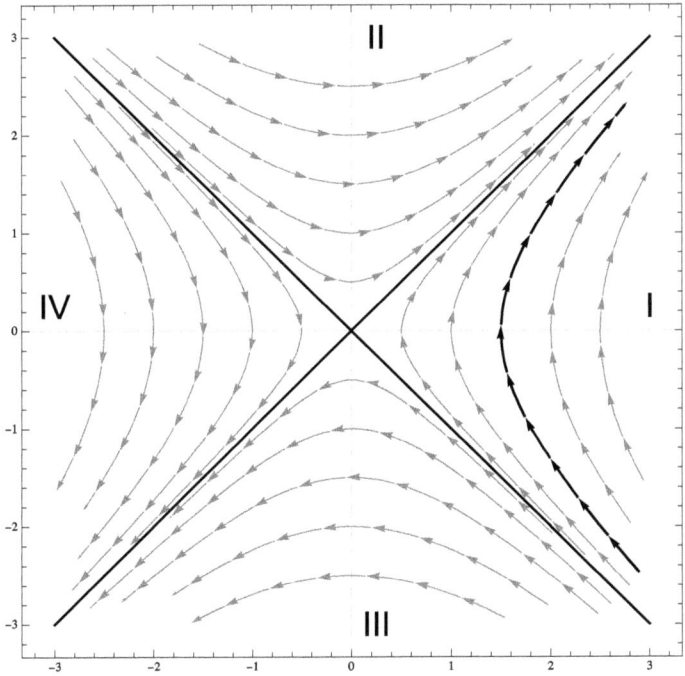

Figure 6.2 A spacetime diagram of a near-horizon geometry, with space on the horizontal axis and time on the vertical axis. Light rays propagate at angles of $\pm\frac{\pi}{4}$, so that the perpendicular direction to a light ray is also null. The thick southwest-to-northeast line is a horizon for regions I and III, in the sense that nothing happening in regions II and IV can influence what is happening in regions I and III. An analogous statement holds for the thick southeast-to-northwest null line. The arrows indicate the directions of the Rindler time η. In region I these lines are also trajectories of constant acceleration, as exemplified by the thicker dashed line. Such a constantly accelerating observer evolves from past null infinity to future null infinity, as the trajectory asymptotically approaches the two horizons. This observer cannot influence what occurs in regions III and IV, and cannot be influenced by regions II and IV.

We can define light cone coordinates $u \equiv t - x$, $v \equiv t + x$, in terms of which the Minkowski metric (6.48) is very simple,

$$ds^2 = -du\,dv \tag{6.51}$$
$$= e^{2a\xi}\left(-d\eta^2 + d\xi^2\right) \tag{6.52}$$
$$= -e^{a(\tilde{v}-\tilde{u})}d\tilde{u}\,d\tilde{v}. \tag{6.53}$$

Here we added two more coordinate systems. The first one, in Eq. (6.52), is defined in order to be adapted to the motion of an accelerated observer, and is obtained via the coordinate transformations

$$t = \frac{1}{a}e^{a\xi}\sinh(a\eta), \qquad x = \frac{1}{a}e^{a\xi}\cosh(a\eta). \tag{6.54}$$

Comparing to (6.50), one can see that an observer undergoing constant acceleration a sits at $\xi = 0$ and experiences proper time η. This is known as the Rindler metric – with the further rescaling $dR \equiv e^{a\xi}d\xi$ it takes the form $ds^2 = -a^2R^2\,d\eta^2 + dR^2$, i.e., it takes the form of the near-horizon metric (6.4), though here this metric is in fact not just an approximation near the horizon, but is exact. From its definition, one can see that the Rindler metric only covers part of Minkowski spacetime, namely the region $x \geq |t|$. This region is known as the *Rindler wedge* and is denoted region I in Fig. 6.2. In the Rindler metric even an infinite coordinate distance (in the direction opposite to the acceleration) only yields a finite proper distance from the observer at $\xi = 0$, namely $\int_{-\infty}^{0} e^{a\xi} = \frac{1}{a}$. This calculation also shows that the horizon (which is at $\xi = -\infty$) is closer to explorers with a higher acceleration.

The coordinates $\tilde{u}, \tilde{v}$ are then the light cone coordinates associated with the Rindler coordinates η, ξ, with $\eta - \xi \equiv \tilde{u}$ and $\eta + \xi \equiv \tilde{v}$, yielding the metric (6.53). The light cone coordinates are related via

$$-au = e^{-a\tilde{u}}, \qquad av = e^{a\tilde{v}}. \tag{6.55}$$

All of these coordinate systems will have their use in our calculation below.

Let us outline the gist of the Unruh effect: In quantum field theory, mode functions are not uniquely defined, as we saw above. Every observer defines their own vacuum state, which is related to the proper time they experience. In particular, a Minkowski observer will define their vacuum with respect to the Minkowski time t, while an accelerated observer will use the proper time η. This leads to different notions of what constitutes a positive-frequency mode, for instance, and leads to different notions of what the vacuum state should be. What the accelerated observer perceives as the vacuum is in fact seen as a state containing particles by the Minkowski observer, and vice versa. And this state turns out to be a thermal state, with temperature $T = a/(2\pi)$, thereby explicitly confirming the heuristic arguments presented in Section 6.1. It is the horizon perceived by the accelerating observer that shuts them off from the rest of Minkowski space, and that leads to this temperature. This is known as the Unruh (sometimes Fulling–Davies–Unruh) effect.

We will illustrate these statements by considering a massless scalar field propagating in Rindler spacetime – analogous calculations can be performed with other fields. The action is given by

$$S = -\frac{1}{2} \int d^2x \sqrt{-g}\, g^{\mu\nu} \partial_\mu \phi \partial_\nu \phi = -2 \int du\, dv\, \partial_u \phi \partial_v \phi \tag{6.56}$$

$$= -2 \int d\tilde{u}\, d\tilde{v}\, \partial_{\tilde{u}} \phi \partial_{\tilde{v}} \phi, \tag{6.57}$$

and thus the equation of motion is

$$\partial_u \partial_v \phi = 0, \qquad \partial_{\tilde{u}} \partial_{\tilde{v}} \phi = 0. \tag{6.58}$$

Here we see the advantage of the two sets of light cone coordinates: The equation of motion becomes extremely simple. The solutions, $\phi(u,v) = f_1(u) + f_2(v)$, are just given by sums of arbitrary differentiable functions of u and of v, respectively, of $\tilde{u}$ and of $\tilde{v}$. For this calculation, we will use a plane-wave basis of solutions. Then, for instance, a right-moving, positive-frequency solution with respect to Minkowski time t is

$$\phi \propto e^{-i\omega u} = e^{-i\omega(t-x)}, \tag{6.59}$$

since $i\partial_t \phi = +\omega\phi$ (cf. the Schrödinger equation $i\partial_t \psi = H\psi$). Here ω is both the frequency and the wavenumber, as the field is massless. Likewise, a right-moving, positive-frequency solution with respect to the Rindler time η is

$$\phi \propto e^{-i\Omega \tilde{u}} = e^{-i\Omega(\eta-\xi)}. \tag{6.60}$$

Analogous expressions exist for left-moving modes, in terms of v and $\tilde{v}$ – these can be treated separately, and we will not write them out explicitly.

On the Rindler wedge, where both sets of light cone coordinates are defined, we can write out the field in terms of two separate sets of operators and mode functions,

$$\hat{\phi} = \int_0^\infty \frac{d\omega}{\sqrt{4\pi\omega}} \left(\hat{a}_\omega^- e^{-i\omega u} + \hat{a}_\omega^+ e^{i\omega u} \right) + \text{left-moving} \tag{6.61}$$

$$= \int_0^\infty \frac{d\Omega}{\sqrt{4\pi\Omega}} \left(\hat{b}_\Omega^- e^{-i\Omega \tilde{u}} + \hat{b}_\Omega^+ e^{i\Omega \tilde{u}} \right) + \text{left-moving}, \tag{6.62}$$

where the operators satisfy the standard commutation relations, adapted here to the case of a continuous-frequency spectrum,

$$[\hat{a}_\omega^-, \hat{a}_{\omega'}^+] = \delta(\omega - \omega'), \quad [\hat{a}_\omega^-, \hat{a}_{\omega'}^-] = 0, \quad [\hat{a}_\omega^+, \hat{a}_{\omega'}^+] = 0, \qquad (6.63)$$

$$[\hat{b}_\Omega^-, \hat{b}_{\Omega'}^+] = \delta(\Omega - \Omega'), \quad [\hat{b}_\Omega^-, \hat{b}_{\Omega'}^-] = 0, \quad [\hat{b}_\Omega^+, \hat{b}_{\Omega'}^+] = 0. \qquad (6.64)$$

A Minkowski observer will define their vacuum as the lowest excitation state with respect to the $\hat{a}$ operators – i.e., the Minkowski vacuum is given by the condition

$$\hat{a}_\omega^- \mid 0\rangle_\text{Mink} = 0, \qquad (6.65)$$

while the vacuum seen by the accelerating observer is defined with respect to Rindler time η and the associated operators $\hat{b}$,

$$\hat{b}_\Omega^- \mid 0\rangle_\text{Acc} = 0. \qquad (6.66)$$

But we can now ask how the accelerated observer perceives the Minkowski vacuum. For this, we would like to relate the two sets of operators to each other, using a Bogolyubov transformation (again adapted to continuous frequencies)

$$\hat{b}_\Omega^- = \int_0^\infty d\omega \left(\alpha_{\Omega\omega} \hat{a}_\omega^- - \beta_{\Omega\omega} \hat{a}_\omega^+\right), \qquad (6.67)$$

where the normalization condition is now given by

$$\int_0^\infty d\omega \left(\alpha_{\Omega\omega} \alpha_{\Omega'\omega}^* - \beta_{\Omega\omega} \beta_{\Omega'\omega}^*\right) = \delta(\Omega - \Omega'). \qquad (6.68)$$

We can substitute the Bogolyubov relation (6.67) into the mode expansion of the accelerating observer (6.62), and equate this expression with the Minkowski mode expansion (6.61) to find

$$\frac{1}{\sqrt{\omega}} e^{-i\omega u} = \int_0^\infty \frac{d\Omega'}{\sqrt{\Omega'}} \left(\alpha_{\Omega'\omega} e^{-i\Omega' \tilde{u}} - \beta_{\Omega'\omega}^* e^{+i\Omega' \tilde{u}}\right). \qquad (6.69)$$

We can turn this expression around by multiplying by factors of $e^{\pm i\Omega \tilde{u}}$ and integrating over $\tilde{u}$ (which leads to a delta function on the right-hand side), obtaining

$$\alpha_{\Omega\omega} = \frac{1}{2\pi} \sqrt{\frac{\Omega}{\omega}} \int_{-\infty}^\infty d\tilde{u} \, e^{-i\omega u + i\Omega \tilde{u}} = \frac{1}{2\pi} \sqrt{\frac{\Omega}{\omega}} \int_{-\infty}^0 \frac{du}{(-au)} e^{-i\omega u - i\frac{\Omega}{a} \ln(-au)}, \qquad (6.70)$$

$$\beta_{\Omega\omega} = -\frac{1}{2\pi} \sqrt{\frac{\Omega}{\omega}} \int_{-\infty}^\infty d\tilde{u} \, e^{i\omega u + i\Omega \tilde{u}} = -\frac{1}{2\pi} \sqrt{\frac{\Omega}{\omega}} \int_{-\infty}^0 \frac{du}{(-au)} e^{i\omega u - i\frac{\Omega}{a} \ln(-au)}. \qquad (6.71)$$

Alternatively, we could have used (6.42) to obtain the Bogolyubov coefficients. One can evaluate the integrals above to yield Gamma functions, but these are not particularly illuminating. Letting $u \to -u$ in the expression for $\beta_{\Omega\omega}$ allows us to relate the two coefficients directly, $|\beta_{\Omega\omega}| = |\alpha_{\Omega\omega}|e^{i\frac{\Omega}{a}\ln(-1)}$, so that we obtain

$$|\beta_{\Omega\omega}|^2 = e^{-\frac{2\pi\Omega}{a}} |\alpha_{\Omega\omega}|^2 . \tag{6.72}$$

We can use this relation to calculate the occupation number of Rindler particles in the Minkowski vacuum, cf. (6.47),

$$\langle N_\Omega \rangle_{\text{Mink}} = \int d\omega \, |\beta_{\Omega\omega}|^2 . \tag{6.73}$$

If we now look at the normalization condition (6.68) at $\Omega = \Omega'$,

$$\int_0^\infty d\omega \left(|\alpha_{\Omega\omega}|^2 - |\beta_{\Omega\omega}|^2 \right) = \delta(0) , \tag{6.74}$$

then we can interpret the right-hand side as being the volume of space $\delta(0) = V$ (had we not used plane-wave mode functions, but more localized wave packets, we could have avoided this divergence). Then the number density of Rindler particles in the Minkowski vacuum can be written as

$$\frac{\langle N_\Omega \rangle_{\text{Mink}}}{V} = \frac{1}{e^{\frac{2\pi\Omega}{a}} - 1} . \tag{6.75}$$

This is precisely the Bose–Einstein distribution (6.12) characterizing a thermal state, in this case with temperature

$$T_{\text{Unruh}} = \frac{1}{\beta} = \frac{2\pi}{a} . \tag{6.76}$$

This is the temperature that an observer undergoing constant acceleration a would measure in the Minkowski vacuum. The notion of vacuum, as well as the concept of a particle, is observer dependent.

An apparent paradox is that the Minkowski vacuum is in fact empty of particles from the point of view of an inertial observer, so how can this inertial observer agree that an accelerating thermometer is measuring a nonzero temperature in the Minkowski vacuum? Note that in order for the thermometer to remain at constant acceleration a rocket is needed, and it must supply a force continuously. Thus, ultimately, the particles that the accelerated thermometer detects get their energy from the rocket on which the thermometer sits.

Let us end this section with a comment on the gravitational calculation of the partition function. As we saw, the near-horizon metric (6.4) does not just approximate the Rindler metric near the horizon, it is in fact valid in the entire Rindler wedge. This metric is thus a (Euclidean) saddle point of the gravitational path integral, if we use boundary conditions in which the Euclidean time is periodic. The correct periodicity in time prevents the appearance of a conical singularity in the metric, which is thus completely smooth and regular, and yields a finite contribution to the partition function. Neglecting backreaction of the matter fields on the geometry (as we have done so far) then implies that all correlation functions of fields on this background space will automatically satisfy the KMS condition (6.13) with the same β. Put differently, all fields that see the same horizon will also experience the same Unruh temperature.

6.4 Hawking radiation

The best-known instance of horizons are black holes, where gravity is so strong that classically nothing can escape from inside the horizon. Famously, Hawking discovered that for quantum fields the situation changes, and in fact field quanta (particles) can be emitted from black holes, thereby allowing isolated black holes to evaporate. Although the focus of this book is on cosmology, the discovery of Hawking radiation was such a seminal result in the development of quantum gravity that it is worthwhile including it. Moreover, as we will see, there is a rather close analogy between Hawking radiation and inflationary particle production.

Historically, Hawking radiation was discovered first and the Unruh effect arose from attempts to understand and simplify Hawking's calculation. This means that from the current perspective, the Hawking effect can be seen as an extension of Unruh's calculation that we reviewed above. We can thus be comparatively brief here.

We will start with the simplest black hole metric, the Schwarzschild metric[1] describing a nonrotating, noncharged black hole of mass M,

$$ds^2 = \left(1 - \frac{2M}{r}\right) dt^2 + \frac{dr^2}{\left(1 - \frac{2M}{r}\right)} + r^2 d\Omega^2 . \tag{6.77}$$

By inspection, one may see that there are potential singularities at radial locations $r = 0$ and $r = 2M$. If one looks at curvature invariants, such as $R_{\mu\nu}R^{\mu\nu}$ or $R_{\mu\nu\rho\sigma}R^{\mu\nu\rho\sigma}$, one finds that they diverge like $1/r^p$ for some positive power p. Hence the center of the black hole, at $r = 0$, is indeed a true curvature singularity in classical general relativity, while the horizon,

---

[1] Schwarzschild in German translates to "black shield" – an astonishing aptronym.

at $r = 2M$, is regular and only appears singular because of our choice of coordinates.

We can look at the horizon region in more detail by studying the form of the metric in its vicinity. To this end, we can define $r \equiv 2M + \frac{\epsilon}{2M}$, allowing us to approximate the metric near $r = 2M$ as

$$ds^2 \approx -\frac{\epsilon}{4M^2}dt^2 + \frac{d\epsilon^2}{\epsilon} + 4M^2 d\Omega^2. \tag{6.78}$$

If we now define $\frac{d\epsilon}{\epsilon^{1/2}} \equiv dR$ and switch to Euclidean time $\sigma = -it$, we obtain

$$ds^2 \approx dR^2 + R^2 \frac{d\sigma^2}{(4M)^2} + 4M^2 d\Omega^2. \tag{6.79}$$

Comparing this metric to the general near-horizon Euclidean metric (6.4), we see that these two metrics agree if we set $a = 1/(4M)$. In other words, the Euclidean metric will cap off smoothly at $R = 0$ if we impose the Euclidean time periodicity $\Delta\sigma = 8\pi M$, thereby avoiding a conical singularity. As a consequence of the general arguments given in Section 6.1, we thus expect the black hole horizon to lead to the emission of radiation with temperature

$$\mathcal{T}_{\text{Hawking}} = \frac{1}{8\pi M}. \tag{6.80}$$

This is the famous Hawking temperature, which scales inversely with the mass M. Thus, an evaporating black hole in fact heats up as it shrinks. Conversely, a black hole that becomes more massive by absorbing matter–energy becomes colder. This is one of the odd thermodynamic properties of black holes.

The analytic continuation of the metric provides the fastest route to deriving thermodynamic properties of black holes, and can be extended to black holes with charge and/or rotation. A more detailed path integral calculation, in the presence of a negative cosmological constant, will be presented in Section 6.5. That calculation will show explicitly how this kind of Euclidean metric can arise as a saddle point of an appropriate path integral.

For the remainder of the present section, we will briefly look at the more explicit calculation of quantum fields on the Schwarzschild background, in analogy with the calculation of the Unruh effect presented above. For simplicity, we will focus on the (1+1)-dimensional time-radial part of the metric; extending to the full $(1+3)$-dimensional case brings in a few modifications, but does not change the core of the argument – interested readers are encouraged to consult books on black holes for further details.

It will be useful to introduce additional coordinate systems. We define the *tortoise* coordinate via

$$r^* \equiv r - 2M + 2M \ln\left(\frac{r}{2M} - 1\right). \tag{6.81}$$

It takes its name from the fact that as $r \to 2M$, the tortoise coordinate diverges $r^* \to -\infty$. This implies that the tortoise coordinate only covers the range $2M \leq r < +\infty$, i.e., it only covers the region outside of the horizon. In terms of this coordinate the metric takes the form

$$ds^2 = \left(1 - \frac{2M}{r}\right)(-dt^2 + dr^{*2}), \tag{6.82}$$

where $r(r^*)$ is implicitly regarded as a function of r^*. We can then define corresponding light cone coordinates $\tilde{u} = t - r^*$ and $\tilde{v} = t + r^*$, in terms of which the metric becomes

$$ds^2 = -\left(1 - \frac{2M}{r}\right)d\tilde{u}\,d\tilde{v}. \tag{6.83}$$

Here $r(\tilde{u}, \tilde{v})$ could in principle be rewritten in terms of the light cone coordinates, though we will not require the explicit expression. Again, these light cone coordinates only cover the outside region of the black hole – they are therefore the equivalent of the Rindler coordinates (6.53) we used when discussing the Unruh effect.

We are once again interested in the quantization of a massless scalar field on this background. An observer far away from the black hole will have the coordinate t coinciding with their proper time, given that in that limit the tortoise metric becomes flat,

$$ds^2 \xrightarrow{r \to \infty} -d\tilde{u}\,d\tilde{v} = -dt^2 + dr^{*2}. \tag{6.84}$$

Following the exact same steps as in Section 6.3, we can infer that to a far-away observer a right-moving, positive-frequency mode will be proportional to $e^{-i\Omega\tilde{u}}$, and for such an observer the scalar field is expanded as

$$\hat{\phi} = \int_0^\infty \frac{d\Omega}{\sqrt{4\pi\Omega}} \left(\hat{b}_\Omega^- e^{-i\Omega\tilde{u}} + \hat{b}_\Omega^+ e^{i\Omega\tilde{u}}\right) + \text{left-moving}. \tag{6.85}$$

Using the annihilation operators $\hat{b}^-$, we can define a vacuum state with these operators,

$$\hat{b}_\Omega^- |0\rangle_B = 0, \tag{6.86}$$

which is known as the *Boulware* vacuum. It is thus empty at large distances from the black hole. However, it turns out that this vacuum state is in fact singular on the horizon, a result that can be guessed from the fact that the tortoise coordinates blow up on the horizon (compared to the Unruh calculation, one may note that the horizon is the locus of an infinitely accelerated observer).

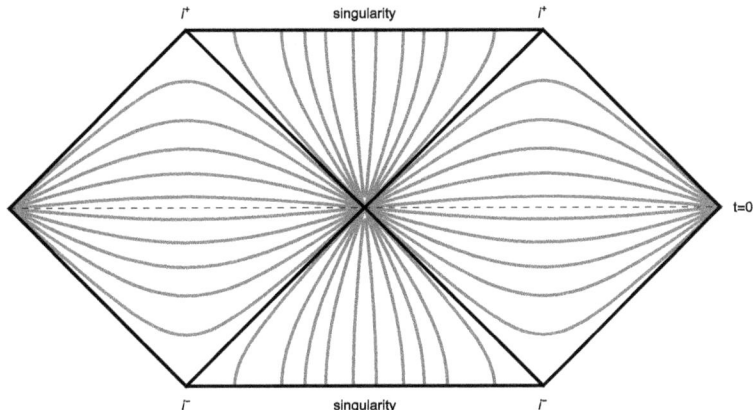

Figure 6.3 A spacetime diagram of the fully extended Schwarzschild solution, with space being horizontal and time vertical. The gray lines show constant t surfaces, with $t = 0$ shown by the dashed horizontal line. Future time-like infinity is located at i^+, and past time-like infinity at i^-. The diagonal black lines linking i^+ and i^- are the horizons, located at $r = 2M$, or equivalently $u = 0$ and $v = 0$. Any time-like curve crossing the horizon will end up at the future singularity (top horizontal line). The fully extended spacetime depicts an eternal black hole (not one formed via gravitational collapse), which is why there is also a past singularity (bottom horizontal line). Note that the constant t surfaces are space-like outside of the horizons, and time-like inside.

To proceed, we can define the *Kruskal* coordinates

$$-\frac{1}{4M}u \equiv e^{-\frac{\tilde{u}}{4M}}, \qquad \frac{1}{4M}v \equiv e^{\frac{\tilde{v}}{4M}}, \qquad (6.87)$$

in terms of which the metric becomes

$$ds^2 = -\frac{2M}{r}e^{(1-\frac{r}{2M})}\,du\,dv, \qquad (6.88)$$

where $r(u,v)$ is left implicit. The advantage of the Kruskal coordinates is that they can be analytically continued beyond the horizons (located at $u = 0, v = 0$), and moreover the metric is nonsingular there – see Fig. 6.3 for a spacetime diagram of the fully extended spacetime. These coordinates can be regarded as the analog of the Minkowski coordinates used in the description of the Unruh effect. They are adapted to an observer near the horizon, where the metric becomes very simple

$$ds^2 \xrightarrow{r \to 2M} -du\,dv. \tag{6.89}$$

To such an observer a positive-frequency, right-moving mode will be proportional to $e^{-i\omega u}$, and the mode expansion reads

$$\hat{\phi} = \int_0^\infty \frac{d\omega}{\sqrt{4\pi\omega}} \left(\hat{a}_\omega^- e^{-i\omega u} + \hat{a}_\omega^+ e^{i\omega u} \right) + \text{left-moving}. \tag{6.90}$$

A corresponding vacuum can be defined as

$$\hat{a}_\omega^- |0\rangle_K = 0. \tag{6.91}$$

This is known as the *Kruskal* vacuum, and it is the natural vacuum to choose for the black hole.

From the point of view of an observer far away from the black hole, the Kruskal vacuum contains particles. If we compare with that calculation, more specifically comparing (6.87) with (6.55), we can see that there is a direct analogy if we set $a = 1/(4M)$. We can then immediately infer that the occupation number of b-particles is given by

$$\frac{\langle N_\Omega \rangle_K}{V} = \frac{1}{e^{8\pi M \Omega} - 1}. \tag{6.92}$$

From this, we conclude that the far-away observer sees a thermal state at the Hawking temperature (6.80), $\mathcal{T} = 1/(8\pi M)$.

A few comments: As one crosses the horizon, time and space change roles, as is evident from the metric (6.77). This property allows for negative frequency modes to arise, a precondition for the detection of particles. A heuristic picture for the extraction of actual particles from the vacuum due to the horizon is as follows: In the vacuum, virtual pairs of particles with their antiparticles constantly form and disappear again. But if such a pair forms near the horizon, it can happen that the positive-energy particle is on the outside, and the negative-energy particle inside. Due to the negative energy, the black hole can lose a tiny bit of mass. Meanwhile, the positive-energy particle, being on the outside, can escape. This process then separates the virtual particles from each other, allowing an actual particle to be detected outside of the black hole.

In closing, let us note that it is clear that the preceding calculations cannot be the end of the story yet. An obvious shortcoming is to assume a fixed background spacetime. Hawking radiation causes a black hole to shrink, and hence backreaction really ought to be included. Also, when radiation is emitted, the black hole will recoil, so a stochastic trajectory of the center of mass might be expected. In addition, there is a deep puzzle that stems from considering the formation of an actual black hole. If we assume that the initial state of the universe, before black hole formation, was a pure state, then according to quantum theory this state should remain pure as long as we keep describing the full universe. But the thermal state we arrived at above is mixed. Hence it looks like information was lost; this is known as the *information paradox*. This is a clear indication that one must go beyond Hawking's calculation. As just argued, backreaction should certainly be included. Moreover, our calculations above suggest that it may be inappropriate to neglect the interior. And in fact, recent calculations (in 2-dimensional toy models) indicate that, inside of the horizon, so-called "island" regions can form that are entangled with the outside Hawking radiation (possibly via wormholes, which thereby would bring the inside into contact with the outside). These calculations indicate that, at late stages in the evaporation process of black holes, the purity of the state may in fact be recovered. This is a topic of intense current research and fast advances – one lesson that appears to come out of it is that this is a further instance in which gravity exhibits surprisingly detailed knowledge about quantum theory.

6.5 Path integrals and thermodynamics: Hawking–Page phase transition

In the calculation above, we assumed without proof that the Euclidean black hole solution was in fact the relevant and dominant contribution to the partition function. We also did not specify the boundary conditions of the partition function very precisely. Here we attempt to go beyond that, and define a gravitational path integral as precisely as is currently feasible, meaning in minisuperspace. We will do this in the particularly interesting setting of a negative cosmological constant, where the empty background solution is provided by Anti-de Sitter space. This does not correspond to our universe, but this setting provides a useful theoretical laboratory for the connections between gravity, quantum theory, and thermodynamics that we are exploring here.

In the presence of a negative cosmological constant, black holes are not expected to dominate at every temperature. Rather, there is the so-called Hawking–Page phase transition, which indicates that a minimum temperature is required for a black hole spacetime to be dominant (below this critical temperature, radiation is not expected to collapse into a black hole). For this calculation, we will make extensive use of the methods of Chapter 5.

Our aim is to calculate the partition function. In order to capture the physics of black holes, we will fix the geometry on an Euclidean "outer" hypersurface of topology $S^1 \times S^2$ (and corresponding radii R_1 and R_2), where we expect the circle periodicity to correspond to inverse temperature. We will perform the calculation at finite radii, but in principle one could imagine letting this hypersurface go off to infinity at the end of the calculation. There will be a radial coordinate r on which the metric will depend. Without loss of generality we will take the range of r to run from 0 to 1, with $r = 1$ representing the hypersurface we just talked about. The idea is then that the spacetime should be filled in smoothly, with the origin residing at $r = 0$ (cf. the cartoon in Fig. 6.1 and imagine holding the outer circle fixed and filling in the rest). One of the tasks of the calculation will be to fix appropriate boundary conditions at this "inner" boundary $r = 0$. The partition function thus reads

$$Z(R_1, R_2) = \sum_B \int_B^{R_1, R_2} d[g_{\mu\nu}] e^{\frac{i}{\hbar}(S_{\text{ND}} - S_{\text{EAdS}})}, \qquad (6.93)$$

where B stands for the explicit boundary conditions at $r = 0$. In principle, we might have to sum over a set of such boundary conditions. Note that we have regulated the partition function by subtracting the action S_{EAdS} of the empty background solution, which will be Euclidean Anti-de Sitter (EAdS) space. The explicit solution will be given below.

We can relate the cosmological constant to the radius of curvature l of EAdS via

$$\Lambda \equiv -\frac{3}{l^2}. \qquad (6.94)$$

Then the action is given by

$$S_{\text{ND}} = \frac{1}{2} \int d^4x \sqrt{-g} \left[R + \frac{6}{l^2} \right] + \int_{\text{outer}} d^3y \sqrt{h} K, \qquad (6.95)$$

where we have added the GHY surface term on the outer boundary, since we want to fix the sizes of the radii $R_{1,2}$ there. We do not add a surface term at $r = 0$, since we do not want to put an explicit boundary there – rather, the idea is that the geometry should be filled in in a smooth manner.

A convenient metric ansatz (belonging to the Kantowski–Sachs class) is the following,

$$ds^2 = -\frac{b(r)}{c(r)} \tilde{N}^2 dr^2 + \frac{c(r)}{b(r)} d\sigma^2 + b(r)^2 d\Omega_2^2, \qquad (6.96)$$

where b, c are scale factors and $\tilde{N}$ is the lapse function, which is defined such

6.5 Path integrals and thermodynamics: Hawking–Page phase transition

that real values of the lapse correspond to Lorentzian metrics. Here σ is a periodic coordinate with period $\Delta\sigma$.

Before continuing with the evaluation of the partition function, let us present the solutions that we expect to be relevant in this calculation. First off, there is the EAdS solution

$$ds^2 = \left(\frac{\rho^2}{l^2} + 1\right) d\sigma^2 + \frac{d\rho^2}{\left(\frac{\rho^2}{l^2} + 1\right)} + \rho^2 d\Omega_2^2, \tag{6.97}$$

where the periodicity $\Delta\sigma$ is arbitrary, as the solution is smooth already. To this solution, one can straightforwardly add a black hole, the metric now becoming

$$ds^2 = \left(\frac{\rho^2}{l^2} + 1 - \frac{2M}{\rho}\right) d\sigma^2 + \frac{d\rho^2}{\left(\frac{\rho^2}{l^2} + 1 - \frac{2M}{\rho}\right)} + \rho^2 d\Omega_2^2, \tag{6.98}$$

with M denoting the black hole mass. The black hole horizon, whose location we will denote by r_+, corresponds to the real root of the cubic

$$\frac{\rho^3}{l^2} + \rho - 2M = 0 \equiv \frac{1}{l^2}(\rho - r_+)(\rho - r_1)(\rho - r_2), \tag{6.99}$$

while the other two roots r_1, r_2 form a complex conjugate pair (as the discriminant of this cubic equation is negative). Turning this relation around, one can express the mass in terms of the horizon radius,

$$M = \frac{1}{2}r_+\left(1 + \frac{r_+^2}{l^2}\right). \tag{6.100}$$

This time, the Euclidean solution (6.98) contains a conical singularity at the location of the horizon, unless we impose the time periodicity

$$\Delta\sigma = \beta = \frac{4\pi l^2 r_+}{3r_+^2 + l^2}. \tag{6.101}$$

Note that for a given periodicity β, the relation (6.101) is satisfied by two values of r_+. Thus, at a given temperature, there can be up to two black holes, one large and one small.

Since we will make use of the metric ansatz (6.96), we must rewrite the EAdS and black hole metrics in this form. This can be done via

$$\rho \equiv b(r) = r(R_2 - r_+) + r_+, \tag{6.102}$$

which allows us to express the black hole metric as

$$\tilde{N} = \pm i(R_2 - r_+), \tag{6.103a}$$

$$c(r) = \frac{1}{l^2}[b^3(r) + l^2 b(r) - r_+^3 - l^2 r_+]. \tag{6.103b}$$

On the outer boundary at $r = 1$ we thus have

$$b(1) = R_2 \quad \text{and} \quad c(1) = \frac{1}{l^2}(R_2^3 + l^2 R_2 - r_+^3 - l^2 r_+). \tag{6.104}$$

The size of the circle direction on the boundary is given by

$$\sqrt{\frac{c(1)}{b(1)}}\,\beta \equiv R_1. \tag{6.105}$$

Fixing R_1 and R_2 thus fixes the size of the outer hypersurface at $r = 1$. At the inner boundary, we have

$$b(0) = r_+ \quad \text{and} \quad c(0) = 0, \tag{6.106}$$

with $b(0)$ specifying the mass of the black hole, using (6.100). In terms of the metric (6.96), a conical singularity is avoided if we impose the periodicity

$$\beta = \frac{4\pi b(0)|\tilde{N}|}{\dot{c}(0)} = \frac{4\pi l^2 r_+}{3r_+^2 + l^2}. \tag{6.107}$$

The EAdS solution can be recovered in the limit of vanishing mass, and then the periodicity β becomes arbitrary.

There is one more subtlety that we must address: The metric (6.96) has a residual gauge invariance, under the transformation

$$\sigma \to \gamma \sigma, \quad c \to \gamma^{-2} c, \quad \text{and} \quad \tilde{N} \to \gamma^{-1} \tilde{N}. \tag{6.108}$$

We will fix this by imposing $\beta = \Delta \sigma$.

Now we are ready to look at the path integral. To this end, we must first evaluate the GHY surface term. The momenta are given by

$$\Pi^{ij} \equiv -\frac{\sqrt{h}}{2}\left(K^{ij} - h^{ij} K\right) \tag{6.109}$$

$$= \text{diag}^{ij}\left[-\frac{b\dot{b}}{\tilde{N}}, -\frac{1}{4}\left(\frac{\dot{c}}{\tilde{N} b} + \frac{c\dot{b}}{\tilde{N} b^2}\right), -\frac{1}{4\sin^2\theta}\left(\frac{\dot{c}}{\tilde{N} b} + \frac{c\dot{b}}{\tilde{N} b^2}\right)\right], \tag{6.110}$$

so that the surface term becomes

$$\Pi^{ij} h_{ij} = \sqrt{h}\, K = -\frac{1}{2}\left(\frac{b\dot{c}}{\tilde{N}} + 3\frac{c\dot{b}}{\tilde{N}}\right). \tag{6.111}$$

6.5 Path integrals and thermodynamics: Hawking–Page phase transition

The action, which has a Neumann boundary condition at $r = 0$ and a Dirichlet condition at $r = 1$, then reduces to

$$S_{\text{ND}} = 4\pi\Delta\sigma \int dr \left[-\frac{b\dot{c}}{\tilde{N}} + \tilde{N}\left(1 + \frac{3b^2}{l^2}\right) \right] - 2\pi\Delta\sigma \left(\frac{b\dot{c}}{\tilde{N}} + 3\frac{c\dot{b}}{\tilde{N}} \right)\bigg|_{r=0}, \tag{6.112}$$

where we have used integration by parts. When we vary the action with respect to the scale factors, we find

$$\delta S_{\text{ND}} = 4\pi\Delta\sigma \int dr \left[\left(\frac{\ddot{c}}{\tilde{N}} + \frac{6\tilde{N}b}{l^2} \right) \delta b + \left(\frac{\ddot{b}}{\tilde{N}} \right) \delta c \right]$$
$$- 4\pi\Delta\sigma \left(\frac{\dot{c}_1}{\tilde{N}}\delta b_1 + \frac{\dot{b}_1}{\tilde{N}}\delta c_1 \right) - 2\pi\Delta\sigma \left(b_0^2 \delta\left(\frac{\dot{c}_0}{\tilde{N}b_0}\right) + \dot{b}_0 \delta c_0 + 3c_0 \delta\dot{b}_0 \right). \tag{6.113}$$

The surface terms will vanish at $r = 1$ if we fix b_1 and c_1. At $r = 0$, we can set $c_0 = 0$, which eliminates the last two terms above. For the remaining term at $r = 0$, we can either set $b_0 = 0$ or fix $\frac{\dot{c}_0 \Delta\sigma}{\tilde{N}b_0}$. We will combine both conditions into the equation

$$\frac{\dot{c}_0 \Delta\sigma}{4\pi i \tilde{N} b_0} \equiv \omega = \text{fixed}, \tag{6.114}$$

where $\omega \to \infty$ corresponds to setting $b_0 = 0$. Note that $\omega = 1$ corresponds to the periodicity required for regularity of the solution. Here we do not wish to include singular solutions, so in fact the partition function will be a sum over only two boundary conditions,

$$Z(R_1, R_2) = \sum_{\omega=1,\infty} \int \mathcal{D}b\,\mathcal{D}c\,d\tilde{N}\, e^{\frac{i}{\hbar}(S_{\text{ND}} - S_{\text{EAdS}})}. \tag{6.115}$$

We can now evaluate the integrals over scale factors by using the same procedure we used in Sections 5.2.1 and 5.2.2: We will shift the integration variable to an (arbitrarily large) fluctuation away from a solution to the classical equations of motion, with the latter given by

$$\ddot{b} = 0, \qquad \ddot{c} = -\frac{6}{l^2}\tilde{N}^2 b. \tag{6.116}$$

The integrals over the fluctuations will simply lead to prefactors, and will be unimportant in a leading-order saddle-point approximation. The important part of the physics resides in the solutions to the equations of motion, which with the given boundary conditions are

$$b(r) = (R_2 - b_0)\, r + b_0,\tag{6.117a}$$

$$c(r) = \left(c_1 + \frac{(2\,b_0 + R_2)\tilde{N}^2}{l^2}\right) r - \frac{3\,b_0 \tilde{N}^2}{l^2} r^2 + \frac{(b_0 - R_2)\tilde{N}^2}{l^2} r^3,\tag{6.117b}$$

with

$$b_0 = \frac{\Delta\sigma \left(l^2 c_1 + \tilde{N}^2 R_2\right)}{4\pi i \omega\, l^2\, \tilde{N} - 2\Delta\sigma\, \tilde{N}^2}, \qquad c_1 = \frac{R_1^2 R_2}{\Delta\sigma^2}.\tag{6.118}$$

The radial integrals over these solutions are straightforward to perform, leaving us with an action that depends solely on the lapse $\tilde{N}$,

$$\frac{l^2}{\pi\Delta\sigma} S_{\text{ND}} = \frac{(3R_2^2 + 4l^2)\Delta\sigma \tilde{N}^2 - 8\pi i\omega l^2 (R_2^2 + l^2)\tilde{N} - 6l^2 R_2 c_1 \Delta\sigma}{\Delta\sigma \tilde{N} - 2\pi i \omega l^2}$$

$$+ \frac{8\pi i \omega l^4 R_2 c_1 \tilde{N} - l^4 c_1^2 \Delta\sigma}{\tilde{N}^2 (\Delta\sigma \tilde{N} - 2\pi i \omega l^2)}.\tag{6.119}$$

The action has a single pole at $\tilde{N} = 2\pi i \omega l^2/\Delta\sigma$ and a double pole at $\tilde{N} = 0$. The saddle points may be found by solving $\partial_{\tilde{N}} S_{\text{ND}} = 0$, which leads to a quintic equation with in general five distinct solutions. One may verify that, for $\omega = 1$, two of the saddle points are given by

$$\tilde{N}_s = -i(R_2 - r_+),\tag{6.120a}$$

$$b(r) = r(R_2 - r_+) + r_+,\tag{6.120b}$$

$$c(r) = \frac{1}{l^2}[b^3(r) + l^2 b(r) - r_+^3 - l^2 r_+],\tag{6.120c}$$

with r_+ given by either of the two solutions of (6.101) (as long as the value of β is such that two real solutions exist). These are the two black hole solutions (6.103a). The on-shell action of these saddle points is given by

$$S^{\text{bh}} = -\frac{2\pi i}{l}\sqrt{\frac{R_1^2 R_2}{(R_2 - r_+)(R_2^2 + l^2 + R_2 r_+ + r_+^2)}}\left(4R_2(R_2^2 + l^2) - 3l^2 r_+ - r_+^3\right).\tag{6.121}$$

The remaining three saddle points must be found numerically.

Meanwhile, the $\omega \to \infty$ limit of the action leads to the simpler expression

$$S_{\text{ND},\omega\to\infty} = \frac{4\pi\Delta\sigma}{l^2}\left((R_2^2 + l^2)\tilde{N} - \frac{R_2 c_1 l^2}{\tilde{N}}\right),\tag{6.122}$$

6.5 Path integrals and thermodynamics: Hawking–Page phase transition

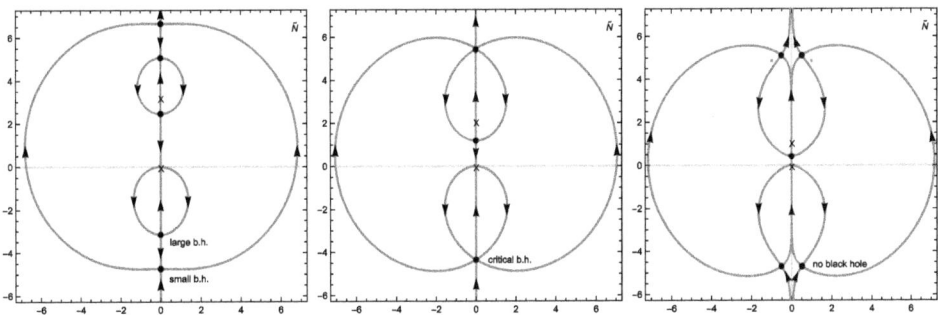

Figure 6.4 The saddle points and associated steepest descent/ascent lines for the action (6.119), shown in the complex plane of the lapse function $\tilde{N}$. Dots mark saddle points, crosses mark poles in the action, and arrows indicate downwards flow of the weighting. The numerical values used are $l = 1, R_2 = 5$, and $R_1 = 10$ (left), $R_1 = 18.5$ (middle), $R_1 = 50$ (right).

with the saddle points corresponding to EAdS space,

$$\tilde{N}_s = \pm i R_2, \quad b(r) = R_2 r, \quad \frac{c(r)}{b(r)} = 1 + \frac{R_2^2 r^2}{l^2}. \qquad (6.123)$$

Here we fix the Euclidean time periodicity such that the circle size on the outer boundary remains R_1, namely

$$\Delta\sigma^{\text{AdS}} = \frac{R_1 l}{\sqrt{R_2^2 + l^2}}. \qquad (6.124)$$

The action for these saddle points is

$$S^{\text{AdS}} = \pm 8\pi i \frac{R_1 R_2}{l} \sqrt{R_2^2 + l^2}, \qquad (6.125)$$

i.e., the two metrics correspond to opposite Wick rotations of the Lorentzian AdS metric. The saddle with enhanced weighting turns out to be the relevant one.

The precise locations of the saddle points, and the associated steepest ascent and descent lines, depends crucially on the boundary values $R_{1,2}$. Its explicit expression is a bit unwieldy, but numerically it is straightforward to handle. (Recall, e.g., from Appendix B, that the steepest ascent/descent contours can be determined by finding the curves of equal phase of the integrand, starting from the saddle points.) We will look at the $\omega = 1$ contributions to the partition function explicitly, leaving the $\omega \to \infty$ part of the calculation as an exercise for the reader (it is very similar to the calculation in Section 5.2.2).

We proceed by fixing the sphere radius R_2, and looking at progressively

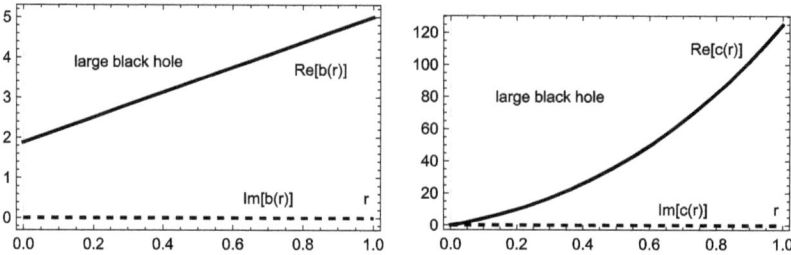

Figure 6.5 Evolution of the scale factors $b(r)$ and $c(r)$ at the large black hole saddle point $\tilde{N} \approx -3.16\,i$, with the same numerical values as those used in the left panel of Fig. 6.4. One can see that the solution is purely Euclidean, starting at the black hole horizon at $r = 0$ and reaching the imposed boundary values $b(1) = R_2$ and $c(1) = R_2^3$.

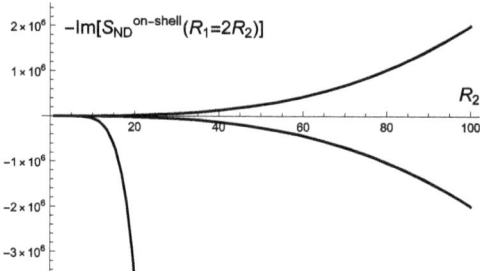

Figure 6.6 The weightings of the saddle points for fixed ratio $R_1/R_2 = 2$ (i.e., fixed temperature) and increasing size of the outer hypersurface. The two black hole weightings are indistinguishable at this scale and represented by the upper curve that continues to grow, while the other saddle points are increasingly suppressed with increasing boundary size (two of those saddles also have almost equal weightings).

larger R_1, and thus progressively lower temperatures. The situation for small R_1 is shown in the left panel in Fig. 6.4. The two black hole saddle points are situated on the Euclidean axis, in the lower half plane. The weighting for the large black hole is higher than that for the smaller black hole, as evidenced by the downwards flow emanating from the larger black hole solution. The scale factors at the large black hole saddle point are shown in Fig. 6.5.

There are three additional saddle points in the upper half plane, that all three have a suppressed weighting – see Fig. 6.6. In fact, the weighting of the additional saddle points becomes increasingly small as the boundary size is increased, and in the infinite radius limit these saddles are infinitely suppressed (which is just as well, as their geometries contain instances where the spatial volume passes through zero – see Fig. 6.7 for an example).

Thus the path integral is dominated by the large black hole solution, as

6.5 Path integrals and thermodynamics: Hawking–Page phase transition   183

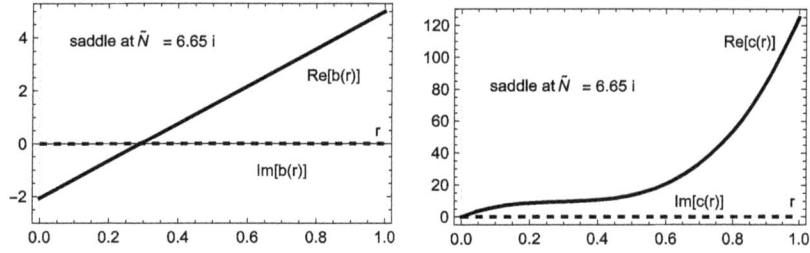

Figure 6.7 Geometry of the singular saddle point located at $\tilde{N} \approx 6.65\,i$ in the left panel of Fig. 6.4. The singularity stems from the fact that the circle radius $b(r)$ passes through zero.

long as the integration contour is chosen such that the corresponding saddle is picked up. As one can see from the figure, in order for this to be the case one needs to choose an integration contour that starts at the singularity at $\tilde{N} = 0$ and leaves the singularity in the negative Euclidean direction. After picking up the large black hole saddle, the steepest descent contour runs to the smaller black hole and from there curves up towards the upper half plane, where it moves off to positive Euclidean infinity (note that the upper half plane is the asymptotic region of convergence of the path integral, as one can see by taking the large $\tilde{N}$ limit of (6.119)). The choice of whether to circumvent the singularities on the left or on the right is arbitrary. In the context of the Anti-de Sitter/Conformal Field Theory (AdS/CFT) correspondence, one expects to be able to calculate the same partition function in a dual conformal field theory, which would be a Euclidean theory with a real-valued partition function. This suggests that one should in fact sum over both choices, i.e., calculate the partition function by summing the results from a contour that goes left and the complex conjugate result obtained by a contour that is reflected across the Euclidean lapse axis. In any case, the partition function can then be approximated to leading order by the large black hole contribution $Z(R_1 < R_{1,\text{limit}}, R_2) \approx e^{i(S^{\text{bh}} - S_{\text{EAdS}})}$.

When we increase the boundary radius R_1, we decrease the temperature. This causes two of the saddle points in the upper half plane, as well as the two black hole saddles, to approach each other and eventually to merge. The limiting case is illustrated in the middle panel in Fig. 6.4. One can obtain the critical value by combining Eqs. (6.120a) and (6.101), and inserting the maximum possible value $\beta_{\max} = \frac{2\pi l}{\sqrt{3}}$, finding

$$R_{1,\text{limit}} = \frac{2\pi}{\sqrt{3}} R_2 \sqrt{1 + \frac{l^2}{R_2^2} - \frac{4l^3}{3\sqrt{3} R_2^3}}. \qquad (6.126)$$

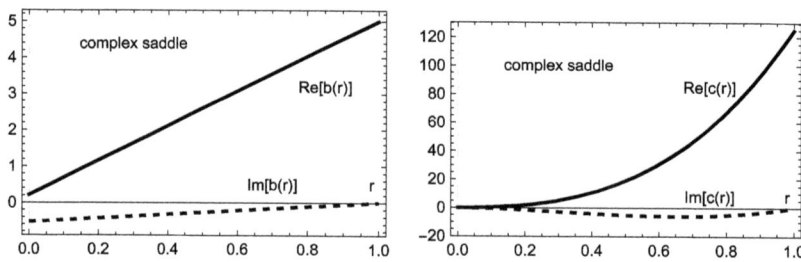

Figure 6.8 The complex geometry at the saddle point $\tilde{N} \approx 0.517 - 4.70\,i$ shown in the right panel of Fig. 6.4. This complex geometry is a deformation of the black hole saddle points when the temperature has become too low for the existence of black holes.

At this value of the outer circle, the path integral is again dominated by a black hole saddle point, which is in fact the black hole at the lowest possible temperature. The integration contour remains unchanged.

When the temperature is lowered further, and R_1 increased beyond $R_{1,\text{limit}}$, a new phenomenon occurs – see the right panel in Fig. 6.4: The two degenerate saddle points on the Euclidean axis split up and *move into the complex domain*. Thus the erstwhile Euclidean black hole saddle points morph into complex solutions that cannot be described as black holes at all. Their geometry is illustrated in Fig. 6.8. This provides a nice illustration of how, at the level of the path integral, black hole solutions cease to exist once the temperature has dropped below the critical value. These complex saddle points have a weighting that is lower than that of the empty EAdS solution at that radius R_1, thus guaranteeing the consistency of the whole picture. The integration contour once again remains unchanged.

We should highlight that we are forced to choose complex contours of integration, thus causing the path integral to sum over complex geometries. The old idea of defining gravitational path integrals as sums over Euclidean metrics simply does not lead to mathematically well-defined (convergent) expressions. However, the sum over the two contours has the Euclidean contour as its "average," which is as close as we can come to utilizing a Euclidean contour. A sketch of the contours of integration is provided in Fig. 6.9.

We can now be more specific about the thermodynamic interpretation, and how the link to standard thermodynamic arguments can be made. The partition function is approximated by the saddle-point contributions, where we must include both the black holes (6.121) and the EAdS space (6.125),

$$\ln Z = i\frac{\Delta S_{\text{ND}}}{\hbar}. \tag{6.127}$$

6.5 Path integrals and thermodynamics: Hawking–Page phase transition 185

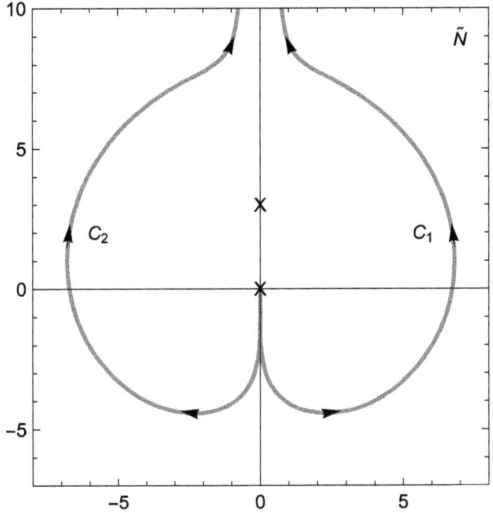

Figure 6.9 A sketch of the required contour of integration in the complex lapse plane. Either of the contours $\mathcal{C}_1$ or $\mathcal{C}_2$ would give a physically acceptable outcome, although if one wants to obtain a real wave function, as expected by AdS/CFT, then one should sum over both since they yield complex conjugate contributions.

The difference in on-shell actions is given by

$$\Delta S_{\text{ND}} = -\frac{2\pi i R_1}{l}\left[\frac{\sqrt{R_2}\left(4R_2^3 + 4l^2 R_2 - 3l^2 r_+ - r_+^3\right)}{\sqrt{R_2^3 + l^2 R_2 - l^2 r_+ - r_+^3}} - 4R_2\sqrt{R_2^2 + l^2}\right] \quad (6.128\text{a})$$

$$= \frac{2\pi i R_1}{l R_2}(l^2 r_+ - r_+^3) - \frac{\pi i l R_1}{R_2^3}(l^2 r_+ - r_+^3) + O(R_2^{-4}), \quad (6.128\text{b})$$

where in the second line we expanded at large R_2. At leading order in the large R_2 expansion, the periodicity is $R_1/R_2 \approx \beta/l$, so that we recover the famous Hawking–Page result

$$\Delta S_{\text{HP}} = -8\pi^2 i\, r_+^2\, \frac{r_+^2 - l^2}{3r_+^2 + l^2} + O(R_2^{-1}). \quad (6.129)$$

When $-\text{Im}[\Delta S_{\text{ND}}] < 0$, then EAdS dominates. At large R_2 this is when the black hole horizon is smaller than the AdS radius of curvature, $r_+ < l$, and at small R_2 there are corrections to this relation encoded in (6.128). The phase transition occurs at $r_+ = l$, which corresponds to the approximate radius $R_{1,\text{HP}} \approx \pi R_2$, as one can see by combining (6.101) and (6.105). An implication of this observation is that the complex saddle points that dominate the $\omega = 1$ contribution at $R_1 > R_{1,\text{limit}} \approx \frac{2\pi}{\sqrt{3}} R_2$ never dominate the full partition function, as the EAdS solution has already taken over by then.

The partition function represents the canonical ensemble, which means that we imagine the system to be kept in equilibrium at a fixed temperature $\mathcal{T}$. This temperature is redshifted as one moves away from the black hole horizon,

$$R_1 = \beta \sqrt{1 + \frac{R_2^2}{l^2} - \frac{2M}{R_2}} = \Delta \sigma^{AdS} \sqrt{1 + \frac{R_2^2}{l^2}} = \frac{1}{\mathcal{T}}, \tag{6.130}$$

where we denoted the Euclidean time periodicity of the EAdS solution by $\Delta \sigma^{\text{AdS}}$. Then, in standard (not reduced) Planck units, we can re-express the partition function as

$$\ln Z = \frac{R_2}{T l_P^2} \left(\sqrt{1 + \frac{R_2^2}{l^2} - \frac{2M}{R_2}} - \sqrt{1 + \frac{R_2^2}{l^2}} \right) + \frac{\pi r_+^2}{l_P^2}, \tag{6.131}$$

where the Planck length is defined via $l_P \equiv \sqrt{\frac{G\hbar}{c^3}}$. From this, we can calculate the expectation value of the energy,

$$\langle E \rangle = k_B T^2 \frac{\partial \ln Z}{\partial T} = \frac{k_B R_2}{l_P^2} \left(\sqrt{1 + \frac{R_2^2}{l^2}} - \sqrt{1 + \frac{R_2^2}{l^2} - \frac{2M}{R_2}} \right) \tag{6.132a}$$

$$= \frac{k_B}{l_P^2} \frac{lM}{R_2} - \frac{k_B}{l_P^2} \frac{Ml^3}{2R_2^3} + O(R_2^{-4}), \tag{6.132b}$$

where we explicitly write out Boltzmann's constant k_B. Meanwhile, the entropy is given by

$$\mathcal{S} = k_B \ln Z + \frac{\langle E \rangle}{T} = \frac{k_B}{l_P^2} \pi r_+^2 = \frac{k_B}{l_P^2} \frac{\text{Area}}{4}. \tag{6.133}$$

As expected, it is given by a quarter of the horizon area in Planck units. As a final consistency check, note that these relations satisfy the quantum statistical relation

$$-k_B T \ln Z = \langle E \rangle - T\mathcal{S}. \tag{6.134}$$

Thus we have recovered, or rather derived, all of the standard thermodynamic relations from a gravitational path integral, thereby very explicitly revealing the intimate connection between gravity, quantum theory, and thermodynamics. We will now switch back to a more cosmological context, and proceed with our exploration of these connections.

6.6 De Sitter thermodynamics and mode functions

The de Sitter spacetime is the maximally symmetric solution of the Einstein equations in the presence of a positive cosmological constant $\Lambda > 0$. There are two phases of evolution of the universe that may be well approximated by a de Sitter solution: the conjectured inflationary phase and the current dark energy phase. We already discussed this spacetime in Section 2.1.1, and you are encouraged to briefly glance back at that discussion. For convenience, we recall the metric, both in the closed and in the flat slicings,

$$ds^2_{\text{closed}} = -d\bar{t}^2 + \frac{1}{\mathsf{H}^2}\cosh^2(\mathsf{H}\bar{t})\, d\Omega_3^2, \quad \mathsf{H} \equiv \sqrt{\frac{\Lambda}{3}}, \tag{6.135}$$

$$ds^2_{\text{flat}} = -dt^2 + \frac{1}{\mathsf{H}^2}e^{2\mathsf{H}t}(dx^2 + dy^2 + dz^2). \tag{6.136}$$

Note that only in the flat slicing is it true that H corresponds to the actual expansion rate $H = \dot{a}/a$. In the closed slicing this is only true asymptotically. A further coordinate system of interest is one in which the metric is static,

$$ds^2_{\text{static}} = -\left(1 - \mathsf{H}^2 r^2\right) dT^2 + \frac{dr^2}{(1 - \mathsf{H}^2 r^2)} + r^2 d\Omega_2^2. \tag{6.137}$$

One may verify that all three metrics lead to the Ricci tensor $R_{\mu\nu} = 3\mathsf{H}^2 g_{\mu\nu} = \Lambda g_{\mu\nu}$. All three of these coordinate systems cover different portions of the full de Sitter spacetime – see Fig. 6.10. The figure shows a Penrose diagram of de Sitter spacetime. This is a conformal diagram, in which light rays propagate at angles $\pm\pi/4$. Space is horizontal and time vertical, with two spatial directions being suppressed. Such a diagram is obtained by performing coordinate transformations such that the ranges of all coordinates are finite. In the present case, a useful transformation is to conformal time τ,

$$\tan\left(\frac{\pi}{4} + \frac{\tau}{2}\right) = \tanh\left(\frac{\mathsf{H}\bar{t}}{2}\right), \quad -\pi < \tau < 0, \tag{6.138}$$

which puts the closed metric into the form

$$ds^2_{\text{closed}} = \frac{\mathsf{H}^2}{\sin^2\tau}\left(-d\tau^2 + d\Omega_3^2\right). \tag{6.139}$$

The diagram allows one to see causal relations in the spacetime, and because of this conformal factors in the metric are unimportant. The Penrose diagram is thus given by a square, with time coordinate τ and one spatial coordinate on the sphere, say $\chi \in [0, \pi]$. The left and right boundaries represent the north and south poles of the spatial 3-sphere, and are thus regular. The closed metric covers this entire manifold, with constant time slices (simply horizontal lines) being Cauchy surfaces.

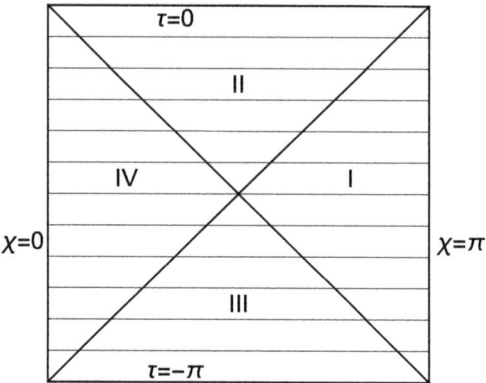

Figure 6.10 A spacetime diagram of the de Sitter spacetime (with two angular coordinates suppressed), which simply maps to a square. Space is horizontal and time vertical, as described in the main text. The bottom line represents the infinite past and the top line the infinite future. Horizontal lines have equal time τ or $\bar{t}$, with a horizontal slice through the middle describing the moment of minimal radius reached at $\bar{t} = 0$). The closed metric covers the entire manifold, while the flat slicing covers only regions II and IV, or II and I. The diagonal lines are horizons, with the static patch describing region I. An observer in region I cannot see what happens in regions II and IV, and cannot influence what happens in regions III and IV.

The static coordinates only cover the so-called *static patch*, which, as one can see from the metric (6.137), is the range $0 \leq r < 1/\mathsf{H}$. The horizon is located at $r = 1/\mathsf{H}$, and thus an observer in the static patch resides inside of the horizon. This corresponds to region I or region IV in the figure. The presence of a horizon indicates that we should expect the presence of thermal radiation. Comparing to our discussion of the Unruh effect, we see that the closed slicing metric (6.135) plays a role analogous to Minkowski space, while the static patch is the analog of the Rindler wedge. Thus, an observer in the static patch would see the vacuum of the full de Sitter spacetime as being filled with particles.

If we define a new radial coordinate via $r = \frac{1}{\mathsf{H}}(1 - \epsilon^2)$, then we can approximate the metric (6.137) near the horizon as

$$\mathrm{d}s^2 \approx \frac{2}{\mathsf{H}^2}\left(-\mathrm{d}\epsilon^2 + \mathsf{H}^2\epsilon^2\mathrm{d}T^2\right) + \frac{1}{\mathsf{H}^2}\mathrm{d}\Omega_2^2. \tag{6.140}$$

The ϵ–T part of the metric again looks like Rindler spacetime (6.4), with the Hubble rate H playing the role of the acceleration. This implies that the

6.6 De Sitter thermodynamics and mode functions

analytic continuation to imaginary time will be smooth as long as the time coordinate has imaginary time periodicity $\beta = \frac{2\pi}{H}$, leading to the (Gibbons–Hawking) de Sitter temperature

$$T_{\text{dS}} = \frac{H}{2\pi}. \tag{6.141}$$

We could also have looked directly at the Euclidean version of the full closed slicing metric, presented in (2.12) and which is just a 4-sphere, precisely with Euclidean time coordinate periodicity $2\pi/H$. This temperature is interpreted as particle production in the de Sitter spacetime, very much in the spirit of the cosmological perturbation theory that we discussed in Chapter 3. In one of the exercises at the end of this chapter, we will make this connection more explicit.

A few more comments are in order: Along with a temperature, one may try to define additional thermodynamic quantities, especially an entropy. In analogy with black holes, this is usually taken to be given by a quarter of the horizon area. Note however that currently no physical understanding of such an entropy exists. In statistical physics, the entropy depends on the number of microstates of the system (it is the logarithm of that number). But here, as we saw, each observer has their own horizon, which is thus position dependent (in contrast to the black hole case, where the black hole singularity provides a reference location). At face value, this implies that for each observer a different description in terms of microstates is in principle required. What these would be made of is however entirely unclear at present. A further consequence of such an entropy would be that the number of states is finite, and thus the Hilbert space would be finite dimensional (of dimension $\sim e^{\text{Area}/4}$). If this is the case, if one waits long enough then history will repeat itself in any universe containing a positive cosmological constant. This might well be one reason for vacuum energy to eventually decay.

6.6.1 Mode functions on de Sitter spacetime

For completeness, let us also write out the actual perturbation modes on de Sitter spacetime. These modes are often used as a first approximation of inflationary perturbations, and thus some familiarity with them is advisable. We will recover some of the results encountered before, e.g., in Sections 3.2.4 and 3.2.5, but from a complementary perspective.

To this end, consider again a massless scalar living on the de Sitter background. The scalar field will be treated as a probe field, enacting no backreaction on the geometry (to be clear: Here the accelerated expansion is caused by the cosmological constant Λ, and not by the scalar).

Let us start with an abridged review of the calculation in the flat slicing. The calculation of the perturbations is most conveniently done in (Lorentzian) conformal time $d\tau = \frac{dt}{a}$, in terms of which the action for the scalar (in Fourier space) reduces to

$$S_\phi = -\frac{1}{2}\int d^4x \sqrt{-g}\, g^{\mu\nu}\partial_\mu\phi\partial_\nu\phi = \frac{1}{2}\int d\tau\, a^2\left((\phi_{,\tau})^2 - k^2\phi^2\right), \quad (6.142)$$

where k is the comoving wave number (the physical wavelength being a/k). A sum over Fourier modes is implicit. The scale factor is given by $a(t) = \frac{1}{H}e^{Ht}$, which translates to $a(\tau) = -\frac{1}{H\tau}$ in conformal time. The Fourier space equation of motion is thus

$$\phi_{,\tau\tau} + 2\frac{a_{,\tau}}{a}\phi_{,\tau} + k^2\phi = 0. \quad (6.143)$$

This can be solved by linear combinations of the two complex conjugate solutions $e^{-ik\tau}(1 + ik\tau)$, $e^{ik\tau}(1 - ik\tau)$. To find the appropriate combination, and to fix the normalization, it is useful to switch to the canonically normalized Mukhanov–Sasaki variable $v \equiv a\phi$, in terms of which the action becomes

$$S_\phi = \frac{1}{2}\int d\tau \left[(v_{,\tau})^2 + \left(\frac{a_{,\tau\tau}}{a} - k^2\right)v^2\right], \quad (6.144)$$

so that the equation of motion is

$$v_{,\tau\tau} + \left(\frac{a_{,\tau\tau}}{a} - k^2\right)v = v_{,\tau\tau} + \left(\frac{2}{\tau^2} - k^2\right)v = 0. \quad (6.145)$$

Now the solutions are linear combinations of $e^{-ik\tau}(1 - i/(k\tau))$, $e^{ik\tau}(1 + i/(k\tau))$. The Klein–Gordon normalization (6.28) then reduces to the Wronskian condition (3.88), requiring $vv^*_{,\tau} - v_{,\tau}v^* = i$. A positive-frequency solution must satisfy $i\partial_\tau\phi = +k\phi$ in the far past. Combining these requirements, we see that the normalized vacuum states are given (up to a phase) by

$$v = \frac{1}{\sqrt{2k}}e^{-ik\tau}\left(1 - \frac{i}{k\tau}\right), \quad \phi = \frac{H}{\sqrt{2k^3}}e^{-ik\tau}\left(1 + ik\tau\right). \quad (6.146)$$

This choice of mode functions is known as the *Bunch–Davies* state, as we saw before. In the far past $\tau \to -\infty$, the v modes agree with the standard Minkowski vacuum states $\frac{1}{\sqrt{2k}}e^{-ik\tau}$, which therefore complements (and justifies) the reasoning we gave in Chapter 3 when choosing a vacuum state for the curvature perturbations.

In the closed slicing, the expressions are slightly more complicated. Now

6.6 De Sitter thermodynamics and mode functions

the Fourier decomposition must be done with spherical harmonics; see Appendix C for some of their properties. We will also go back to ordinary, nonconformal time, as conformal time is of no great help in this case. The action, in physical time, is now ($n > 1$)

$$S_\phi = \frac{1}{2} \int d\bar{t} \left(a^3 \dot{\phi}^2 - a(n^2 - 1)\phi^2 \right), \tag{6.147}$$

and the equation of motion that follows from it reads

$$\ddot{\phi} + 3\frac{\dot{a}}{a}\dot{\phi} + \frac{n^2 - 1}{a^2}\phi = 0. \tag{6.148}$$

With $a(\bar{t}) = \frac{1}{H}\cosh(H\bar{t})$, cf. (2.10), the equation of motion becomes

$$\ddot{\phi} + 3H\coth(H\sigma)\dot{\phi} - \frac{H^2(n^2 - 1)}{(\cosh(H\bar{t}))^2}\phi = 0. \tag{6.149}$$

This equation possesses two solutions,

$$F_1(\bar{t}) = \left(1 + \frac{i}{\sinh(H\bar{t})}\right)^{\frac{n-1}{2}} \left(1 - \frac{i}{\sinh(H\bar{t})}\right)^{-\frac{n+1}{2}} \left(1 - \frac{in}{\sinh(H\bar{t})}\right), \tag{6.150}$$

$$F_2(\bar{t}) = \left(1 - \frac{i}{\sinh(H\bar{t})}\right)^{\frac{n-1}{2}} \left(1 + \frac{i}{\sinh(H\bar{t})}\right)^{-\frac{n+1}{2}} \left(1 + \frac{in}{\sinh(H\bar{t})}\right), \tag{6.151}$$

which are complex conjugates of each other. The general solution will be a linear combination of these. Note that $F_{1,2} \to 1$ at late times.

To compare with the solutions in the flat slicing, we can now transform to conformal time. Using the definition (6.138), we arrive at the relation

$$\sinh(H\bar{t}) = 2\frac{\tan(\frac{\pi}{4} + \frac{\tau}{2})}{1 - \tan^2(\frac{\pi}{4} + \frac{\tau}{2})} = -\frac{1}{\tau} + \frac{\tau}{3} + \frac{\tau^3}{45} + \cdots, \tag{6.152}$$

where we included the late-time expansion. Plugging this expansion into the solution (6.150), we obtain

$$F_1 = 1 + \frac{n^2 - 1}{2}\tau^2 - i\frac{n^3 - n}{3}\tau^3 + \cdots. \tag{6.153}$$

This is precisely the same expansion as that of the Bunch–Davies vacuum mode (6.146), in the large n limit, since

$$e^{-ik\tau}(1 + ik\tau) = 1 + \frac{k^2}{2}\tau^2 - i\frac{k^3}{3}\tau^3 + \cdots, \tag{6.154}$$

where we have to allow for the fact that (6.153) involves spherical Fourier wave numbers, while (6.154) involves flat/toroidal wave numbers. If we now impose a final condition that at some specified time $\bar{t}_1$ the perturbation reaches the value ϕ_1, then the appropriate perturbative, positive-frequency solution is

$$\phi(\bar{t}) = \frac{F_1(\bar{t})}{F_1(\bar{t}_1)} \phi_1 \,. \tag{6.155}$$

This corresponds to the Bunch–Davies vacuum.

Finally, we can work out the wave function of the perturbations. We will do this for the flat slicing and work in conformal time, with the action given above in (6.142). Starting from the mode expansion (6.27) and focusing on a single frequency, we have

$$\hat{\phi} = \hat{a}^- f + \hat{a}^+ f^*, \quad \dot{\hat{\phi}} = \hat{a}^- \dot{f} + \hat{a}^+ \dot{f}^*, \tag{6.156}$$

which, making use of the orthonormality condition (6.28), can be combined to yield

$$i\hat{a}^- = \dot{f}^* \hat{\phi} - f^* \dot{\hat{\phi}} \,. \tag{6.157}$$

In the field basis, we can use the replacements $\hat{\phi} \to \phi$, $\hat{\pi} = a^2 \dot{\hat{\phi}} \to -i \partial_\phi$, so that the vacuum satisfies

$$\hat{a}^- |\Psi\rangle = 0 \quad \Rightarrow \quad \left(\dot{f}^* \phi_1 + i \frac{f^*}{a^2} \partial_{\phi_1} \right) |\Psi\rangle = 0, \tag{6.158}$$

where we denote the ϕ value (which is the argument of the wave function) by ϕ_1. This is a simple differential equation that is solved by a Gaussian state (as expected, since we are considering a free field),

$$|\Psi\rangle = e^{\frac{1}{2} i a^2 \frac{\dot{f}^*}{f^*} \phi_1^2} \,. \tag{6.159}$$

Note that the wave function is calculated using the complex conjugate mode function.[2] Plugging in the de Sitter vacuum mode (6.146), recalling $a = -1/(H\tau)$, and taking the late-time limit $\tau \to 0^-$, we find

$$|\Psi\rangle \approx e^{i \frac{k^2}{2\hbar H} a_1 \phi_1^2 - \frac{k^3}{2\hbar H^2} \phi_1^2} \,. \tag{6.160}$$

This is the wave function for a single-frequency mode; the general expression is given by a product of such terms, one for each frequency. There are two contributions at each frequency: The first is a phase, which grows in proportion to the size a_1 of the universe. By contrast, the weighting is independent

---

[2] This property lets different people use different assignments of the term "positive-frequency" mode. Here we assign this term with respect to the sign of the energy eigenvalue. Thus we call the mode in (6.146), which satisfies $i\partial_\tau v = +kv$ at early times, a positive-frequency mode.

of the scale factor, and thus remains constant as the universe expands. This is therefore a JWKB wave function, in which the phase varies rapidly compared to the amplitude. Thus we confirm that on a de Sitter background the scalar perturbations behave classically at late times.

The weighting in (6.160) is Gaussian, meaning that large perturbations ϕ_1 are suppressed, and the most likely configuration is one where the perturbations vanish. This is a reflection of the stability of these perturbations. The weighting also contains the factor k^3/H^2, which implies that the amplitude of the perturbations is determined by the Hubble rate (which is in accord with the de Sitter temperature formula (6.141)), while the spectrum is exactly scale invariant.

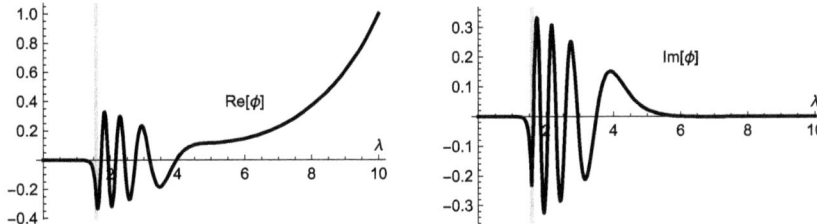

Figure 6.11 The real and imaginary parts of a perturbation mode in the closed slicing, prepared by evolution on the Euclidean semi-4-sphere. The time coordinate evolves from $\sigma = 0$ up to $\sigma = \pi/(2\mathsf{H})$ and from there in the Lorentzian time direction $\bar{t}$. This is represented here by having an affine parameter λ that follows the Euclidean evolution until $\pi/(2\mathsf{H})$, a moment marked by a gray line in the figures, and then continues increasing in the Lorentzian direction until reaching $\lambda = 10$. Here $\mathsf{H} = 1, l = 15, \phi_1 = 1$.

An entirely analogous calculation can be performed for the closed slicing, with mode functions (6.150)–(6.151); it is just a little lengthier. One then finds an analogous stable, Gaussian, scale-invariant wave function. Note that while (6.150) represents the Heisenberg picture positive-frequency vacuum mode, the wave function is actually calculated with the complex conjugate, which is (6.151).

If we think back to the embedding of de Sitter spacetime as a hyperboloid (see Chapter 2), then we realize that the minimum radius $a = \sqrt{3/\Lambda}$ coincides with the radius of the Euclidean 4-sphere with equal curvature. This 4-sphere, given in Eq. (2.12), is the imaginary time version of the Lorentzian de Sitter solution. The equality of radii means that onto the waist of the de Sitter hyperboloid one can glue half of the 4-sphere, with the gluing being done at the equator from the point of view of the sphere. This setup is often described by saying that the Euclidean evolution *prepares* the vacuum state, which then evolves in physical time from the waist of the hyperboloid onward. In other words, the time coordinate would run in the Euclidean first, from $\sigma = 0$ up to $\sigma = \pi/(2\mathsf{H})$, and from then on in the physical, Lorentzian time direction $\bar{t}$, according to $\sigma = \pi/(2\mathsf{H}) + i\bar{t}$. An example of a perturbation mode (6.151), evaluated precisely along this time contour, is shown in Fig. 6.11. Note the property that the mode function starts at zero. This is not accidental, and we will discuss the fundamental meaning of this construction when discussing the no-boundary proposal in Chapter 7. For now, we may just consider it as a useful prescription for preparing the Bunch–Davies vacuum state via Euclidean evolution.

Exercises

6.1 Go back and actually do the calculation of the $\omega \to \infty$ contribution to the partition function in Section 6.5. Determine the required contour of integration for the lapse.

6.2 Express the entropy in terms of the mass of a black hole. By taking a derivative, recover the first law of thermodynamics.

6.3 Putting in the numbers: 1. How quickly would one have to accelerate to feel an ambient temperature of 300 K? 2. What size black hole would one need to put in an oven to bake a cake? 3. What are the entropies of solar mass and supermassive (say, of 1 billion solar masses) black holes? 4. What is the temperature associated with the current dark energy phase? In light of these calculations, comment on the observability of the horizon-induced radiation temperatures.

6.4 Assume, for the sake of argument, that black holes were created around the time of the hot big bang. If they have been evaporating continuously, then how massive must these "primordial" black holes have been to still exist today?

6.5 Consider a massless scalar field living on a fixed de Sitter background spacetime (you may take the flat slicing). Calculate its variance at early and late times, and compare it to the de Sitter temperature. Now assume that the spacetime is also dynamical and relate the curvature perturbation to the de Sitter temperature.

7
Creating space and time: the no-boundary proposal

For a long time it was thought that the universe was infinite, both in extent and in age, even though an infinite universe leads to all sorts of paradoxes (for instance, anything that can happen with nonzero probability will happen an infinite number of times, so there must be an infinite number of copies of us, etc.). The discovery of the expansion of space changed that view. Space and time are changing, thus they may have had a beginning and they may have an end. In classical general relativity, Penrose and Hawking then proved that an expanding universe filled with ordinary matter must have had a singular beginning – the big bang. But a singularity is another kind of infinity.

Meanwhile, the development of quantum theory showed that new kinds of effects could happen. For instance, a particle can tunnel into existence where none was before. This led to the idea (already discussed by Lemaître in the 1930s, and proposed more precisely by E. Tryon in 1973) that perhaps space and time can also tunnel into existence, from nothing. In the early 1980s, J. Hartle and S. Hawking started developing a framework for calculating such a creation event, in a way that avoids the big bang singularity and relies explicitly on principles of quantum gravity – their framework is known as the *no-boundary proposal*.

7.1 Motivation and examples of no-boundary instantons

7.1.1 Heuristic arguments

One can subdivide physics into dynamics and initial/boundary conditions. The dynamics tells us all of the things that could possibly happen, while the initial conditions specify which of these possibilities are actualized. Put succinctly, the no-boundary proposal is a theory for the initial conditions of the universe.

Given a theory for the dynamics of the universe, the no-boundary proposal provides probabilities for different histories to occur by specifying how one should calculate the wave function of the universe. But it is not completely independent of the dynamics: As we will see, not all physical theories allow for no-boundary solutions, and not all physical theories lead to the prediction of a classical, viable universe. This interplay between dynamics and initial conditions enhances the predictive and explanatory power of the no-boundary proposal.

Avoiding an infinite regression

When we describe the history of the universe, we see a succession of phases: currently dark energy dominated, before that a matter phase, and before a radiation phase. No matter how far back we go, we always have to specify initial conditions for the evolution – see the left panel in Fig. 7.1. If we now add an inflationary phase, we again must specify initial conditions at the beginning of that phase. This process could go on forever – in standard theories one eventually ends up with a big bang singularity in the past, but this does not explain the initial conditions (middle panel). What would be much preferable would be to have an actual explanation of initial conditions.

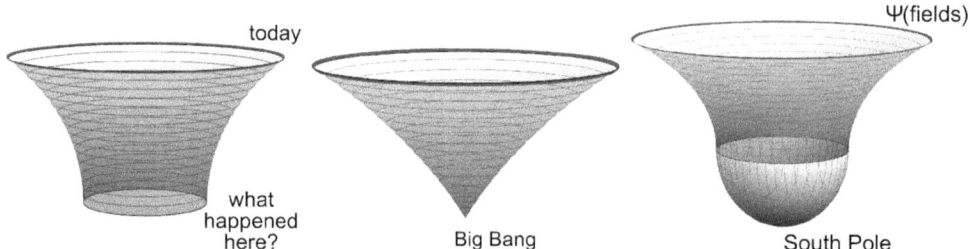

Figure 7.1 *Left panel:* Each time we model a prior cosmological phase, we must specify initial conditions for it. *Middle panel:* Going back in time classically, the expected outcome is a curvature singularity, commonly called the big bang. *Right panel:* The no-boundary proposal ends a potentially infinite regression by positing that the geometry of spacetime is rounded off in the past, and that there is thus no boundary in the past on which one might have to impose boundary conditions. The point on the "bottom" of this geometry (which is a fully regular point) is often referred to as the *South Pole*.

The no-boundary proposal achieves this by proposing that, when calculating the wave function of the universe, one consider geometries that are smoothly rounded off in the past (right panel). Note that we are talking about the entire spacetime here, including time. In classical Lorentzian geometry, such a rounding off is not possible, but in Euclidean geometry it is.

7.1 Motivation and examples of no-boundary instantons

Thus we require an extension of classical gravity to include Euclidean, and in general complex, geometries. As we will see, it is precisely the departure from classical Lorentzian geometry that is associated with the quantum nature of the spacetime creation event. The rounding off of the geometry then also implies that the universe is entirely self-contained, both in space and in time.

Ground state of the universe and quantum creation

Let us start by considering the amplitude for a particle to propagate from $(x=0, t=0)$ to (x,t),

$$\langle x, t \mid 0, 0 \rangle = \sum_n \psi_n(x) \bar{\psi}_n(0) \, e^{-iE_n t} \tag{7.1}$$

$$= \sum_n \psi_n(x) \bar{\psi}_n(0) \, e^{-E_n \sigma'} \, . \tag{7.2}$$

In the first line we inserted a complete set of energy eigenstates ψ_n with energy E_n, and in the second line we Wick rotated to Euclidean time $\sigma = it$. Now, when the Euclidean time becomes large, only the lowest-energy eigenstate will contribute significantly. In this way, evolution in Euclidean time can be seen as a means of preparing the ground state of the system (cf. our analogous discussion for perturbations in Section 6.6, and also compare to the low-temperature limit of the arguments in Section 6.1).

One of the motivations that Hartle and Hawking highlighted for the no-boundary proposal was precisely the extension of this idea to include gravity – namely, that an initial evolution in Euclidean time can prepare the universe in its ground state. When gravity is included, the question still remains then of which kinds of asymptotic Euclidean manifolds one should consider. Hartle and Hawking argued that there are only two clear options: asymptotic flat space, and compact Euclidean manifolds. But asymptotic flat space is more appropriate for scattering calculations, where particles come in from infinity and fly off to infinity again. There is also an obvious tension with the expansion of the universe. This leaves compact Euclidean metrics, which are indeed more appropriate, given that we observe at a finite time, from inside the universe. Compactness facilitates having a finite action, which is required to obtain sensible amplitudes, and moreover it provides a pathway for avoiding a big bang singularity. This brings us to the original proposal of Hartle and Hawking, namely that the wave function of the universe should be calculated by summing over compact, Euclidean metrics that interpolate to the spatial metric h_{ij}, schematically

$$\Psi_{\text{HH}}[h_{ij}] = \mathcal{N}_{\text{norm}} \int^{h_{ij}} \mathcal{D}g_{\mu\nu} \, e^{-S_E[g_{\mu\nu}]} \, , \tag{7.3}$$

where S_E is the Euclidean action, $\mathcal{N}_{\text{norm}}$ a normalization factor, and where the lapse integration contour should run over Euclidean times. Such a path integral then has the interpretation of calculating amplitudes to create the spatial hypersurface h_{ij} from nothing – i.e., it describes the creation of the universe. Making this definition precise, and in fact seeing whether it is meaningful, will be the central aim of this chapter.

A few comments: Strictly speaking, a closed universe does not have a well-defined energy, as there is no asymptotic region in which one could be defined. Thus, saying that the universe is in its ground state is really only an analogy. That said, as we saw when discussing de Sitter mode functions in Section 6.6, requiring regularity on a compact Euclidean manifold indeed describes a state of minimal quantum excitation. This also fits very nicely with a description that P. A. M. Dirac gave of the potential for quantum theory to explain the initial conditions of the universe. Here is what Dirac said, in a speech delivered upon receiving the James Scott Prize in 1939 (reproduced from the *Proceedings of the Royal Society of Edinburgh (1940), Vol. 59, pp. 122-129)*:

"Let us now return to dynamical questions. With the new cosmology the universe must have been started off in some very simple way. What, then, becomes of the initial conditions required by dynamical theory? Plainly there cannot be any, or they must be trivial. We are left in a situation which would be untenable with the old mechanics. If the universe were simply the motion which follows from a given scheme of equations of motion with trivial initial conditions, it could not contain the complexity we observe. Quantum mechanics provides an escape from the difficulty. It enables us to ascribe the complexity to the quantum jumps, lying outside the scheme of equations of motion. The quantum jumps now form the uncalculable part of natural phenomena, to replace the initial conditions of the old mechanistic view."

We should also mention that the idea of the universe tunneling out of nothing was independently proposed by A. Vilenkin, and goes by the name of *tunneling proposal*. There is a close conceptual analogy with the no-boundary proposal, though there exist important technical differences, which we will discuss below.

Before continuing, let us point out an immediate implication of the definition (7.3). Since the integral is defined as a sum over real Euclidean metrics only, the wave function is necessarily real. Nevertheless, a late-time Lorentzian universe must come out. Thus we would expect such an integral to contain complex saddle points, if it is to describe our universe. This is also necessary for a probabilistic interpretation, as a real wave function leads to a vanishing Klein–Gordon current.

7.1 Motivation and examples of no-boundary instantons

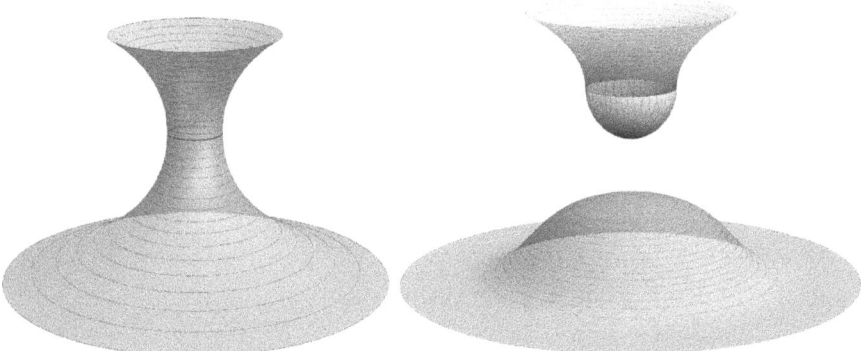

Figure 7.2 If we choose asymptotic flat space as the initial condition, then the path integral will contain bouncing configurations that are not saddle points (left panel) as well as topologically nontrivial contributions (right panel).

The path integral might effectively be of no-boundary type anyway

What if, after all, we had chosen asymptotically flat space as the boundary condition in the far past? Then there would be configurations interpolating between flat space and an expanding universe – see the left panel in Fig. 7.2. But such configurations would not be saddle points, as bounces require a violation of the null energy condition (see the discussion in Section 2.2.2). But there would be other, topologically nontrivial contributions, in which the original space pinches off and a new space is created smoothly, before reaching the final hypersurface with metric h_{ij}. The pinched-off manifolds would not really affect the later nucleation event, and thus it might well be that with such a definition one would effectively recover the "standard" no-boundary proposal.

7.1.2 Basic examples and stability of perturbations

De Sitter solutions

The simplest example of a no-boundary *instanton*[1] is just given by the de Sitter solution, or, more precisely, a slice through the analytic continuation of de Sitter spacetime in the closed slicing. For an illustration, see Fig. 7.3. Recall that the de Sitter solution (2.10) is a solution of the Einstein equations in the presence of a positive cosmological constant Λ and with no other matter present. The trick is to extend the Lorentzian solution into the imaginary time direction, where the solution can smoothly round off.

---

[1] Finite action solutions, especially Euclidean solutions, are often referred to as instantons.

Creating space and time: the no-boundary proposal

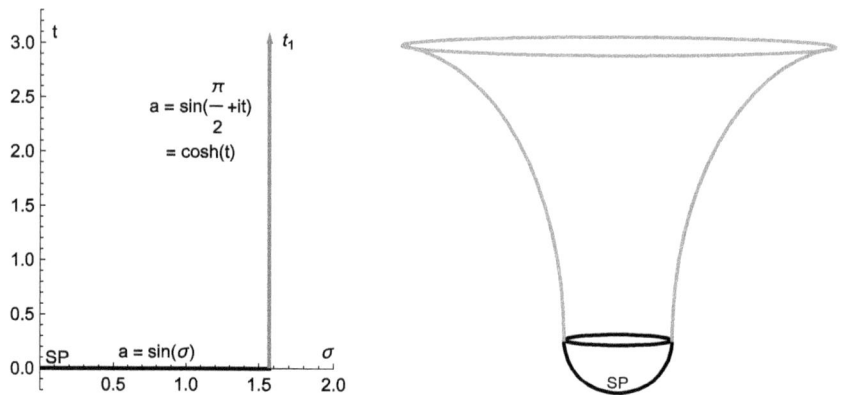

Figure 7.3 The prototype no-boundary solution is simply a section of complexified de Sitter spacetime. Onto the Lorentzian hyperboloid (in gray) one glues a Euclidean 4-hemisphere (in black). Then both the metric and its first derivative are continuous at the gluing. The solution may be seen as the path in the complexified (Euclidean) time plane shown on the left, with SP denoting the South Pole. In the figure we use $\Lambda = 3$.

So let us start with the Lorentzian solution (2.10), with scale factor

$$a = \sqrt{\frac{3}{\Lambda}} \cosh\left(\sqrt{\frac{\Lambda}{3}} t\right). \tag{7.4}$$

Into the future, one extends the solution in t until the desired final scale factor a_1 is reached at some t_1. At the minimum radius, $t = 0$ and $a_{\min} = \sqrt{\frac{3}{\Lambda}}$, one extends the solution into the imaginary time direction

$$\sigma = \sqrt{\frac{3}{\Lambda}} \frac{\pi}{2} + it, \tag{7.5}$$

whereupon the scale factor becomes

$$a = \sqrt{\frac{3}{\Lambda}} \sin\left(\sqrt{\frac{\Lambda}{3}} \sigma\right). \tag{7.6}$$

The solution can then be continued to zero size at $\sigma = 0$, which corresponds to the (smooth) South Pole of a 4-sphere. The action can be evaluated by recalling that for this maximally symmetric spacetime $R = 4\Lambda$, so that

$$S = \frac{1}{2} \int d^4x \sqrt{-g}(R - 2\Lambda)$$

$$= 2\pi^2 \Lambda \left(-i \int_0^{\sqrt{\frac{3}{\Lambda}}\frac{\pi}{2}} d\sigma\, a^3 + \int_0^{t_1} dt\, a^3\right)$$

7.1 Motivation and examples of no-boundary instantons

$$= \frac{12\pi^2}{\Lambda}\left[-i - \left(\frac{\Lambda}{3}a_1^2 - 1\right)^{3/2}\right], \tag{7.7}$$

where in the second line we performed the integral over the spatial 3-sphere, yielding a factor of $2\pi^2$, and in the final line we performed the integration over the complex time contour. Note that the end result does not depend on the particular time contour chosen, as long as it interpolates between the South Pole and t_1, given that there are no singularities in the integrand here. In particular, one could also choose a contour that does not perform any sharp turns. Thus the particular illustration shown in Fig. 7.3 should really be seen as just one possible realization of the instanton solution.

Assuming that this solution is relevant to the path integral, it will contribute to the no-boundary wave function an approximate term

$$\Psi \approx e^{\frac{12\pi^2}{\hbar\Lambda}\left[1-i\left(\frac{\Lambda}{3}a_1^2-1\right)^{3/2}\right]}, \tag{7.8}$$

where the smallness of $\hbar$ (or, better said, the large size of $1/(\hbar\Lambda) \gg 1$) guarantees that the saddle-point approximation will be good. This term is of JWKB form, with the amplitude being constant and the phase growing roughly as the volume of space a_1^3. Thus we may expect such a contribution to describe the emergence of a classical de Sitter universe of size a_1, out of nothing. The relative probability,

$$\Psi^\star \Psi \approx e^{\frac{24\pi^2}{\hbar\Lambda}}, \tag{7.9}$$

is somewhat meaningless here, as there exists only a single possible history. We will have more to say about the amplitude when considering less trivial examples below.

Note that the wave function (7.8) only starts oscillating when the scale factor has exceeded the Hubble radius associated with the vacuum energy, $a_1 \geq \sqrt{3/\Lambda}$. In other words, we can only start talking about a classical spacetime when the scale factor is larger than the de Sitter radius. This is a concrete illustration of the more heuristic arguments to this effect that we gave in Exercises 4.1 and 5.4, and that were based on the uncertainty principle (a yet more precise version will be given in Chapter 8). Here we see that the no-boundary proposal supports that interpretation, and that the uncertainty principle may well limit how far back in time we can go in our exploration of the universe.

Before continuing, we must deal with an important ambiguity, which has to do with the nonuniqueness of the background. Our boundary conditions include the condition that the scale factor should reach some specified value

a_1. But if we find a value t_1 for which this is the case, then we immediately see that $-t_1$ will also satisfy this condition, as $a_1 = \frac{1}{H}\cosh(Ht_1) = \frac{1}{H}\cosh(-Ht_1)$. And at early times, we want to find an analytic continuation such that the solution rounds off smoothly. But in the imaginary time direction, the scale factor is given by $\frac{1}{H}\cosh(iHt) = \frac{1}{H}\cos(Ht)$. This is zero, and moreover rounds off smoothly, at $it = \pm\frac{\pi}{2H}, \pm\frac{3\pi}{2H}, \pm\frac{5\pi}{2H}, \ldots$. Thus a priori we have an infinity of possible de Sitter sections that can all be no-boundary solutions – see Fig. 7.4. To assess their reasonableness we must include perturbations.

In Section 6.6 we had already started exploring small scalar perturbations (without backreaction) on de Sitter spacetime, by adding the perturbative action (6.147) and its associated equation of motion (6.148) (one could equally well look at gravitational wave perturbations, with analogous results). There we found that in the closed slicing a general solution is given by a linear combination of the modes (6.150) and (6.151).

For our present purposes, it is useful to rewrite these in terms of Euclidean time, defined via $\sigma = \frac{\pi}{2H} + i\bar{t}$. The shift by $\frac{\pi}{2H}$ is done in analogy with the time contour used above (and in Fig. 7.3). It has the consequence that the scale factor is now more straightforwardly written in terms of the sine function, namely $a(\sigma) = \frac{1}{H}\sin(H\sigma)$, with the zeros located at $\sigma = z\pi/H$ with $z \in \mathbb{Z}$. In Euclidean time the perturbation modes then read (up to a phase)

$$F_1(\sigma) = \frac{(1+\cos(H\sigma))^{(n-1)/2}(\cos(H\sigma) - n)}{(1-\cos(H\sigma))^{(n+1)/2}}, \qquad (7.10)$$

$$F_2(\sigma) = \frac{(1-\cos(H\sigma))^{(n-1)/2}(\cos(H\sigma) + n)}{(1+\cos(H\sigma))^{(n+1)/2}}. \qquad (7.11)$$

There is an important distinction between the two mode functions, and that is the location where they blow up and become singular. For F_1, this can be seen to happen at $H\sigma = 2\pi z$, while for F_2 it happens at $H\sigma = (2z+1)\pi$, still with $z \in \mathbb{Z}$. This allows us to eliminate all but two South Pole locations: Whenever the scale factor $a(\sigma)$ passes through two or more zeros, then it will necessarily include at least one location where the perturbations blow up. Put differently, the perturbations cannot be regular at the North and South Poles simultaneously. Thus "necklace" configurations, where one adds 4-spheres to the standard no-boundary geometry – see Fig. 7.4 – must be considered unphysical and discarded.[2]

This leaves two possibilities for the South Pole, namely $\sigma = 0$ (or $H\bar{t} = i\frac{\pi}{2}$) and $H\sigma = \pi$ (or $H\bar{t} = -i\frac{\pi}{2}$). Together with the two possibilities for choosing

---

[2] One could consider deforming the time contour so as to avoid the zeros of the scale factor. As we will discuss in Section 7.3.2, this does not solve the problem.

7.1 Motivation and examples of no-boundary instantons

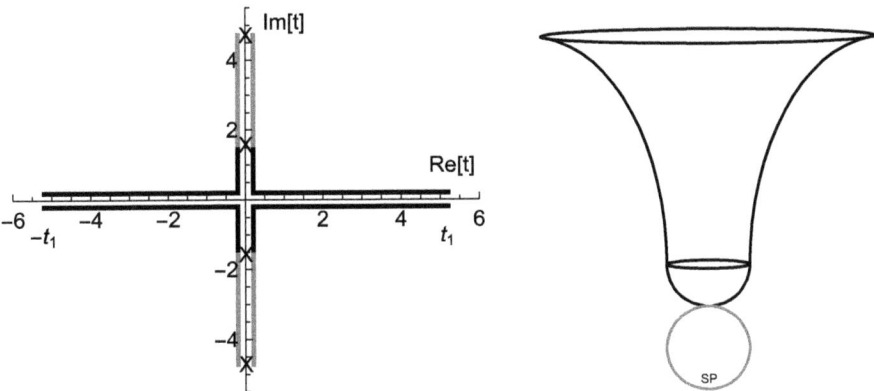

Figure 7.4 The four black complex time contours (slightly shifted away from the axes for better visibility) all superficially yield the same no-boundary geometry, though a closer analysis reveals that only the two contours that start at $H\bar{t} = i\frac{\pi}{2}$ support stable perturbations, while those starting at $H\bar{t} = -i\frac{\pi}{2}$ are unphysical. The zeros of the scale factor are indicated by crosses. If one were to continue the integration contours to the next zero, as shown in gray, then one would obtain a geometry containing a further 4-sphere – this geometry, and analogous ones with additional spheres attached, do not admit regular perturbations on them and must also be discarded.

the final time coordinate location, this gives us a priori four possible saddle points, which may be seen as time reverses and/or complex conjugates of each other – see again Fig. 7.4.

For the two solutions starting at $\sigma = 0$, the regular mode function is $F_2(\sigma)$. We can evaluate the on-shell action of the perturbations by making use of the equation of motion (6.148), finding

$$S_\phi^{\text{on-shell}} = 2\pi^2 \int d\bar{t} \frac{1}{2} a^3 \left(\dot{\phi}^2 + \phi\ddot{\phi} + 3\frac{\dot{a}}{a}\phi \right)$$
$$= \pi^2 a^3 \phi \dot{\phi} \, |_{\text{SP}}^{\bar{t}=t_1} \, . \tag{7.12}$$

At the South Pole there is no contribution, while at the final time t_1, where the fluctuation reaches the value ϕ_1, we get

$$S_\phi^{\text{on-shell}}(\phi_1) = \pi^2 a_1^3 \frac{\dot{F}_2(t_1)}{F_2(t_1)} \phi_1^2$$
$$= i\pi^2 a_1^2 \frac{n^2 - 1}{n - i\sqrt{H^2 a_1^2 - 1}} \phi_1^2 \tag{7.13}$$
$$\approx -\frac{2\pi^2 (n^2 - 1) a_1}{H} \phi_1^2 + i\frac{\pi^2 (n^3 - n)}{H^2} \phi_1^2 + \mathcal{O}(a_1^{-1}), \tag{7.14}$$

where we expanded at large final scale factor a_1 in the last line. If we had used the contour ending at $-t_1$ instead, the real part above would have had the opposite sign. The on-shell action can thus be seen to have a fast-growing (in magnitude) real part and a constant imaginary part at late times, which is appropriate for perturbations that behave classically. Combining with the background weighting in (7.8) and for $a_1 \gg n/\mathsf{H}$, we thus see that the combination of background and perturbations, for contours starting at $\sigma = 0$, leads to the weighting

$$|\Psi_{\text{no-boundary}}| \approx e^{\frac{12\pi^2}{\hbar\Lambda} - \frac{3\pi^2(n^3-n)}{\hbar\Lambda}\phi_1^2} . \tag{7.15}$$

We can see that the perturbations follow a Gaussian distribution, with larger fluctuations being strongly suppressed. This shows that these saddle points are stable, and represent legitimate contributions to the no-boundary path integral. The perturbations are in the Bunch–Davies ground state, and this can be seen as a consequence of requiring regularity at the South Pole. Thus the no-boundary condition of having a smooth, regular background geometry has automatically selected the Bunch–Davies ground state for the perturbations. In this sense, the no-boundary proposal selects and explains the Bunch–Davies vacuum. This is also in line with the heuristic statement made above, that the no-boundary state can be regarded as a state of minimum excitation.

We can perform a similar calculation of the action for the solutions with time contours starting at $\mathsf{H}\sigma = \pi$, and which are the complex conjugate solutions to the solutions starting at $\sigma = 0$. Since the calculation proceeds along exactly the same lines, we will not write out the details here. In the end one finds, for background and perturbations, and again for $a_1 \gg n/\mathsf{H}$, the weighting

$$|\Psi_{\text{tunneling}}| \approx e^{-\frac{12\pi^2}{\hbar\Lambda} + \frac{3\pi^2(n^3-n)}{\hbar\Lambda}\phi_1^2} . \tag{7.16}$$

The weighting is thus exactly the inverse of the one in (7.15). We labeled the associated contribution to the wave function by the term "tunneling," as these are the saddle points that are typically associated with the tunneling proposal. They are characterized by having a suppressed weighting for the background (note that it is not so clear what this implies in the present context, as for the creation of the universe there is no classical background we could compare to). However, the weighting of the perturbations is problematic, as it is given by an *inverse* Gaussian. This means that such saddle points lead to unphysical consequences, with large perturbations being favored over small ones. There are two possibilities here: Either the wave function follows this inverse Gaussian behavior to arbitrarily large fluctuations, in which case the wave function is mathematically ill-defined, or the enhanced perturbations are only predicted for small fluctuations, and then the predictions of the tunneling proposal are in conflict with CMB measurements. Either way, we see that such saddle points must be discarded. We will see how this can be done when looking at minisuperspace implementations of the no-boundary path integral.

No-boundary solutions with a massive scalar field

Having looked rather thoroughly at the simplest de Sitter solutions, we are ready to extend our analysis to include a massive scalar field. This case is not entirely realistic, in the sense that CMB observations disfavor quadratic inflation (as we discussed in Chapter 3), but it has the advantage that the corresponding no-boundary solutions can be approximated analytically. This will provide us with qualitatively new insights regarding the background instanton solutions, and these insights carry over to general inflationary no-boundary solutions.

Our framework is thus gravity coupled to a scalar field ϕ with mass m, which can be implemented via a potential $V(\phi) = \frac{1}{2}m^2\phi^2$. It proves useful to work in Euclidean time σ, as this facilitates solving the equations of motion near the South Pole. We will only consider the simplest possible metric, namely that of a closed FLRW universe,

$$\mathrm{d}s^2 = \mathrm{d}\sigma^2 + a^2(\sigma)\,\mathrm{d}\Omega_3^2 \,, \tag{7.17}$$

with the spatial sections being 3-spheres. Having positively curved hypersurfaces at constant σ is required for regularity, as will become obvious shortly. The scalar field is then also only a function of time, $\phi(\sigma)$. Note that ϕ is not considered to be a small perturbation here. Rather, ϕ will drive the dynamical evolution. At first, we will keep the scalar potential $V(\phi)$ general. The Euclidean action $S_E = -iS$ then reads

$$S_E = 2\pi^2 \int \mathrm{d}\sigma \left(-3aa'^2 - 3a + a^3 \left(\frac{1}{2}\phi'^2 + V(\phi) \right) \right) \,, \tag{7.18}$$

where $' \equiv \mathrm{d}/\mathrm{d}\sigma$. Here we have dropped surface terms, as they turn out not to play an important role for our present purposes – we will be more careful in Section 7.2, where surface terms will be crucial. The equations of motion are

$$a'' + \frac{a}{3}\left(\phi'^2 + V\right) = 0 \,, \tag{7.19}$$

$$\phi'' + 3\frac{a'}{a}\phi' - V_{,\phi} = 0 , \qquad (7.20)$$

while the constraint (Friedmann equation), arising from the time-time component of the Einstein equations, is

$$a'^2 - 1 = \frac{a^2}{3}\left(\frac{1}{2}\phi'^2 - V(\phi)\right) . \qquad (7.21)$$

We can use the constraint to simplify the on-shell action, i.e., the action evaluated on a solution of the equations of motion,

$$S_E^{\text{on-shell}} = 4\pi^2 \int d\sigma \left(-3a + a^3 V(\phi)\right) . \qquad (7.22)$$

Now we will specialize to a mass term $V(\phi) = \frac{1}{2}m^2\phi^2$. We start by solving the equations of motion as a Taylor series near the South Pole. Since we are simply looking for a solution of the classical field equations (albeit a complex solution thereof), we are free to impose a condition on both the scale factor and its derivative (which is related to the momentum). Our first requirement is that the solution should start at zero size, $a(0) = 0$. But then the constraint (7.21) reduces to

$$a'^2 - 1 = 0 \rightarrow a' = \pm 1 \quad \text{when} \quad a = 0 . \qquad (7.23)$$

Two remarks: First, the term -1 in (7.21) and in the line above results from the positive curvature of the 3-sphere part of the metric. This is the reason why positive curvature is required – otherwise we could not obtain a regular solution when the scale factor goes to zero. And second, the choice of sign for a' is in fact the same choice of sign we had to make when extending the time contours in the de Sitter case into the Euclidean time direction. It corresponds to the choice between no-boundary and tunneling instantons, i.e., the choice between the cases where perturbations are stable (which corresponds to $a' = +1$ – to see this compare with (7.6)) and the cases where perturbations are unphysical ($a' = -1$). Thus we must choose the positive sign,

$$\frac{da}{d\sigma}\bigg|_{a=0} = +1 \quad \text{(regularity and stability)} . \qquad (7.24)$$

With these boundary conditions, the Euclidean time-series solution of the equations of motion near the South Pole becomes

$$a(\sigma) = \sigma - \frac{1}{36}m^2\phi_{\text{SP}}^2 \sigma^3 + \mathcal{O}(\sigma^5) , \qquad (7.25)$$

$$\phi(\sigma) = \phi_{\text{SP}} + \frac{1}{8}m^2\phi_{\text{SP}} \sigma^2 + \mathcal{O}(\sigma^4) , \qquad (7.26)$$

where ϕ_{SP} denotes the value of the scalar field at the South Pole. We can see that the entire solution depends on that single (in general complex-valued) parameter, which remains to be determined. The analytic approximation turns out to be most accurate when ϕ_{SP} is almost real and large. Hence we will assume henceforth that $|\phi_{SP}^I| \ll 1 \ll |\phi_{SP}^R|$, where it will be convenient to denote real and imaginary parts by the superscripts R, I.

Near the South Pole, ϕ remains approximately constant, and Eq. (7.19) implies

$$a \approx \frac{\sqrt{6}}{m\,\phi_{SP}^R} \sin\left(\frac{m\,\phi_{SP}^R}{\sqrt{6}}\sigma\right), \qquad \phi \approx \phi_{SP}^R. \tag{7.27}$$

This solution describes a 4-sphere with radius determined by the location of the scalar field on the potential. We expect this solution to only be valid up to the equator, which is reached at Euclidean time $\sigma_{max}^R = \frac{\sqrt{6}}{m\,\phi_{SP}^R}\frac{\pi}{2}$. The series expansion (7.26) confirms that ϕ remains constant up to this point with fractional error $1/(\phi_{SP}^R)^2$.

At large scale factor, we can find a solution by making use of the slow-roll approximation for inflationary models, developed in Section 3.2.5. In Euclidean time, we then obtain

$$a(\sigma) \approx a_0 e^{-i\frac{m\,\phi_{SP}}{\sqrt{6}}\sigma + \frac{m^2}{6}\sigma^2}, \tag{7.28}$$

$$\phi(\sigma) \approx \phi_{SP} + i\sqrt{\frac{2}{3}}m\sigma. \tag{7.29}$$

The scalar field rolls down the potential linearly in time, and the scale factor expands exponentially in the imaginary σ direction, which is to say in Lorentzian time. The approximation breaks down once the second term in the exponent for a overtakes the first one, i.e., once $\sigma^I \sim \phi_{SP}^R/m$.

So much for the solutions at large and small a. Now we would like to link these two regimes, and also satisfy the appropriate boundary conditions. At the South Pole, we imposed the no-boundary conditions of compactness ($a = 0$) and regularity ($a' = 1$), but this left the value of the scalar field ϕ_{SP} undetermined. At late times, we would like the solution to reach specified real values. More precisely, at some final time $\sigma_1 \in \mathbb{C}$, we require

$$a_1 = a(\sigma_1) \in \mathbb{R}, \quad \phi_1 = \phi(\sigma_1) \in \mathbb{R}, \quad \text{with} \quad \Psi = \Psi(a_1, \phi_1). \tag{7.30}$$

These final, real values are the arguments of the wave function.

Looking again at (7.29), we can notice that the imaginary part of the scalar field does not change if we move in the imaginary σ direction. Hence, if we follow the Euclidean time direction from the South Pole up to the equator at σ_{max}^R, and then continue in the Lorentzian direction, the scalar field will have a fixed imaginary part given by $\phi_{\text{SP}}^I + i\sqrt{\frac{2}{3}}m\sigma_{\text{max}}^R$. Thus the late time scalar field can be made real if we choose ϕ_{SP}^I accordingly. Moreover, the scale factor (7.28) will also be real valued if we choose the phase of a_0 appropriately. Combining these requirements, we can match the early and late solutions with

$$a_0 \approx \frac{i\sqrt{6}}{m\phi_{\text{SP}}^R}, \quad \phi_{\text{SP}}^R \approx \phi_1, \quad \phi_{\text{SP}}^I \approx -\frac{\pi}{\phi_{\text{SP}}^R}, \quad \sigma_1^I \approx \frac{\sqrt{6}}{m\phi_{\text{SP}}^R}\ln\left(\frac{m\phi_{\text{SP}}^R}{\sqrt{6}}a_1\right). \tag{7.31}$$

We can draw two immediate conclusions from this analysis: Once a scalar field is added, the no-boundary instantons become generally complex valued. And along a time contour very similar to the one shown for the de Sitter solution in Fig. 7.3, the fields of a slow-roll solution only have a small imaginary part.

Using the values just derived, we can also estimate the action (7.22) of these solutions. Integrating from $\sigma = 0$ to σ_{max}^R gives a real part $S_E^R \approx -\frac{24\pi^2}{m^2\phi_1^2}$. Meanwhile, we can approximate the integral along the Lorentzian direction by neglecting the term linear in a in (7.22),

$$\frac{S_E^I}{2\pi^2} \approx \int_{\sigma_{\text{max}}^R}^{\sigma_1^I} \mathrm{d}\sigma\, m^2\phi^2 a^3 = \int \mathrm{d}\sigma\, m\phi a^2\sqrt{6}a' = \sqrt{\frac{2}{3}}\int \mathrm{d}\sigma\, m\phi\frac{\mathrm{d}}{\mathrm{d}\sigma}(a^3)$$

$$\approx \sqrt{\frac{2}{3}}m\phi_1 a_1^3, \tag{7.32}$$

where the last step was obtained by assuming slow-roll in the scalar.

This no-boundary instanton would thus contribute the following term to the wave function,

$$\Psi(a_1,\phi_1) \approx e^{\frac{12\pi^2}{\hbar V(\phi_1)} - i2\pi^2\sqrt{\frac{2}{3}}m\phi_1 a_1^3}. \tag{7.33}$$

The weighting implies that low values of the potential come out as preferred. This means that the no-boundary wave function assigns a high probability to short inflationary histories. This may be problematic, as a short inflationary phase may not be enough to resolve the flatness and horizon puzzles, and may not be able to amplify primordial density fluctuations over the relevant scales seen in the CMB. As we will discuss in Section 7.3.2, this problem might be overcome by looking more carefully at which metrics should be (and, mainly, which should not be) included in the gravitational path integral.

7.1 Motivation and examples of no-boundary instantons

Keeping this potential problem for later, we may focus on one truly remarkable feature of this calculation, which is the approach to classicality. We may straightforwardly notice that the wave function is of JWKB form, which is both an indication for classicality and a prerequisite for a probabilistic interpretation, as we discussed in Section 4.2. More precisely, using $(\nabla S)^2 = G^{AB}\partial_A S \partial_B S$ and $G_{a_1 a_1} = -12\pi^2 a_1$, $G_{\phi_1 \phi_1} = 2\pi^2 a_1^3$, we can compare the rates of change of both scale factor and scalar field using the explicit form (7.33),

$$\frac{(\nabla S_E^R)^2}{(\nabla S_E^I)^2} \sim \frac{1}{m^6 \phi_1^8 a_1^6}, \tag{7.34}$$

which is indeed driven to tiny values as a_1 grows.

When the wave function has achieved this semiclassical form, the physical spacetime and the actual dynamical evolution of the scalar field correspond to those series of (a_1, ϕ_1) values that arise with approximately constant weighting. Here, we can infer their explicit expression by eliminating σ from (7.28) and (7.29), resulting in

$$a_1 \approx \frac{i\sqrt{6}}{m\phi_{\text{SP}}^R} e^{\frac{1}{4}(\phi_{\text{SP}}^2 - \phi_1^2)}, \tag{7.35}$$

and they have an approximately fixed scalar field value $\phi_{\text{SP}} = \phi_{\text{SP}}^R - i\frac{\pi}{\phi_{\text{SP}}^R}$ at the South Pole. These are all attractor solutions of the present model. An interesting fact is that these histories are parametrized by a single real parameter, ϕ_{SP}^R, whereas standard classical solutions require two integration constants. This illustrates one additional aspect of the predictivity of the no-boundary proposal, namely that not all classical histories are included in the no-boundary wave function.

A final point worthy of note is that this scalar field model shows that the no-boundary instantons and the physical spacetime (plus matter evolutions) are separate concepts. Strictly speaking, the no-boundary instantons are used to calculate the wave function, and then one infers the physical evolution of the fields from the wave function, as we just did. However, when the solutions are attractors, as they are here, then for large instantons the late-time behavior is very similar to the late-time behavior of the associated classical solution.

7.1.3 General prescription, numerical examples, classical histories

In the previous section, we presented simple examples of no-boundary solutions that can be described (at least approximately) analytically. For a general scalar potential, this is unfortunately not possible and one must revert to numerical methods. However, the general prescription for identifying no-boundary saddles can be obtained by a direct generalization of the case of a massive scalar treated above, the leading principles still being compactness and regularity. We will work once more in the setting of gravity with a scalar field in a potential, in the simple context of spatially homogeneous and isotropic spacetimes. The action and equations of motion were given in Eqs. (7.17)–(7.22).

In the next section, we will aim to define the path integral more precisely. For now, we will take the saddle-point approximation for granted, with the no-boundary wave function loosely defined as a sum over an appropriate class of metrics,

$$\Psi(a_1, \phi_1) = \int_\mathcal{C} \mathcal{D}a \mathcal{D}\phi \, e^{\frac{i}{\hbar} S(a,\phi)}$$
$$\sim \sum e^{\frac{i}{\hbar} S^{\text{saddle}}(a_1,\phi_1)} = \sum e^{-\frac{1}{\hbar} S_E^{\text{saddle}}(a_1,\phi_1)}, \qquad (7.36)$$

where the saddle-point actions are complex in general, and thus it does not matter at this point whether we consider the fundamental action to be the Lorentzian (S) or Euclidean ($S_E = -iS$). For definiteness, we will work with the Euclidean action here, as this was traditionally used in the literature on the no-boundary proposal. The saddle points must then satisfy the following conditions:

- Since the saddles extremize the action, the fields a, ϕ must be solutions of the equations of motion. In general, they will be complex solutions – with the imaginary part describing the quantum part of the nucleation process.
- The solutions must be compact, so somewhere we must have $a(0) = 0$, where the origin of the Euclidean time coordinate σ has been shifted accordingly. This point is called the South Pole.
- At the South Pole, the solution must also be regular, by which we mean that the fields and their derivatives must take finite values. The Friedmann equation (7.21) then implies $a'(0) = \pm 1$, expressing the fact that the geometry must be Euclidean at the South Pole. As we saw, requiring

perturbations to be stable fixes the sign, and we must impose

$$a'(0) = +1, \quad \text{or} \quad \dot{a}(0) = +i, \tag{7.37}$$

where we recall the relation between Euclidean and Lorentzian times, $dt = -i\, d\sigma$.

- The equation of motion for ϕ, Eq. (7.20), shows that regularity implies the condition $\phi'(0) = 0$. This means that no-boundary solutions are labeled by the (complex) value of the scalar field at the South Pole, which we write as

$$\phi(0) \equiv \phi_{\rm SP} = \phi_{\rm SP}^R + i\phi_{\rm SP}^I. \tag{7.38}$$

- The South Pole is a regular singular point of the equations of motion. This means that we cannot integrate numerically directly from the South Pole, but rather we must use a Taylor series expansion near it. This Taylor series follows directly from the equations of motion (7.19)–(7.20),

$$a(\sigma) = \sigma - \frac{V}{18}\sigma^3 + \frac{8V^2 - 27V_{,\phi}^2}{8640}\sigma^5 + \mathcal{O}(\sigma^7), \tag{7.39}$$

$$\phi(\sigma) = \phi_{\rm SP} + \frac{V_{,\phi}}{8}\sigma^2 + \frac{2VV_{,\phi} + 3V_{,\phi}V_{,\phi\phi}}{576}\sigma^4 + \mathcal{O}(\sigma^6). \tag{7.40}$$

In numerical applications, it is typically useful to start the integration from $\sigma = 10^{-5}$ or even smaller.

- The solution must reach final real values of the scale factor (a_1) and scalar field (ϕ_1). In other words, there must exist a point $\sigma_1 \in \mathbb{C}$ in the complex σ plane where the fields take their specified values $a_1, \phi_1 \in \mathbb{R}$ simultaneously,

$$a(\sigma_1) = a_1 \quad \text{and} \quad \phi(\sigma_1) = \phi_1. \tag{7.41}$$

The values a_1, ϕ_1 are then also the arguments of the wave function. If such a point σ_1 does not exist, then there is no no-boundary solution for these final values.

Let us add four remarks: First, the action $S_E(a_1, \phi_1)$ will be evaluated along a path starting at $\sigma = 0$ and ending at $\sigma = \sigma_1$. The choice of path is irrelevant, due to Cauchy's theorem, as long as there are no singularities or branch points present in the complex σ plane. (In the presence of singularities, the "maximal curve" described in Section 7.3.2 generically seems to be the best choice of path.)

Second, the integration contour in the complex time plane that we are considering here has nothing to do with the complex integration contours (which are contours over the fields) used to define path integrals using the Picard–Lefschetz method that we discussed in Chapter 5 and that we will describe in Section 7.2. Here we are talking about saddle points of such integrals. But these saddle points are complex, and can be represented in different ways using different complex time paths. The different representations are physically equivalent, since the action is invariant under changes of path.

Third, we can give an argument why this setting is relevant. Imagine having a general perfect fluid, also homogeneous and isotropic, with energy density ρ and pressure p, as we considered in Chapter 1. The equation of continuity (1.34) can then be rewritten as

$$a\dot{\rho} + 3\dot{a}(\rho + p) = 0 \quad \Rightarrow \quad \rho + p \mid_{a=0} = 0. \qquad (7.42)$$

Regularity implies that $a\dot{\rho} \to 0$ as $a \to 0$, and also that $\dot{a} \mid_{a=0} \neq 0$. Then one can conclude immediately that, at the South Pole ($a = 0$), the perfect fluid must act like vacuum energy with $p = -\rho$. There are two ways that this relation can be satisfied: One can either have a cosmological constant, as we considered for the prototypical no-boundary instanton in Fig. 7.3; or one can have a scalar field that is momentarily at rest somewhere on its potential, cf. (2.20) and (2.21). Since a cosmological constant can be included in the scalar potential, we can see that this setting is appropriately general for now. (Of course, other types of matter can form later on. It is only at the nucleation of the universe that all other matter types are required to be zero.)

As a final comment, we should stress once more that the saddle-point solutions should not be regarded as the physical spacetime. They allow one to calculate the wave function with specified boundary conditions, and the physical spacetime must then be deduced from the wave function. After all, it is the wave function that contains all of the physics (to be fully precise: all of the physics accessible to a particular observer).

Let us now look at an example, which will also allow us to illustrate the numerical techniques required to find no-boundary solutions. We will consider an inflationary potential of exponential form,

$$V(\phi) = e^{\frac{\phi}{3}}, \qquad (7.43)$$

so that the (constant) slow-roll parameter is given by $\epsilon = \frac{V_\phi^2}{2V^2} = \frac{1}{18}$. For spatially flat FLRW metric, the classical solutions can be found explicitly and were given in (2.26), where the present case corresponds to $c = 1/3$. With the inclusion of perturbations, such solutions lead to a big bang sin-

7.1 Motivation and examples of no-boundary instantons

gularity as the scale factor tends to zero. Here, however, we would like to find solutions that are compact and regular in the past.

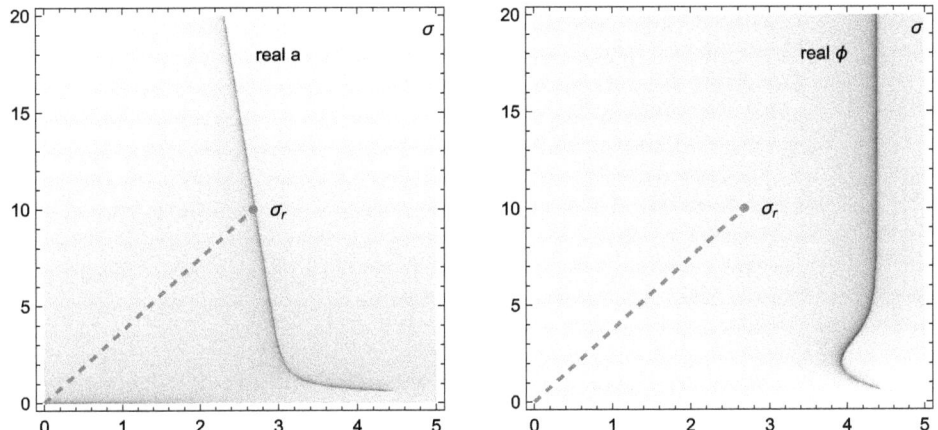

Figure 7.5 This graph shows the complexified Euclidean time (σ) plane, with dark lines indicating the locus where field values are real (more precisely, the figures show a density plot of $\log|\text{Im}(a(\sigma), \phi(\sigma))|$ for a dense grid of values of σ). The left panel shows this locus for the scale factor a, while the right panel is for the scalar field ϕ. White regions correspond to complex values of the fields. Along the dashed contour starting near the South Pole, we reach a real value of a, but at the corresponding point on the right the scalar field still takes complex values. One can then tune the values of ϕ_{SP} and σ_r until both fields are simultaneously real, as shown in Fig. 7.6. In the present graph, ϕ_{SP} is detuned by a factor $e^{i\pi/10}$ from the optimized value.

To this end, we can proceed as follows: First we choose a complex value ϕ_{SP} for the scalar field at the South Pole. It is useful to pick a value of ϕ_{SP}^R that lies in a region of the potential where inflationary solutions may be expected to be found, i.e., in a relatively flat region of the potential. For the present potential, any value will do. Then one integrates the equations of motion (7.19) and (7.20) from very close to the South Pole, say from $\sigma = 10^{-5} e^{i\pi/4}$, making use of the expansions (7.39)–(7.40). The integration is usefully done in a direction with angle between 0 and $\pi/2$ radians, until one reaches a real value of the scale factor. Let us call that location σ_r. Generically, the scalar field will not also be real at σ_r – see Fig. 7.5.

At this point, the idea is to *tune* the values of ϕ_{SP} and σ_r (corresponding to four real parameters) until one satisfies the four conditions $\text{Re}(a) = a_1$, $\text{Im}(a) = 0$, $\text{Re}(\phi) = \phi_1$, and $\text{Im}(\phi) = 0$ to the desired accuracy. The tuning can be done by using a Newtonian algorithm, which can straightforwardly be implemented in such a higher-dimensional parameter space.

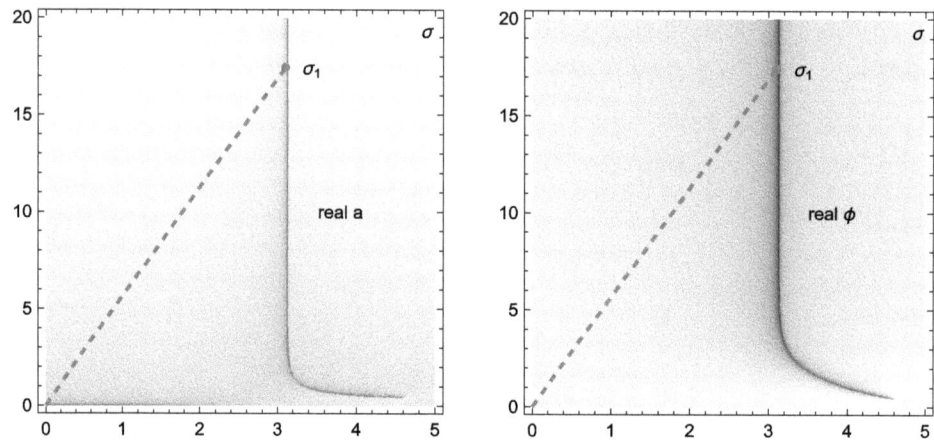

Figure 7.6 Same as Fig. 7.5, but now with the optimized value $\phi_{\rm SP} \approx -0.897 - 0.554\,i$. At time $\sigma_1 \approx 3.14 + 17.4\,i$, the scale factor reaches the value $a_1 = 1577$, while the scalar field is $\phi_1 = -3.046$. See also Fig. 7.7.

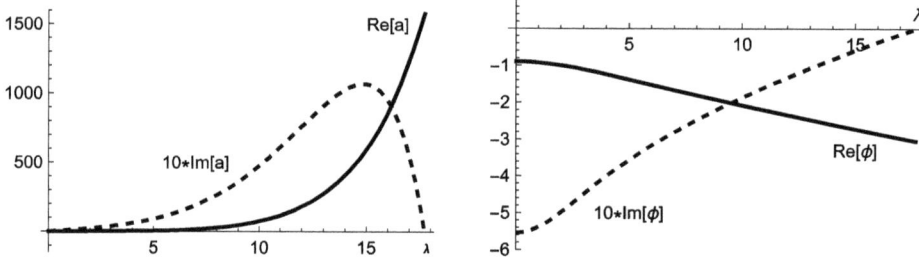

Figure 7.7 The field values along the dashed path shown in Fig. 7.6, for the optimized South Pole value $\phi_{\rm SP} \approx -0.897 - 0.554\,i$. The integration path is parametrized by a parameter λ. As one can see, the fields become real valued at the final time, and reach the values $a_1 = 1577, \phi_1 = -3.046$. The solution is a complex perturbation of an inflationary solution, with the scale factor expanding roughly exponentially, and the scalar field slowly rolling down the potential. However, the imaginary parts of the fields are necessary to allow the solution to round off smoothly at the origin.

However, some trial and error is typically required at first, in order to find oneself in the basin of attraction of the solution. Here it is useful to choose target values a_1 and ϕ_1 that are close to the values $a(\sigma_r)$ and $\text{Re}[\phi(\sigma_r)]$. The result of a successful optimization is shown in Fig. 7.6, where the endpoint, at which the fields reach the real values a_1, ϕ_1, is now denoted as σ_1. The corresponding field values along the dashed contour are shown in Fig. 7.7. Thus we have now found our first numerical no-boundary instanton! This is in fact the hardest part – successive instantons can be found by changing

7.1 Motivation and examples of no-boundary instantons 215

the final values a_1, ϕ_1 in small steps, with small changes in the optimized South Pole and σ_1 values.

A striking feature of the optimized solution in Fig. 7.6 are the two vertical dark lines, which would largely overlap if one were to superpose the left and right graphs. These are a result of the inflationary attractor, which pulls the solution close to a classical one (in the vertical, Lorentzian time direction). In fact, it is because of the attractor nature of inflation that the optimization is feasible. These vertical dark lines also imply that evolving along a classical solution will result in instantons that are changed very little. Put differently, we could move up the vertical dark lines (with the fields staying approximately real) and have an essentially unchanged bottom part of the instanton.

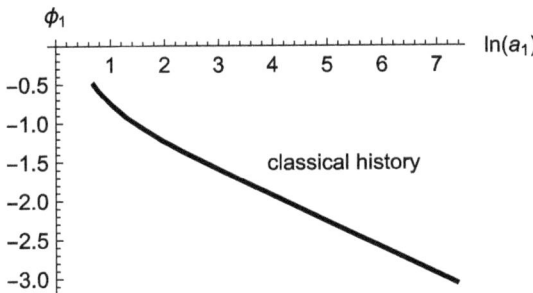

Figure 7.8 Values of the scale factor and scalar field for a classical inflationary solution to the equations of motion in the potential $V(\phi) = e^{\phi/3}$. These values are used as arguments of the wave function $\Psi(a_1, \phi_1)$.

Let us make this more precise, by considering an actual classical solution. An example is shown in Fig. 7.8. It depicts a standard inflationary solution, with the scale factor expanding quasi-exponentially (note the horizontal axis shows the logarithm of a), and the scalar field slowly rolling down the potential. We can now take this classical solution, or rather the sequence of paired $[a_1, \phi_1]$ values, as argument for the wave function. Numerically speaking, the classical solution depends on a parameter λ (that one can often take to be classical time), so we may write $[a_1(\lambda), \phi_1(\lambda)]$. We can then discretize by moving forward with small steps in λ. In this way we approximate the continuous classical solution by a discrete time-step version. This leads to a sequence of corresponding instantons, whose optimized ϕ_{SP} values are shown in Fig. 7.9. As one can see, the values indeed stabilize very quickly as the universe gets larger.

We can also look at the action of this series of instantons – see Fig. 7.10. The right panel shows the phase of the wave function, which is seen to change

216   *Creating space and time: the no-boundary proposal*

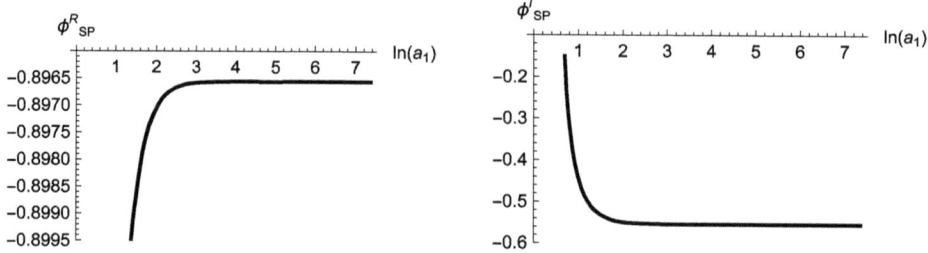

Figure 7.9 The optimized values of ϕ_{SP} as a function of the final scale factor value, for the series of final values given by the classical history shown in Fig. 7.8. The graph shows the values of 200 discrete time steps, joined together by interpolation.

fast. Meanwhile, the left panel shows that the weighting quickly stabilizes, as the universe gets larger. This is important, as it shows that the wave function is semiclassical, and allows us to assign a (constant) relative probability to this history, provided the scale factor is large enough – the rough criterion here being that $a_1 \gtrsim e^2$.

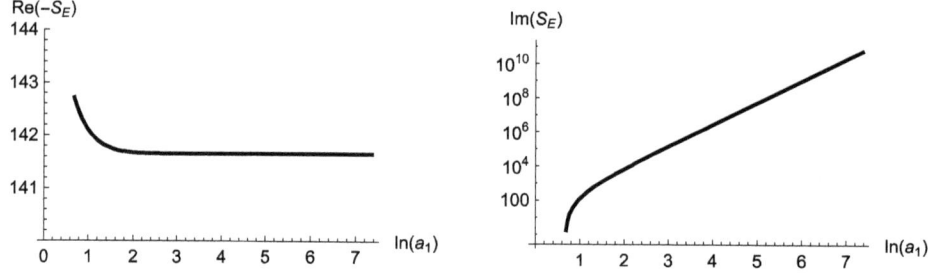

Figure 7.10 The weighting (on the left) and phase (on the right) as a function of the final scale factor, for arguments of the wave function corresponding to the classical history shown in Fig. 7.8. Again, the plot shows the values of 200 discrete time steps, joined by interpolation.

In fact, we can be a bit more precise regarding the approach to JWKB classicality. We may recall from (4.35) that the actual criterion for semiclassicality is that the ratio, given by

$$\frac{(\nabla S_E^R)^2}{(\nabla S_E^I)^2}, \tag{7.44}$$

must be small, where ∇_A is the covariant derivative on the (a, ϕ) field space. We performed just this calculation in (7.34) in the previous section. Here, we will have to work a little harder to obtain an analytic estimate of this ratio.

7.1 Motivation and examples of no-boundary instantons

Here we can make use of the fact that, for an exponential potential, the action transforms in a simple way under shifts of the scalar field. For this calculation, we will write the potential again in more general form as $V(\phi) = e^{-c\phi}$, since this allows us to model any inflationary model with a constant equation of state $\epsilon = c^2/2$. In the numerical calculations shown above we had set $c = -1/3$.

Taking the Euclidean action as the starting point,

$$S_E = -\int d^4x \sqrt{g} \left(\frac{R}{2} - \frac{1}{2} g^{\mu\nu} \partial_\mu \phi \, \partial_\nu \phi - e^{-c\phi} \right), \tag{7.45}$$

one can shift the scalar and perform a related scaling of the metric,

$$\phi \equiv \bar{\phi} + \Delta\phi, \quad g_{\mu\nu} \equiv e^{c\Delta\phi} \bar{g}_{\mu\nu}, \tag{7.46}$$

to find that the action has transformed into

$$S_E = -e^{c\Delta\phi} \int d^4x \sqrt{\bar{g}} \left(\frac{\bar{R}}{2} - \frac{1}{2} \bar{g}^{\mu\nu} \partial_\mu \bar{\phi} \partial_\nu \bar{\phi} - e^{-c\bar{\phi}} \right). \tag{7.47}$$

These relations imply that the field equations are invariant under

$$\bar{a}(\bar{t}) = e^{c\Delta\phi/2} a\left(e^{-c\Delta\phi/2} \bar{t} \right), \tag{7.48}$$

$$\bar{\phi}(\bar{t}) = \phi\left(e^{-c\Delta\phi/2} \bar{t} \right) + \Delta\phi, \tag{7.49}$$

where overbars denote the transformed quantities. Under this transformation, the scaling solution (2.26) morphs into

$$\bar{a} = \bar{a}_0 \left(\bar{t}\right)^{2/c^2}, \quad \bar{a}_0 = \exp\left(\frac{(c^2 - 2)\Delta\phi}{2c} \right) a_0, \quad V(\bar{\phi}) = \frac{12 - 2c^2}{c^4} \frac{1}{\bar{t}^2}, \tag{7.50}$$

which shows that a_0 is a constant of motion,

$$a_0 = a \left(\frac{c^4}{12 - 2c^2} V \right)^{1/c^2}. \tag{7.51}$$

This means that we can label different solutions by their value of a_0.

As shown in Section 2.1.3 and as can clearly be seen in Fig. 7.6, at late Lorentzian times the attractor pulls the solution close to a real classical solution, and thus the imaginary part of the Euclidean action scales as (with $d\sigma = idt$, and see also Exercise 4.3)

$$S_E^I \sim i \int dt \, a^3 V \sim i a_0^3 \, t^{-1 + \frac{6}{c^2}} \sim i a_0^3 \, V^{\frac{1}{2} - \frac{3}{c^2}}. \tag{7.52}$$

Using the constant of motion (7.51), one can thus determine the dependence on the final values a_1, ϕ_1 to be

$$S_E^I \sim i a_1^3 V(\phi_1)^{1/2} . \qquad (7.53)$$

Meanwhile, the scaling of the real part of the Euclidean action is governed by the scaling/shift symmetry. Since we are interested in changes in the final values a_1, ϕ_1, we must take the shift $\Delta\phi$ to be real valued. Then from (7.47) we obtain

$$\bar{S}_E^R = e^{c\Delta\phi} S_R = \left(\frac{\bar{a}_0}{a_0}\right)^{\frac{2c^2}{c^2-2}} S_E^R , \qquad (7.54)$$

so that

$$S_E^R \sim a_0^{\frac{2c^2}{c^2-2}} \sim a_1^{\frac{2c^2}{c^2-2}} V(\phi_1)^{\frac{2}{c^2-2}} . \qquad (7.55)$$

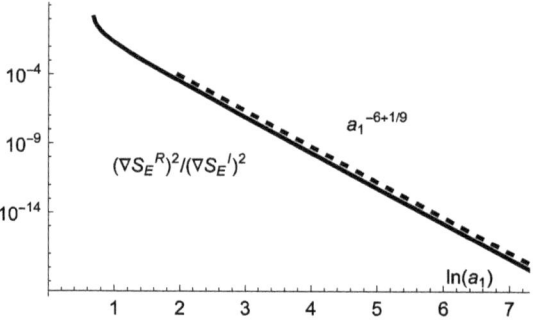

Figure 7.11 The approach to classicality as a function of the final scale factor, for the wave function containing as its argument the classical history shown in Fig. 7.8, with $c = -1/3$. Here we show the numerically calculated values with the continuous interpolating curve, and compare to the theoretical expectation depicted by the dashed line. The scalings match to great precision.

Now it becomes straightforward to work out the JWKB condition (4.35), which says that the amplitude of the wave function should vary slowly compared to its phase. With $(\nabla I)^2 = G^{AB}\partial_A I \partial_B I$ with $G_{aa} = -12\pi^2 a_1$, $G_{\phi\phi} = 2\pi^2 a_1^3$, one finds

$$\frac{(\nabla S_E^R)^2}{(\nabla S_E^I)^2} \sim \frac{a_1^{\frac{4c^2}{c^2-2}-3} V^{\frac{4}{c^2-2}}}{a_1^3 V} \sim a_1^{c^2-6} . \qquad (7.56)$$

Thus we see a confirmation that classicality is reached under the same conditions under which we had found a dynamical attractor, namely for $c^2 < 6$.

7.1 Motivation and examples of no-boundary instantons   219

For our numerical example, the scaling implied by (7.56) is obeyed to great precision, as shown in Fig. 7.11. Thus the inflationary phase automatically drives the wave function to JWKB semiclassicality. This explains how one obtains a classical universe (on large scales) shortly after nucleation.[3]

Let us elaborate a little on this important feature. The interesting point is that we have a purely dynamical way of obtaining classicalization here. The usual procedure in quantum mechanics is to invoke decoherence, i.e., the loss of quantum coherence due to interactions of a system with its environment. Such a process is highly relevant on Earth, where interactions are extremely common. But it is not available at the creation of the universe, for two reasons: On the one hand, there is no environment as the universe contains everything by definition and is still empty. And on the other hand, decoherence is a process that happens over time, meaning that once time has become classical, decoherence can take place. Here we have just seen how time (and space) can become classical in the first place, purely due to cosmological dynamics. In this way, the no-boundary proposal can explain the classicality of the early universe.

Now that we have thoroughly discussed a single possible classical history, one can of course repeat these calculations for different histories. In this way one obtains relative probabilities for different classical evolutions, as long as the corresponding no-boundary instantons exist, and as long as the wave function for the corresponding series of saddles becomes JWKB semiclassical. When these conditions are met, what one finds is that the weighting is well approximated by (cf. also (7.33))

$$|\Psi(a_1, \phi_1)| \sim e^{\frac{12\pi^2}{\hbar V(\phi_{\rm SP}^R)}} . \tag{7.57}$$

This implies that inflationary histories that start lower on the potential come out as preferred. Therefore, shorter inflationary evolutions, with fewer e-folds, are more likely than longer ones. And inflation with a smaller energy scale (i.e., small-field models) is more likely than inflation at a high energy scale (large-field inflation). As we saw in Chapter 3, the latter property agrees well with observations – the current nondetection of primordial B-mode polarization implies a small tensor-to-scalar ratio, and thus favors inflationary models that take place at a lower energy scale. However, one still needs a sizeable amount of inflation, typically at least 50 e-folds, to explain the flatness of the universe (this is important here, as the no-boundary

---

[3] Only one other mechanism is currently known to achieve this, and that is an ekpyrotic phase; in that case, after the universe nucleates it starts contracting, and this contraction is an attractor, much in the same way as inflation. During an ekpyrotic contraction phase, which would correspond to $c^2 > 6$ in (7.56), the wave function is also driven to JWKB classicality.

solution is spatially closed) and to explain the temperature fluctuations seen on the CMB sky (if inflation is too short, then no modes are amplified on the scales that we see in the CMB). Thus, as it stands, the no-boundary proposal favors inflationary histories that are too short to agree with observations.

It remains unclear at present if and how this discrepancy can be resolved. One possibility goes by the name of *anthropic principle*, which states that certain features of the universe must be as they are because otherwise they would not be compatible with us living in the universe. Put differently, the fact that we exist requires a rather large and old universe, and thus it requires a rather long inflationary phase. Even if the a priori probability for a short inflationary phase is higher, this is not meaningful if it only produces a universe that we cannot inhabit. This line of reasoning would suggest that we must live in the universe with the shortest possible inflationary phase, at the lowest possible energy scale, while still being compatible with the development of life. Not enough is currently known about either the development of life in general, nor about the fundamental scalar potential, to be able to support or rule out this reasoning.

A second possibility, which we will discuss in more detail in Section 7.3.2, is that certain classes of spacetimes are not allowed to be summed over in the gravitational path integral. Surprisingly, what one finds is that a reasonable criterion for which spacetimes should be allowed also requires inflation to be rather long. There are indications that this might be stringent enough to offer an explanation for the length of an early inflationary phase. But in that case, it comes down to having a precise definition of gravitational path integrals. In the next section, we will embark on this quest, starting with simple models and building up complexity gradually.

7.2 Minisuperspace implementations

So far, the no-boundary proposal was based very loosely on a path integral summing over compact, regular metrics and matter configurations. Now, our aim is to make this definition more precise. After all, it is not enough to have some saddle-point solutions. These saddle points must be saddles of an actual integral. We would like to know what exactly they are approximating, and which saddle points are relevant to that approximation. We will be able to make good use of the results of Chapter 5 on path integrals, but there are a number of issues that must be faced.

- There is one immediate conundrum: The no-boundary condition that we imposed for saddle points cannot be imposed as a condition on the full

7.2 Minisuperspace implementations

path integral, as it is in conflict with the uncertainty principle. This is because it is a condition on both the field value $a(\sigma = 0) = 0$, ensuring compactness, and on the expansion rate $a'(\sigma = 0) = 1$, ensuring regularity. Thus we cannot define a path integral that sums over metrics that are both compact and regular. It has to be one or the other, or perhaps a condition on a linear combination of field value and momentum. The (somewhat Zen-flavored) question then is: Which boundary conditions correspond best to no-boundary conditions? Note that imposing compactness will not guarantee that the saddle points turn out to also be regular, and vice versa. We will have to check this at the end of the calculation.

- In the action for gravity coupled to matter, the kinetic term for the scale factor of the universe enters with a different sign than all of the other kinetic terms, both those of anisotropic components of the metric and those of matter fields,

$$S \sim \int \mathrm{d}t \left[-3a\dot{a}^2 + \frac{1}{2}a^3\dot{\phi}^2 + \cdots \right]. \tag{7.58}$$

This is known as the *conformal mode problem*, so-called because the scale factor can be seen as the conformal mode of spatial sections. If the path integral is defined as a Euclidean integral, then the integrand will thus be unbounded above and below, regardless of the overall choice of sign. Hence it is doubtful that the Euclidean path integral might make sense. A Lorentzian path integral seems more promising as it would not have this problem (by being a sum over phases), but there we have the issue that the integral is only conditionally convergent and it has to be defined carefully to make its meaning unambiguous.

- Gravity is not renormalizable, which means that we expect an infinite number of correction terms involving ever higher powers of the Riemann tensor and ever higher numbers of derivatives. This might however still be fine, as long as the curvature remains well below the Planck scale. At least for the saddle points we looked at so far, this happened to be true: They all had curvatures on the order of the Hubble scale in the early universe, which for inflationary examples is constrained by observations to be $H/M_P \lesssim 10^{-5}$. Under such circumstances, we may expect the path integral to yield reliable semiclassical results. We will analyze this in more detail in Section 7.3.1.

- It should be expected that the general sum over 4-manifolds is very difficult to make precise. One has the freedom to sum over metrics that can differ at all spacetime points. Moreover, in analogy with the paths integrated over in quantum mechanics, one might expect the required manifolds in general not to be differentiable anywhere. What is more, even the topology of 4-manifolds remains ill-understood. Hence it seems difficult at present to properly define a sum over 4-manifolds. We will take a more pragmatic approach, and restrict to metrics that have certain symmetries, in particular cosmologically relevant symmetries. This is the framework of minisuperspace, where the metric is parametrized by a finite number of functions of time. One may object to this on the basis that we are neglecting infinitely many degrees of freedom, and, what is worse, we are setting both these degrees of freedom and their conjugate momenta to zero simultaneously. Nevertheless, we know from observations that the early universe was highly symmetric. Hence, we should expect a realistic theory of initial conditions to predict a high probability for precisely these minisuperspace kinds of metrics. What is crucial then is to check at the end of our calculations whether or not metrics that are perturbed around the minisuperspace saddle points come out as suppressed. If so, then we may have some confidence in our calculations.

Dirichlet boundary condition

In their original paper, Hartle and Hawking envisioned the no-boundary proposal as corresponding to a Euclidean path integral over compact metrics. We will interpret the compactness condition as implying that we fix the initial size to zero, which corresponds to a Dirichlet boundary condition. We will try to see to what extent such an integral can be realized in the simple setting of having only gravity and a cosmological constant, with action

$$S = \int_{\mathcal{M}} d^4x \sqrt{-g} \left(\frac{R}{2} - \Lambda\right) + \sum \gamma \int_{\partial \mathcal{M}_{0,1}} d^3y \sqrt{h} K, \quad (7.59)$$

and path integral

$$\Psi_{\rm DD}(q_1) = \int_{\mathcal{C}} \int_{q_0=0}^{q_1} d\tilde{N} \mathcal{D}q \, e^{\frac{i}{\hbar} S}. \quad (7.60)$$

The GHY surface terms are essential as we intend to keep the field values fixed both on the initial ($\partial \mathcal{M}_0$, with $\gamma = -1$) and on the final ($\partial \mathcal{M}_1$, with $\gamma = +1$) hypersurfaces. We will make extensive use of the results of Section 5.2.1, where we looked at transition amplitudes with Dirichlet boundary conditions. Hence we can be brief here. With $q_0 = 0$, the solution to the equation of motion (though not necessarily the constraint) is given by

$$\bar{q}(t_q) = \frac{\Lambda}{3} \tilde{N}^2 t_q^2 + \left(-\frac{\Lambda}{3} \tilde{N}^2 + q_1\right) t_q. \quad (7.61)$$

Shifting the integration variable to $q(t_q) = \bar{q}(t_q) + Q(t_q)$ and performing the integral over Q then results in the lapse integral

$$\Psi_{\mathrm{DD}}(q_1) = \sqrt{\frac{3\pi i}{2\hbar}} \int_{\mathcal{C}} \frac{\mathrm{d}\tilde{N}}{\tilde{N}^{1/2}} e^{\frac{i}{\hbar} 2\pi^2 S_0}, \qquad (7.62)$$

with

$$S_0 = \tilde{N}^3 \frac{\Lambda^2}{36} + \tilde{N}\left(3 - \frac{\Lambda}{2} q_1\right) - \frac{3}{4\tilde{N}} q_1^2. \qquad (7.63)$$

The lapse integral contains an essential singularity at $\tilde{N} = 0$, which can be understood physically as the impossibility of evolving from size zero to size $q_1 \neq 0$ in vanishing time.

The four saddle points of the integral are located at

$$\tilde{N}^{\mathrm{saddle}} = c_1 \frac{3}{\Lambda} \left[i + c_2 \left(\frac{\Lambda}{3} q_1 - 1\right)^{1/2} \right], \qquad (7.64)$$

with $c_1, c_2 \in \{-1, 1\}$, while the action at the saddle points is given by

$$S_0^{\mathrm{saddle}} = c_1 \frac{6}{\Lambda} \left[i - c_2 \left(\frac{\Lambda}{3} q_1 - 1\right)^{3/2} \right]. \qquad (7.65)$$

The saddle points split into two pairs, with an important distinction between them. This can be appreciated by calculating the expansion rate at the South Poles of the four saddle-point solutions. From (7.61), we have

$$\frac{\mathrm{d}a}{\mathrm{d}t}(t=0) = \frac{1}{2\tilde{N}} \frac{\mathrm{d}\bar{q}}{\mathrm{d}t_q}(t_q = 0) = \frac{q_1 - \frac{\Lambda}{3}\tilde{N}^2}{2\tilde{N}} \qquad (7.66)$$

$$= -\frac{i}{c_1} \quad \text{at the saddles}. \qquad (7.67)$$

A shown in Section 7.1.2, this means that the saddle points with $c_1 = +1$ are in fact unstable (if one adds perturbations, these inherit an inverse Gaussian probability distribution), while the two saddles in the lower half $\tilde{N}$ plane ($c_1 = -1$) are stable upon inclusion of perturbations. Put differently, for the saddles in the lower half plane the minisuperspace restriction makes sense, while for those in the upper half plane it does not (and the latter do not stand a chance of explaining our universe).[4]

The crucial question then is what integration contour $\mathcal{C}$ over the lapse should be chosen. The flow lines are identical to those shown in Fig. 5.5, hence we do not need to reproduce the figure here. As discussed there, the

---

[4] In Fig. 7.4, the stable integration contours appear in the upper half plane. This is because, compared to the current calculation, the origin of the lapse is shifted. If one translates the contours in Fig. 7.4 such that they start at the origin, then the stable contours indeed move to the lower half plane, and the unstable ones into the upper half plane.

Euclidean path integral (i.e., choosing $\mathcal{C}$ to run over imaginary $\tilde{N}$ values) simply does not exist, because of the divergences associated with the conformal mode problem. Hence we must abandon the idea of a purely Euclidean definition. As for integrals over Lorentzian metrics (real $\tilde{N}$ values), we found that they always pick up the unstable saddles in the upper half plane.

In fact, only a contour that starts at $-i\infty$ and reaches the singularity at the origin $\tilde{N} = 0$ by approaching it from the upper half plane results in a path integral that solely picks up either of the two stable saddle points with $c_1 = -1$. This is an inherently complex contour of integration, since the region adjacent to the origin and situated in the lower half plane is a region of divergent weighting, which must be circumvented. Moreover, such an integration contour results in a propagator (Green's function) rather than a wave function, because it is only half-infinite – i.e., it satisfies the inhomogeneous WDW equation (5.70) rather than the standard WDW equation (5.69). This is a subtle but important distinction as the propagator describes how states evolve, while a wave function describes the actual state itself. The aim of the no-boundary proposal is clearly to explain the state of the universe itself, and thus we must discard this half-infinite contour in the present context.

For completeness, let us mention one other complex contour, which asymptotically approaches the Lorentzian line, but circumvents the singularity by passing through the high-weighting region below the origin. This is an infinite contour, leading to a wave function that picks up all four saddles. To date, it has not been possible to conclusively establish what happens when (large) perturbations are included, as now the wave function is approximated by a combination of stable and unstable saddle points. To settle this issue, one would need to extend the calculation beyond minisuperspace, and this has not been possible so far.

We conclude that an implementation of the no-boundary proposal as a Euclidean path integral over compact metrics is not possible, and that the imposition of a Dirichlet boundary condition seems problematic in general. We will now see how this situation can be improved significantly by adopting a better-suited boundary condition.

Neumann boundary condition

Above we attempted to define the path integral with fixed (zero) initial size. For consistency this meant that we had to include the GHY surface term on the initial hypersurface in (7.59). But the spirit of the no-boundary proposal is that there should be no boundary. Hence why should one include a boundary term? And where should it be placed, if the intention is that there should be no boundary? It seems much more natural not to include a boundary term. If we do this, it changes the variational problem, and in fact leads to a Neumann boundary condition (see Section 5.2.2), allowing us to fix the initial momentum $\dot{q}/\tilde{N}$.

The regularity requirement of the no-boundary proposal implies that we need the initial expansion rate to be Euclidean. Imposing it also requires a choice of sign, cf. (7.37). This choice of sign is precisely what distinguishes the stable from the unstable saddle points. One may also think of this choice of sign as the choice of Wick rotation – i.e., do we define $t = +i\sigma$ or $t = -i\sigma$? Only one sign assignment leads to stable, Gaussian-distributed fluctuations, and this is the choice $a'(0) = +1$. So can we define a path integral with this boundary condition?

The action will now include a GHY surface term only on the final hypersurface,

$$S = \int_{\mathcal{M}} d^4x \sqrt{-g}\left(\frac{R}{2} - \Lambda\right) + \int_{\partial\mathcal{M}_1} d^3y \sqrt{h} K. \tag{7.68}$$

Again working with closed FLRW universes, and using integration by parts, the surface term at $t_q = 1$ is eliminated by the GHY term, while a surface term is then generated at $t_q = 0$,

$$S = 2\pi^2 \int_0^1 dt_q \left[-\frac{3}{4N}\dot{q}^2 + N(3 - \Lambda q)\right] - \frac{3\pi^2}{N} q\dot{q}|_{t_q=0}. \tag{7.69}$$

As discussed in Section 5.2.2, variation of this action leads to the equation of motion $\ddot{q} = \frac{2\Lambda}{3}N^2$ and the boundary condition that we can fix $\dot{q}/\tilde{N}$ at $t_q = 0$. Recall that $\tilde{N} dt_q/q^{1/2} = dt$, so that consequently $\dot{q}/\tilde{N} = 2da/dt$, implying that we should fix

$$\frac{\dot{q}}{2N} = +i. \tag{7.70}$$

This is the no-boundary regularity condition (7.37) expressed in our currently used variables. On the final hypersurface we will again set $q(t_q = 1) = q_1 > 3/\Lambda$. With these boundary conditions, the solution to the equation of motion reads

$$\bar{q}(t_q) = \frac{\Lambda}{3}\tilde{N}^2 t_q^2 + 2\tilde{N}it_q + q_1 - \frac{\Lambda}{3}\tilde{N}^2 - 2\tilde{N}i. \tag{7.71}$$

Evaluating the path integral over q in the same manner as before, we then obtain the following expression for the Neumann–Dirichlet wave function,

$$\Psi_{ND}(q_1) = \int_{\mathcal{C}} d\tilde{N}\, e^{\frac{i}{\hbar}2\pi^2 \left[\frac{\Lambda^2}{9}\tilde{N}^3 + i\Lambda\tilde{N}^2 - q_1\Lambda\tilde{N} - 3iq_1\right]}, \tag{7.72}$$

where once again the contour of integration $\mathcal{C}$ for the lapse integral remains to be specified.

But before discussing contours, we may already notice some important differences with the Dirichlet action (7.63). There is no singularity at $\tilde{N} = 0$, since it is now possible (even if unlikely) that the initial geometry, satisfying $\dot{q} = 2\tilde{N}i$, already coincides with the final geometry, satisfying $q = q_1$. Also, the action now contains explicit factors of i, due to the boundary condition (7.70). Thus we do not expect saddle points to come in complex conjugate pairs anymore. In fact, there are only two saddle points this time, and they are located at

$$\tilde{N}_\pm = \frac{3}{\Lambda}\left[-i \pm \frac{3}{\Lambda}\sqrt{\frac{\Lambda}{3}q_1 - 1}\right]. \qquad (7.73)$$

The corresponding field evolutions are shown for an example in Fig. 7.12. Note that the saddle points are not only regular, which they are by design, but also compact: At the saddle points, one may see and verify that $\bar{q}(t_q = 0) = 0$. Also note that the fields are only becoming real right at the end of the time evolution, as this now does not correspond to a Euclidean-plus-Lorentzian contour in the complex time plane, but rather to a solution with fixed (complex) lapse. Comparing to (7.64), we see that in fact only the two saddle points in the lower half plane are left. That is to say, the unstable saddle points have been eliminated, and we are left purely with the stable saddle points!

We still have to figure out which integration contour to take – see the right graph in Fig. 7.12, in which the saddle points and their steepest ascent/descent paths are shown. Asymptotic regions of convergence are indicated by the shading.[5] We may notice right away that the Euclidean contour once again does not work. Since there is no singularity at $\tilde{N} = 0$, it would have to be defined over the entire imaginary lapse line in order to yield an invariant result. However, we can see that there is again a divergence at large positive imaginary values, and thus the Euclidean definition does not make sense.

In fact, all convergent integration contours must interpolate between two of the three asymptotic regions of convergence. This includes the Lorentzian contour, which can be deformed into the sum over both thimbles $\mathcal{C} = \mathcal{J}_+ + \mathcal{J}_-$, with orientations chosen in the direction of increasing real parts of the

---

[5] Comparing with (7.72), we can see that these are obtained by finding the regions for which at large $|\tilde{N}|$ we have $\mathrm{Re}\left[i\left(\frac{\Lambda^2}{9}\tilde{N}^3 + i\Lambda\tilde{N}^2 - q_1\Lambda\tilde{N} - 3iq_1\right)\right] \to -\infty$.

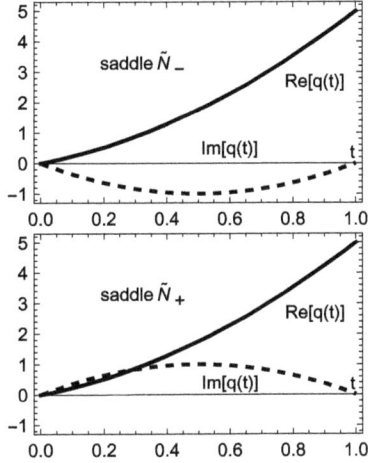 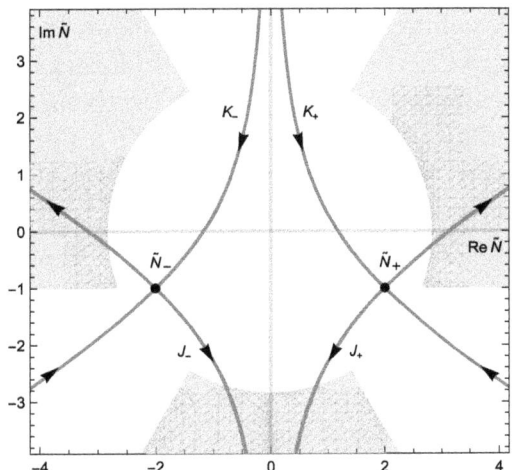

Figure 7.12 *Left:* An example of the evolution of the scale factor squared for the saddle points of the no-boundary wave function defined with an initial Neumann/momentum condition, with $\Lambda = 3$, $q_1 = 5$, so that the saddle points are at $N_\pm = \pm 2 - i$. One can see that the saddle points are compact, as they start at zero size. *Right:* the saddle points and their associated steepest ascent/descent lines in the complex plane of the lapse function (arrows indicate downward flow). Convergent integration contours interpolate between two of the three asymptotic regions of convergence (indicated by the shaded regions) and can be deformed into one or a combination of both thimbles $\mathcal{J}_\pm$.

lapse. The wave function then becomes approximated by

$$\Psi_{\rm ND}(q_1) \approx e^{+\frac{12\pi^2}{\hbar\Lambda} - i\frac{12\pi^2}{\hbar\Lambda}\left(\frac{\Lambda}{3}q_1 - 1\right)^{3/2}} + e^{+\frac{12\pi^2}{\hbar\Lambda} + i\frac{12\pi^2}{\hbar\Lambda}\left(\frac{\Lambda}{3}q_1 - 1\right)^{3/2}}. \quad (7.74)$$

Note that it is real, as it is the sum of two complex conjugate contributions, even though the integral is not defined over a Euclidean contour. This is due to the symmetry between negative and positive real parts of the lapse. One further remark: Picard–Lefschetz theory implies that relevant saddle points must have a lower weighting than that of the defining contour. One may thus wonder how it is possible to obtain an enhanced weighting from a Lorentzian integral. This is because, even though the starting action (7.68) is real, the boundary condition (7.70) is complex, and this results in a positive weighting in (7.72), even at real values of the lapse.

In fact, there are only a few other contours we could contemplate, including starting at $-i\infty$ (where the geometries that are summed over essentially correspond to very large Euclidean 4-spheres with their North Pole cap removed at radius-squared q_1) and ending in one of the other regions of convergence, thereby picking up only one of the two saddle points. One further possibility is to sum over the two thimbles, but with opposite orientations $\mathcal{C} = \mathcal{J}_+ - \mathcal{J}_-$, i.e., to sum from negative imaginary infinity up to negative real infinity, plus an integral from negative imaginary infinity to positive real infinity. This choice would give a pure imaginary wave function – however, since we are ignoring the prefactor, which could be imaginary too, this must be seen as equivalent to a real wave function. At the semiclassical level, the implications are however largely unaffected by this choice of orientation of the thimbles. This is because the two saddle points behave effectively independently, as soon as perturbations and the resulting decoherence is taken into account. We will discuss decoherence in more detail in Chapter 8.

Having found a minisuperspace path integral implementation of the no-boundary wave function, we may also study its relation to the WDW equation. Recall from (5.69) that the Hamiltonian is given by

$$H = -\frac{\tilde{N}}{6\pi^2}\left[p^2 + 12\pi^4(3 - \Lambda q)\right] = \tilde{N}\hat{H}, \quad (7.75)$$

with the canonical momentum $p = -\frac{3\pi^2}{N}\dot{q}$. The WDW equation then corresponds to the operator version of this equation,

$$\hat{H}\Psi = 0, \quad (7.76)$$

with Ψ being the wave function of the universe. Now we have to pay attention to the boundary conditions we imposed, namely Dirichlet on the final hypersurface and Neumann on the initial one. The canonical commutation relation $[q, p] = i$ must be implemented correspondingly. On the final hypersurface, where we work in field space, we replace the momentum by a derivative operator $p \to \hat{p} = -i\frac{\partial}{\partial q}$, leading to

$$\hat{H}_{(q)}\Psi = 0 \to \frac{\partial^2 \Psi}{\partial q^2} + 12\pi^4(\Lambda q - 3)\Psi = 0. \quad (7.77)$$

By contrast, on the initial hypersurface we impose a momentum condition. In order to obtain the WDW equation in momentum space we therefore substitute $q \to \hat{q} = i\frac{\partial}{\partial p}$, leading to

$$\hat{H}_{(p)}\Psi = 0 \to (p^2 + 36\pi^4)\Psi + 12\pi^4\Lambda i\frac{\partial \Psi}{\partial p} = 0. \quad (7.78)$$

There is one subtlety here: Because we are imposing this equation on the initial boundary, we have to flip the sign $\frac{\partial}{\partial p} \to -\frac{\partial}{\partial p}$.

The momentum space equation (7.78) is of first order and yields an essentially unique solution, the exponential of a cubic in p_0. Meanwhile, the position space equation (7.77) can be identified as an Airy equation, with

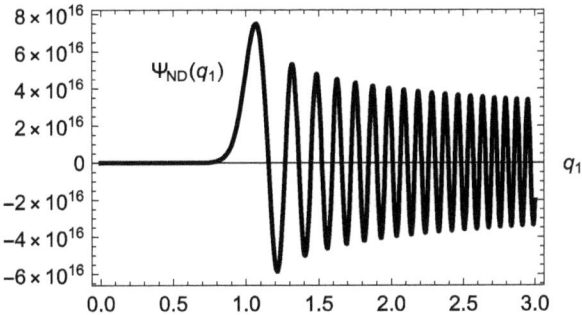

Figure 7.13 The no-boundary wave function as an Airy function, see (7.79), and plotted as a function of the final radius squared q_1, with $\Lambda = 3$. The wave function grows monotonically during nucleation, and then oscillates when the universe is larger than the Hubble radius.

two linearly independent solutions, the Ai and Bi functions. Choosing a particular linear combination is directly related to the choice of contour in the path integral approach. For instance, the Lorentzian contour yields the Ai function, and the contour summing both thimbles from negative imaginary infinity yields the Bi function. Explicitly, if we stick to the Lorentzian integration contour, then the equivalent – exact – solution to the WDW equation is

$$\Psi(p_0, q_1) = e^{\frac{3}{\hbar \Lambda} i p_0 + \frac{1}{36\pi^4 \hbar \Lambda} i p_0^3} \text{Ai}\left[\left(\frac{18\pi^2}{\hbar \Lambda}\right)^{2/3} \left(1 - \frac{\Lambda}{3} q_1\right)\right], \quad (7.79)$$

with initial momentum given by (7.70) – that is to say,

$$p_0 = -\frac{3\pi^2}{\tilde{N}} \dot{q}(t_q = 0) = -6\pi^2 i\,. \quad (7.80)$$

The wave function is shown in Fig. 7.13 as a function of the final size $q_1 = a_1^2$. As one can see there, the wave function rises exponentially (from a tiny, though nonzero value) at $q_1 = 0$ and then starts to oscillate once the universe has become larger than the Hubble radius. These two regimes correspond to quantum tunneling from nothing and subsequent classical evolution, respectively. It remains unclear how to interpret the fact that the wave function is nonzero at zero size. One suggestion is that this could be due to the contribution of nontrivial topologies, but the present calculation did not contain additional topologies. This interesting question deserves further study.

A further comment: a priori, one might think that field space and momentum space definitions might be equivalent, as they can be Fourier transformed into each other. However, the Fourier transform would sum over all possible initial momenta, and would thus also include momenta that correspond to unstable Wick rotations. This argument implies that the discordant results obtained with Dirichlet versus Neumann initial conditions are not mutually inconsistent.

For the particular value of the initial momentum (7.80), which was chosen to ensure regularity, we find from (7.78) that at the no-boundary point, the wave function satisfies the additional relation

$$i\frac{\partial}{\partial p}\Psi = \hat{q}\,\Psi = 0 \qquad \text{(WDW no-boundary condition)}. \qquad (7.81)$$

This is particularly suggestive: It says that the no-boundary wave function satisfies the momentum space equivalent of the zero-size condition. That is to say, the regularity condition we imposed on the wave function turns out to be equivalent to the operator expression for the zero-size condition! This certainly conforms well with the spirit of the no-boundary proposal. It also means that at the nucleation of the universe, there is no momentum transfer into the universe. In other words, this condition also expresses the notion that the universe is self-contained.

Biaxial Bianchi IX – a model with anisotropies

There are only a handful of (at least partially) analytically tractable minisuperspace models. One model of particular interest is based on the Bianchi IX metric, which can be thought of as having spatial 3-spheres that can be squashed in three directions. The full Bianchi IX model cannot be treated analytically, but a subset of this model, with two squashings set equal and dubbed the biaxial Bianchi IX (BBIX) spacetime, has equations of motion that can be solved analytically. This model has the advantage that it can describe anisotropic deformations of unlimited magnitude (albeit of a prescribed shape). This allows us to go beyond the perturbative treatment of perturbations around the isotropic saddle points on which we relied up to now. In addition, the model provides a neat example that shows that one should include a sum over topologies, and thus also a sum over appropriate boundary conditions, in the gravitational path integral.

If one thinks of a 3-sphere as a fibration of a circle S^1 over a sphere S^2, then the BBIX metric can be pictured as giving different radii to the S^1 and S^2. In other words, it describes a squashed 3-sphere,

$$\mathrm{d}s^2 = -\frac{\tilde{N}^2}{q}\mathrm{d}t^2 + \frac{p}{4}(\sigma_1^2 + \sigma_2^2) + \frac{q}{4}\sigma_3^2, \qquad (7.82)$$

where $q(t), p(t)$ are the time-dependent radii of the circle and sphere, respectively. The spatial metric is written in terms of $\sigma_1 = \sin\psi d\theta - \cos\psi \sin\theta d\varphi$, $\sigma_2 = \cos\psi d\theta + \sin\psi \sin\theta d\varphi$, and $\sigma_3 = d\psi + \cos\theta d\varphi$, which are differential forms on the 3-sphere with coordinate ranges $0 \le \psi \le 4\pi$, $0 \le \theta \le \pi$, and $0 \le \varphi \le 2\pi$. Sometimes, this is also called the Taub spacetime.

We will be interested in two different versions, with differing topologies: the Taub-NUT space in which both q and p reach zero simultaneously, and the Taub-Bolt space in which the circle radius q reaches zero while the 2-sphere radius p remains nonzero. Both versions have zero spatial volume $2\pi^2 q^{1/2} p$ at the origin, and both are regular there, thus providing suitable saddle points for the no-boundary wave function.

We will pursue the philosophy that there should be no surface term on the initial hypersurface, and that one should thus impose Neumann boundary conditions there. The minisuperspace no-boundary wave function is then defined via

$$\Psi(q_1, p_1) = \sum_{\text{Neumann}} \int^{q_1} \mathcal{D}q \int^{p_1} \mathcal{D}p \int_\mathcal{C} d\tilde{N} \, e^{\frac{i}{\hbar}S}, \qquad (7.83)$$

where we leave the room open for summing over inequivalent Neumann boundary conditions. When evaluated on the metric (7.82), the action (7.68) reduces to

$$S = 2\pi^2 \int_0^1 dt \left[\tilde{N}\left(4 - \frac{q}{p} - p\Lambda\right) + \frac{1}{4\tilde{N}}\left(-\frac{q}{p}\dot{p}^2 + 4q\ddot{p} + 2p\ddot{q} + 4\dot{p}\dot{q}\right)\right]$$
$$+ \frac{\pi^2}{\tilde{N}}(-p\dot{q} - 2q\dot{p})|_{t=1}, \qquad (7.84)$$

where the second line consists of the boundary contribution on the final hypersurface at $t=1$.

Varying the action with respect to the fields yields the equations of motion

$$2p\ddot{p} - \dot{p}^2 = 4\tilde{N}^2, \qquad (7.85)$$

$$\ddot{q} + \frac{\dot{p}}{p}\dot{q} = \tilde{N}^2\left(2\Lambda - 4\frac{q}{p}\right), \qquad (7.86)$$

supplemented by the boundary variations

$$(\Pi_q \delta q + \Pi_p \delta p)|_{t=1} + (q\delta\Pi_q + p\delta\Pi_p)|_{t=0}, \qquad (7.87)$$

which we have written in terms of the canonical momenta

$$\frac{1}{2\pi^2}\Pi_q = -\frac{1}{2\tilde{N}}\dot{p}, \qquad \frac{1}{2\pi^2}\Pi_p = -\frac{1}{2\tilde{N}}\left(\dot{q} + \frac{q}{p}\dot{p}\right). \qquad (7.88)$$

Before discussing boundary conditions let us point out a peculiarity of this model, which is that the action is linear in q. Thus q acts as a Lagrange multiplier, imposing the equation of motion Eq. (7.85), which simplifies the action to

$$S = 2\pi^2 \int_0^1 dt \tilde{N} \left(4 - p\Lambda\right) + \frac{\pi^2}{\tilde{N}} \left[-q\dot{p}\,|_{t=1} - (q\dot{p} + p\dot{q})\,|_{t=0}\right]. \quad (7.89)$$

The action now depends on q only via its boundary values.

To proceed, we must impose appropriate boundary conditions. On the final hypersurface we will simply impose Dirichlet conditions $q(t = 1) \equiv q_1$ and $p(t = 1) \equiv p_1$. On the initial hypersurface, Eq. (7.87) allows for four possible choices a priori: 1. fixing $q_0 = p_0 = 0$, 2. fixing $p_0 = 0$ and $\Pi_{q,0} = -2\pi^2 i$ (cf. Eq. (7.70)), 3. fixing $q_0 = 0$ and $\Pi_{p,0} = -2\pi^2 i$, and 4. fixing both momenta $\Pi_{q,0} = \Pi_{p,0} = -2\pi^2 i$. Choice 1 would eliminate Taub-Bolt metrics. For choices 2 and 3, the sign of the Wick rotation is fixed by the sign of the momentum. Meanwhile, choice 4 turns out to be overconstraining. Thus the two interesting choices are 2 and 3, and it seems natural that one should sum over both. The results below will support this point of view.

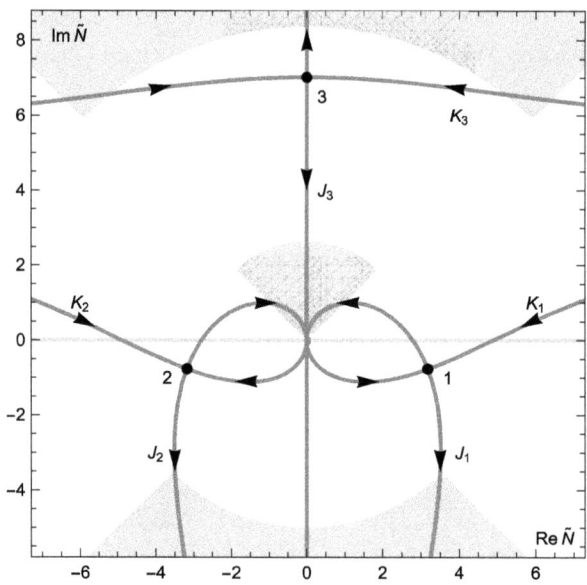

Figure 7.14 The three saddle points of the Taub-NUT action (7.91) and the associated steepest ascent ($\mathcal{K}_{1,2,3}$) and descent lines ($\mathcal{J}_{1,2,3}$), in the complex plane of the lapse function $\tilde{N}$. Downwards flow is indicated by the arrows. The asymptotic regions of convergence, both at large $|\tilde{N}|$ and near the singularity at the origin, are indicated by shading. For this example, $\Lambda = 3$, $q_1 = 10$, and $p_1 = 15$.

7.2 Minisuperspace implementations

Choice 2 of the boundary conditions leads to a Taub-NUT-dS spacetime, in which the 2-sphere is shrunk to zero size initially. The explicit solution to Eq. (7.85) is given by

$$p(t) = p_0 + 2\left(\sqrt{p_0 p_1 - \tilde{N}^2} - p_0\right) t + \left(p_0 + p_1 - 2\sqrt{p_0 p_1 - \tilde{N}^2}\right) t^2, \tag{7.90}$$

where we fix a branch by taking $\sqrt{p_0 p_1 - \tilde{N}^2} \to +i\tilde{N}$ as $p_0 \to 0$. After substitution, one is left with an action that depends purely on the lapse

$$\frac{1}{2\pi^2} S_{\text{NUT}}(\tilde{N}) = -\frac{p_1 q_1}{\tilde{N}} + i q_1 + \tilde{N}\left(4 - \frac{\Lambda}{3} p_1\right) - i\frac{\Lambda}{3}\tilde{N}^2. \tag{7.91}$$

This action admits three saddle points – see Fig. 7.14. One of these is purely Euclidean, and does not lead to a classical spacetime. The other two saddle points are of physical interest (they yield a complex action, with a real part that grows rapidly with increasing p_1, q_1), and are simply related by a reflection in the real part of $\tilde{N}$. They are picked up by the path integral if one chooses a contour of integration for the lapse that contains their thimbles. Note that a Lorentzian integration contour, as we had in the isotropic case, is not possible this time as the integral simply diverges along such a contour. If we insist on obtaining a wave function (rather than a propagator), then we must start and end in the region of convergence towards $-i\infty$ and loop around the singularity at the origin, thus summing over $\mathcal{J}_1 + \mathcal{J}_2$. Alternatively, one can also choose contours that run from $-i\infty$ to $+i\infty$, either passing to the left or right of the singularity at the origin. In such a case, saddle 3 contributes too, but its weighting is exponentially suppressed compared to the weightings of saddles 1 and 2.

The weighting of the relevant saddles 1 and 2 is equal and shown in the left panel of Fig. 7.15. As one can see from this example, the isotropic configuration is most probable and anisotropic configurations are suppressed. So far this validates the minisuperspace approach, and also confirms that the no-boundary wave function puts the highest weight on isotropic configurations.

But this is not the full story yet, as we have not analyzed the second possible boundary condition, called choice 3 above. That choice amounts to imposing $q_0 = 0$ and $\Pi_{p,0} = -2\pi^2 i$, which implies setting $\dot{q}(0) = 2\tilde{N}i$. Obtaining the latter condition fixes the initial value of p to be

$$p_0 = \frac{p_1(\Lambda \tilde{N}^2 - 3q_1)^2}{4\tilde{N}^2(\Lambda \tilde{N} + 3i)^2} + \frac{\tilde{N}^2}{p_1}. \tag{7.92}$$

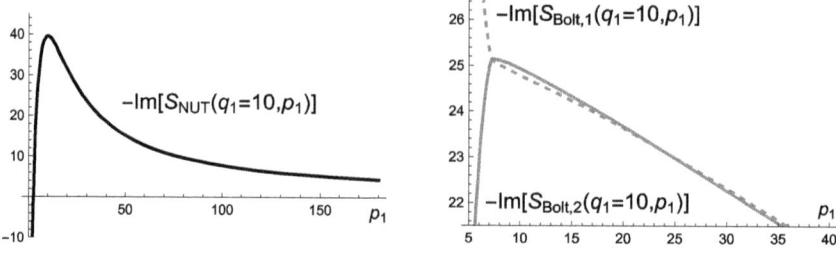

Figure 7.15 The weightings of different saddle points of interest for Taub metrics. The NUT case is shown in the left panel and the Bolt case on the right, with $\Lambda = 3$ and $q_1 = 10$. Note that for the NUT case, the isotropic configuration is most likely, while for the Bolt saddles this is not the case. Fig. 7.16 shows how the two cases compare.

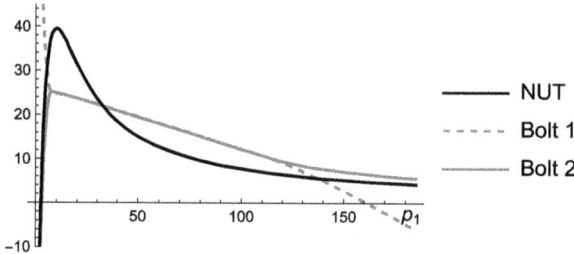

Figure 7.16 A comparison of the NUT and Bolt weightings. A Stokes phenomenon ensures that the Bolt saddle numbered 1 does not come to dominate at small p_1 values. This is crucial for the overall consistency of the model.

In general, this will be nonzero. After solving the equations of motion with the required boundary conditions[6] and integrating over time, one obtains the lapse action

$$\frac{1}{2\pi^2}S_{\text{Bolt}}(\tilde{N}) = \left(4 - \frac{\Lambda}{3}\right)\tilde{N} - \frac{p_1 q_1}{\tilde{N}} - \frac{\tilde{N}^2(\Lambda\tilde{N} + 3i)}{3p_1} + \frac{p_1(\Lambda\tilde{N}^2 - 3q_1)^2}{12\tilde{N}^2(\Lambda\tilde{N} + 3i)}.$$
(7.93)

There is a simple pole located at $\tilde{N} = -3i/\Lambda$ and a double pole at $\tilde{N} = 0$.

This action admits seven saddle points. Examples of the associated thimbles are shown in the left panel of Fig. 7.17. Three saddles are Euclidean and not of physical interest. The other four are however of potential physical relevance. They arise again in pairs, with equal weighting. The imaginary parts of the actions of saddles 1 and 2 are shown in the right panel in Fig. 7.15. We

---

[6] The solution to the equation of motion for q is a quartic in t, divided by $p(t)$.

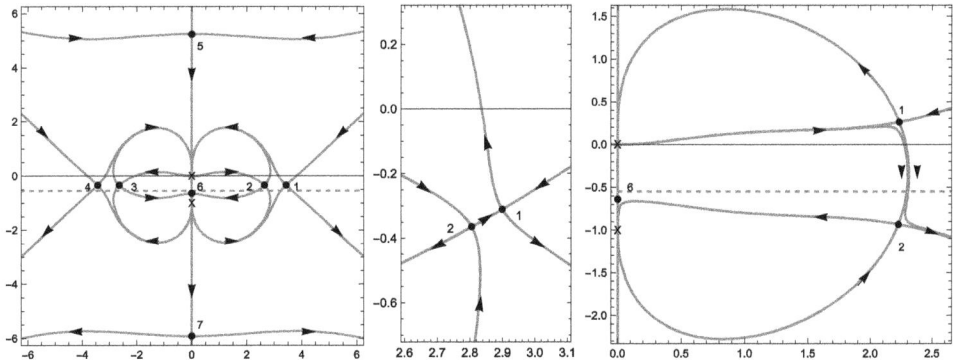

Figure 7.17 Saddle points and associated steepest ascent/descent contours for Bolt boundary conditions, for the action (7.93), in the complex plane of the lapse function. Crosses indicate singularities, dots saddles, and arrows downwards flow. The left panel shows an isotropic example ($p_1 = q_1 = 10$, $\Lambda = 3$), and the right panel one with smaller $p_1 = 3$. The middle panel shows the cross-over between these two regimes, at $p_1 \approx 7.2798$. This is described by a Stokes phenomenon, where the downwards flow from saddle 2 simultaneously becomes the upwards flow from saddle 1. The dashed line is located parallel to the real $\tilde{N}$ line in the lower half plane, and represents the required integration contour. The middle and right panels show only the relevant regions. Note that in the right panel, saddles 1 and 2 are not connected by steepest ascent or descent lines, they just happen to evolve close to each other in some locations.

can see that there is a crossover in likelihoods, with the transition between dominance of the two saddles taking place when p_1 is a little smaller than q_1. Also, the weighting diverges at small p_1 for saddle number 1. Thus, the predictions depend rather crucially on which saddles are picked up in the path integral.

The thimbles are shown in Fig. 7.17, both for q_1 fixed and various values of p_1. Let us first start with the case where the scale factors are equal – see the left panel in the figure. The obvious contour of interest is the one running parallel to the real $\tilde{N}$ line, in between the two singularities (in fact, for convergence, the contour must run below $\tilde{N} = -i/\Lambda$). This contour can be deformed into a sum of the four thimbles associated with saddles 1 to 4, and thus all four saddle points are picked up. The ones with the highest weighting will dominate the path integral, in this case saddles 2 and 3.

Now comes a crucial point: If it remains true that all four complex saddles contribute to the path integral, then at small p_1 (for fixed q_1) saddles 1 and 4 will dominate, and in fact their weighting grows without bound at small p_1 – cf. again Fig. 7.15. This would favor highly anisotropic configurations, and might well render the wave function nonnormalizable – we would have to conclude that the no-boundary wave function is either in conflict with observations or simply does not exist. However, a topological change in the steepest ascent/descent paths occurs as p_1 shrinks below a critical value (in our numerical example, for $p_1 \approx 7.2798$), which is near the point where the dominance of the two saddles switches. This is an example of a Stokes phenomenon.

At the critical p_1 value, the steepest descent contour from saddle 2 coincides with the steepest ascent contour passing through saddle 1 – i.e., the actions at both saddles have the same real part (and similarly for saddles 3 and 4), though the weightings of saddles 2 and 3 are still higher than those of saddles 1 and 4. Below this critical value, the thimbles of saddles 2 and 3 run directly to infinity, and the integration contour for the lapse that we considered before can now simply be rewritten as the sum of the thimbles associated with the saddles 2 and 3. In other words, saddles 1 and 4 do not contribute anymore (their steepest ascent paths do not cross the integration contour – see the right panel in Fig. 7.17) after this Stokes phenomenon has occurred. At small p_1, the NUT geometry will then dominate – see Fig. 7.16 – thereby ensuring the consistency of the model.

We can draw three lessons from this model: First, it verifies at the nonlinear level that the no-boundary proposal assigns the highest weighting to isotropic configurations, and thus provides an encouraging verification of the minisuperspace approach.[7] Second, it shows that the gravitational path integral must include a sum over topologies, enacted here by a sum over inequivalent no-boundary conditions. And third, a proper definition of integration contours is crucial for consistency, as exemplified by the Stokes phenomenon occurring in this model.

7.3 Further developments

7.3.1 Quantum gravity and higher curvature corrections

When gravity is quantized, effective terms with higher derivatives are generated from graviton loops at higher orders in $\hbar$. These terms typically arrange themselves as higher powers of the Riemann tensor and derivatives thereof. A pertinent question, even when sticking to an effective 4-dimensional description of gravity, is thus whether no-boundary solutions are still acceptable (regular, finite action) solutions of the corrected theory. In other words, are

---

[7] If we had assigned the opposite signs in the momentum conditions termed choices 2 and 3, then anisotropic configurations would have come out as preferred, thereby demonstrating at the nonlinear level that the tunneling wave function is unstable.

no-boundary solutions robust to the inclusion of quantum corrections? In string theory, such higher-order terms also arise, now accompanied by coefficients that contain powers of the string tension α'. Thus, the analogous question arises there too.

The most crucial aspect of this question is to see whether the rounded-off region near the South Pole of no-boundary instantons remains a good solution. Since perturbations go to zero near the South Pole, it is in fact sufficient to focus on homogeneous and isotropic backgrounds. In that case, i.e., for metrics of the standard closed FLRW form, $\mathrm{d}s^2 = -N^2 \mathrm{d}t^2 + a^2 \mathrm{d}\Omega_3^2$, the Riemann tensor depends on only two combinations, namely

$$A_1 \equiv \frac{\dot{a}^2 + N^2}{a^2 N^2}, \qquad A_2 \equiv \frac{\ddot{a}}{aN^2}. \qquad (7.94)$$

An action that is a function of the Riemann tensor may be expanded in terms of these combinations as

$$S = \int \mathrm{d}^4 x \sqrt{-g}\, f(\text{Riemann}) = 2\pi^2 \int \mathrm{d}t N a^3 \sum_{p_1, p_2} c_{p_1, p_2} A_1^{p_1} A_2^{p_2}, \qquad (7.95)$$

where c_{p_1, p_2} are coefficients, and the power of the Riemann terms is given by $P = p_1 + p_2$. The constraint equation can be found by taking a derivative of the action with respect to the lapse function, with the result

$$0 = \frac{\delta S}{\delta N} = 2\pi^2 \sum_{p_1, p_2} c_{p_1, p_2} \left[2p_1(p_2-1)\frac{a\dot{a}^2}{N^2} A_2^{p_2} A_1^{p_1-1} + (1-p_2) a^3 A_2^{p_2} A_1^{p_1} \right.$$
$$\left. + p_2(p_2-1)\frac{a\dot{a}a^{(3)}}{N^4} A_2^{p_2-2} A_1^{p_1} - p_2(2p_1 + p_2 - 3)\frac{a\dot{a}^2}{N^2} A_2^{p_2-1} A_1^{p_1} \right]. \qquad (7.96)$$

The equation of motion for the scale factor follows from taking a time derivative of the constraint, and hence we do not need to consider it separately. The constraint is in fact invariant under the following transformation,

$$\begin{cases} t \to -t, \\ a \to -a, \end{cases} \Rightarrow \quad a(-t) = -a(t), \qquad (7.97)$$

which implies that the scale factor is odd in t. The no-boundary ansatz, with its Euclidean rounding off at $a = 0$, then corresponds to a series expansion

$$\begin{cases} a(t) = a_1 t + \dfrac{a_3}{6} t^3 + \dfrac{a_5}{120} t^5 + O(t^7); \\ a_1^2 = -N^2. \end{cases} \qquad (7.98)$$

In physical time, we have $a_1^2 = -1$, i.e., $a_1 = \pm i$. For any solution, there exists a time-reversed solution obtained by sending $t \to -t$, and moreover the solutions come in complex conjugate pairs. Hence there are always four related solutions (two of which can be eliminated by fixing the initial expansion rate, as we have seen). With this ansatz, the components of the Riemann tensor read

$$\begin{cases} A_1 = -\dfrac{a_3}{a_1^3} + \dfrac{(a_3^2 - a_1 a_5)}{12 a_1^4} t^2 + \dfrac{(a_3 a_5 - a_1 a_7)}{360 a_1^4} t^4 + O(t^6); \\ A_2 = -\dfrac{a_3}{a_1^3} + \dfrac{(a_3^2 - a_1 a_5)}{6 a_1^4} t^2 - \dfrac{(10 a_3^3 - 13 a_1 a_3 a_5 + 3 a_1^2 a_7)}{360 a_1^5} t^4 + O(t^6). \end{cases}$$

(7.99)

Importantly, the series expansions start at order t^0 and not t^{-2}, as naively expected, and this is the main reason why no-boundary solutions can have finite action. Solving the constraint equation leads to the following expressions at the leading orders,

$$\mathcal{O}(t) : \quad \sum_{p_1, p_2} \frac{c_{p_1, p_2}}{N^{2P}} a_1^{4-P} a_3^{P-1} (p_2 - p_1) = 0, \qquad (7.100)$$

$$\mathcal{O}(t^3) : \quad \sum_{p_1, p_2} \frac{c_{p_1, p_2}}{N^{2P}} a_1^{3-P} a_3^{P-2} \left(a_3^2 \cdot G_3[p_1, p_2] + a_1 a_5 \cdot G_5[p_1, p_2] \right) = 0, \qquad (7.101)$$

with

$$G_3[p_1, p_2] = \frac{1}{6} \left(p_1^2 - 15 p_1 + 6 - 4 p_2^2 + 12 p_2 \right), \qquad (7.102)$$

$$G_5[p_1, p_2] = \frac{p_1(1 - p_1)}{6} - \frac{2 p_2(1 - p_2)}{3}. \qquad (7.103)$$

Again it is a consequence of the no-boundary ansatz that there are no non-trivial conditions at negative powers of t. The order t condition is most easily solved by $c_{p_1, p_2} = c_{p_2, p_1}$, which turns out to be satisfied quite generally, in particular for all terms that are powers of the Ricci scalar $R = 6(A_1 + A_2)$ and all terms involving the quadratic combinations $R_{\mu\nu\rho\sigma} R^{\mu\nu\rho\sigma} = 6 \left(A_1^2 + A_2^2 \right)$ and $R_{\mu\nu} R^{\mu\nu} = 12 \left(A_1^2 + A_1 A_2 + A_2^2 \right)$. The condition at order t^3 then fixes a_5 in terms of a_3,

$$a_5 \cdot \sum_{p_1, p_2} \frac{c_{p_1, p_2}}{N^{2P}} a_1^{3-P} a_3^{P-2} G_5[p_1, p_2] = -\frac{a_3^2}{a_1} \cdot \sum_{p_1, p_2} \frac{c_{p_1, p_2}}{N^{2P}} a_1^{3-P} a_3^{P-2} G_3[p_1, p_2]. \qquad (7.104)$$

At higher orders, the next coefficients a_7, a_9, etc. are then fixed in turn. Thus, a full solution, with manifestly finite action, is obtained, with free parameter a_3. This parameter specifies the early expansion rate and, in the case where a scalar field is added, is linked to the initial value of the scalar field.

In string theory, additional terms appear, in particular with derivatives acting on Riemann tensors. It is not possible to analyze such terms in full generality, but it turns out that the first few correction terms, both in heterotic and type IIB string theory, still admit no-boundary solutions with finite action. This suggests a basic compatibility of string theory with the no-boundary proposal.

7.3.2 Allowable metrics

The usefulness of complex metrics became apparent through the study of complex black hole metrics, which provide the quickest way of deriving (and to some extent, understanding) the thermodynamic properties of black holes, as we saw in Chapter 6. However, not all complexified metrics make physical sense.

E. Witten gave a simple counterexample: Start with flat space in polar coordinates $ds^2 = dR^2 + R^2 d\Omega_3^2$ and promote $R \to R(u)$, for a real parameter u. Now if $R(u)$ interpolates between asymptotic regions $R \to \pm\infty$ while avoiding $R = 0$ by passing around the origin in the complexified R plane, then this solution interpolates between two asymptotically flat regions of spacetime and describes a complex wormhole in between – see Fig. 7.18. But since the metric is simply obtained by a coordinate change from flat space, the Ricci curvature remains zero and so does the action. If such solutions were allowed, they would have the same likelihood of occurring as classical evolution (which also proceeds without suppression, with zero weighting). Such wormholes should thus surround us everywhere. Since this is manifestly not the case, there must be some criterion that eliminates these from gravitational path integrals.

In order to review a promising criterion that has been suggested, let us note first that locally, i.e., at any point, one can write the metric in diagonal form by finding suitable variables,

$$g_{\mu\nu} = \delta_{\mu\nu} \lambda_\mu. \tag{7.105}$$

We will allow the diagonal elements to be complex, $\lambda_\mu \in \mathbb{C}$. Note that when all λ_μ are real and positive, the metric is Euclidean.

J. Louko and R. Sorkin analyzed the question of allowability in a simplified 2-dimensional context, where they proposed to allow complex metrics only when they admit a well-defined (that is to say, convergent) scalar field theory on them. In fact, even a simple mass term already illustrates the concept.

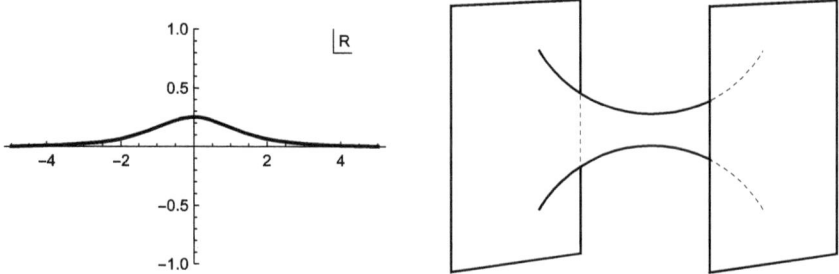

Figure 7.18 A complex wormhole with zero action. *Left panel:* the deformed path of the radial coordinate R. *Right panel:* The wormhole metric is complex and links two flat regions of spacetime.

So, assume that the theory we study contains a massive scalar field, with mass term

$$\int \cdots \int \mathcal{D}\phi e^{-\int d^4 x \sqrt{g}\frac{1}{2}m^2\phi^2} \cdots \qquad (7.106)$$

In order for the theory to be well defined, one would require the path integral over ϕ to be convergent. If the integration domain is over real values $\phi \in \mathbb{R}$ then convergence is assured only if we require

$$\text{Re}[\sqrt{g}] > 0 \quad \rightarrow \quad -\frac{\pi}{2} < \text{Arg}[\sqrt{g}] < \frac{\pi}{2}$$
$$\rightarrow \quad -\pi < \sum_\mu \text{Arg}[\lambda_\mu] < \pi, \qquad (7.107)$$

which must be satisfied at every point. If this inequality is not satisfied, the path integral diverges and the theory is quite simply not properly defined. One could say that a violation of the inequality implies the presence of an instability, as if the field had a negative mass squared (this is usually called a tachyonic instability).

M. Kontsevich and G. Segal have generalized this criterion to p-forms, i.e., fields that are represented by antisymmetric tensors. Requiring the integral over their kinetic terms to be convergent leads to the analogous requirement

$$|e^{\frac{i}{\hbar}S}| < 1 \text{ or } |e^{-\frac{1}{\hbar}S_E}| < 1, \text{ implying}$$
$$\text{Re}\left[\sqrt{g}g^{\mu_1\nu_1}\cdots g^{\mu_{p+1}\nu_{p+1}}F_{\mu_1\cdots\mu_{p+1}}F_{\nu_1\cdots\nu_{p+1}}\right] > 0, \qquad (7.108)$$

where the fields strengths $F_{\mu_1\cdots\mu_{p+1}}$ are antisymmetric and real-valued. If one writes out all implied inequalities, for all p, on the arguments of the λ_μ diagonal metric elements, one finds that all conditions are simultaneously satisfied if the following inequality is satisfied,

$$\Sigma \equiv \sum_\mu |\mathrm{Arg}[\lambda_\mu]| < \pi. \tag{7.109}$$

A few comments: Why should one define the matter integrals to integrate over real values? For one, path integrals over p-forms with $p \geq 1$ contain sums over quantized integer fluxes (sometimes called monopole numbers, or instanton numbers), which cannot be analytically continued. But the main reason is that real integrals are required to define matter states in Hilbert space via path integrals. A Euclidean transition amplitude, $\Psi[0 \to 1] = \langle 1|e^{-\sigma \hat{H}}|0\rangle$ can be written as a path integral fixed at its two ends. Then a quantum state, e.g., the evolved ket $|\psi_0\rangle = e^{-\sigma \hat{H}}|0\rangle$, can be viewed as a path integral fixed at only one of its ends. If the integration is over real field values, then this can be defined a priori. By contrast, if one were to define such integrals on thimbles, then this would introduce an element of nonlocality, as the thimbles depend both on the initial and final conditions (as we have seen in explicit examples) and thus one could not define Hilbert space states as stand-alone entities. Note that we care about the matter states, as this is what we observe. We never actually observe gravitational degrees of freedom directly (even gravitational waves are inferred from the behavior of laser interferometers).

Why is this criterion constructed using p-form matter? This is simply because the Weinberg–Witten theorem implies that these are the most general forms of matter that have local covariant stress–energy tensors. The idea then is that on allowable backgrounds, one may define general quantum field theories. One additional argument in its favor is that for allowable metrics, the value of topological invariants such as the Gauss–Bonnet integral are invariant, as expected. For nonallowable metrics this property can be violated.

The arguments just provided underline that the condition (7.109) is a reasonable one to try out. However, it should be emphasized that it was derived using mainly quantum field theory considerations – quantum gravity may ultimately provide a deeper criterion. For now, let us note that the bound (7.109) implies that metrics are indeed allowed to become complex, but not too much. In fact, a Lorentzian metric with one time direction ($\mathrm{Arg}[\lambda_0] = \pi$) already has $\Sigma = \pi$ and thus lies on the boundary of the allowed domain. This can be straightforwardly regularized – for instance, one may deform an FLRW metric to

$$\mathrm{d}s^2 = -(1 \mp i\epsilon)\mathrm{d}t^2 + a(t)^2 \mathrm{d}\Omega^2 \tag{7.110}$$

for a small real number ϵ. In such a case, one must stick to a particular sign of ϵ, and cannot cross from positive to negative ϵ, or vice versa. Another immediate consequence is that metrics with two time directions are ruled out.

Most importantly, the allowability criterion (7.109) rules out known pathological metrics. For the complex wormholes discussed above, this is easy to see: When the curve $R(u)$ crosses the imaginary axis, then momentarily R^2 turns negative, and hence each spatial direction contributes $\mathrm{Arg}[R^2] = \pi$ to the sum. Thus $\Sigma \geq (D-1)\pi$, if we are in a D-dimensional spacetime, and the inequality is drastically violated. We will see other examples of disallowed metrics momentarily.

In the context of this chapter, the interesting point is that the allowability criterion restricts no-boundary solutions. It is instructive to start with the pure cosmological constant case, where the saddle points are sections of complexified de Sitter space. In Section 7.2 we obtained the saddle points in a gauge where the lapse is constant, cf. Eq. (7.73). One can transform the associated solutions to ordinary (nonrescaled) Euclidean time by defining $\frac{\tilde{N}}{\sqrt{q}}dt_q \equiv -idt_E$, which leads to

$$\sqrt{\frac{\Lambda}{3}}\, t_E(t_q) = 2i\,\mathrm{arsinh}\left(\sqrt{\frac{\Lambda \tilde{N} t_q}{6i}}\right). \quad (7.111)$$

The left panel of Fig. 7.19 shows the resulting path (dashed line) in the complex σ plane. Meanwhile, the right panel in the figure shows the corresponding sum Σ of absolute values of arguments of the metric components, as defined in (7.109). What may come as a surprise is that the bound is seen to be violated. Thus, in the constant lapse gauge, the saddle points appear to violate the allowability bound (7.109). However, as we discussed previously, the action, and thus also the physical consequences, are unchanged when the path is deformed. In fact, the original Euclidean-plus-Lorentzian contour gives $\Sigma = 0$ along the Euclidean segment, and $\Sigma = \pi$ along the Lorentzian one, and thus implies that no-boundary saddle points actually saturate the allowability bound. But one may worry that this contour is nonsmooth. Let us therefore consider a smooth family of paths, obtained by specifying $t_E = \theta(t_q)$ with

$$\theta(t_q) = \frac{\pi}{2} - \frac{\pi}{2}(1-t_q)^n + \left(t_E(1) - \frac{\pi}{2}\right) t_q^n, \quad 0 \leq t_q \leq 1. \quad (7.112)$$

In Fig. 7.19 an example with $n = 3$ is plotted (black line) in the left panel.

7.3 Further developments 243

There we can see that now the bound is indeed satisfied and saturated only at the end point – see also the black line in the right panel of the figure.

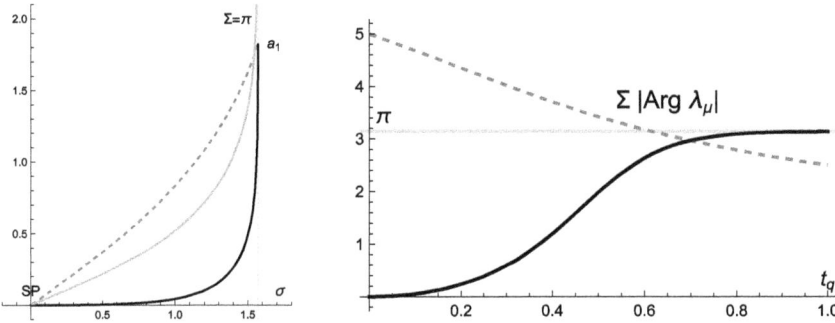

Figure 7.19 *Left panel:* a no-boundary saddle-point solution, interpolating between the South Pole (SP) and a final scale factor value $a_1 = \sqrt{10}$, with $\Lambda = 3$, in the complexified time plane. The horizontal axis corresponds to Euclidean physical time σ, and the vertical axis indicates the Lorentzian time direction. The dashed line corresponds to the path taken in the constant lapse gauge, Eq. (7.73). The black line corresponds to the $n = 3$ path defined in Eq. (7.112). Meanwhile, the gray curve is the "maximal" curve, defined via $\Sigma = \pi$ and lying at the edge of the domain of allowability. *Right panel:* the sum of arguments Σ, as defined in (7.109), along the same paths.

A useful concept here is that of the *maximal curve*, which represents the curve that saturates the bound (7.109), i.e., it is the curve defined by $\Sigma = \pi$. We can find this curve as follows: Let us parametrize a curve $\sigma(u)$ in the complexified Euclidean time plane with a real parameter u. Then the metric along this curve is given by

$$ds^2 = \sigma'^2 du^2 + [a(\sigma(u))]^2 d\Omega_3^2, \tag{7.113}$$

where we keep the scale factor general for now. The maximal curve is then defined via

$$\text{Arg}[\sigma'_x(u)] + 3\,\text{Arg}[a(\sigma_x(u))] = \frac{\pi}{2}, \tag{7.114}$$

where it turns out that in the no-boundary examples of interest we can drop the absolute value signs on the arguments. In other words, for simple no-boundary metrics the maximal curve is specified by (7.107) because the arguments in question turn out to all be positive. This equation can be solved by stipulating

$$\sigma'_x(u) \cdot [a(\sigma_x(u))]^3 = i. \tag{7.115}$$

Other, equally suitable, solutions may be used just as well. In a specific background, one can then integrate the relation above and find the maximal curve. For a solution to be allowable, a path must exist that links the South Pole to the end point while staying below the maximal curve at all times, so that the inequality (7.107) remains satisfied throughout.

For pure de Sitter space, we have $a(\sigma) = \frac{1}{H}\sin(H\sigma)$, and the integral of (7.115) yields

$$\frac{1}{12H^4}\cos(3H\sigma_x(u)) - \frac{3}{4H^4}\cos(H\sigma_x(u)) = iu - \frac{2}{3}, \quad (7.116)$$

which implicitly defines the maximal curve. As one can see in the left panel of Fig. 7.19, the end point marked a_1 is indeed located below the maximal curve (drawn in gray). In fact, the maximal curve asymptotically approaches the line $\mathrm{Re}(\sigma) = \frac{\pi}{2H}$, on which the scale factor takes real values, without crossing it. Thus, we conclude that pure de Sitter no-boundary solutions always satisfy the allowability criterion (7.109).

As should be clear from the preceding discussion, the allowability bound is only narrowly satisfied at large scale factor values. And indeed, as soon as we generalize our treatment to inflationary models with a nonzero slow-roll parameter ϵ, nontrivial conditions emerge. To this end, we have to generalize the treatment above to include a scalar field and its potential. Along a path $\sigma(u)$, the metric is still given by (7.113). But the equations of motion now read

$$a'' - \frac{\sigma''}{\sigma'}a' + \frac{a}{3}\left(\phi'^2 + \sigma'^2 V(\phi)\right) = 0, \quad (7.117a)$$

$$\phi'' - \frac{\sigma''}{\sigma'}\phi' + 3\frac{a'}{a}\phi' - \sigma'^2 V_{,\phi} = 0, \quad (7.117b)$$

where we have used the chain rule. The maximal curve σ_x, defined so as to satisfy $\Sigma = \pi$, is once more specified by (7.115). This time, in view of the fact that the equations below have to be solved numerically and that the scale factor becomes large during inflation, it is advantageous to use the following solution to (7.115),

$$\sigma' = i\left(\frac{a^*}{a}\right)^{3/2}. \quad (7.118)$$

By taking a derivative we obtain

$$\frac{\sigma''}{\sigma'} = \frac{3}{2}\left(\frac{a^{*\prime}}{a^*} - \frac{a'}{a}\right), \quad (7.119)$$

7.3 Further developments

so that the equations of motion end up being given by

$$a'' - \frac{3}{2}\left(\frac{a^{*\prime}}{a^*} - \frac{a'}{a}\right)a' + \frac{a}{3}\left(\phi'^2 - \left(\frac{a^*}{a}\right)^3 V(\phi)\right) = 0, \quad (7.120a)$$

$$\phi'' - \frac{3}{2}\left(\frac{a^{*\prime}}{a^*} - \frac{a'}{a}\right)\phi' + 3\frac{a'}{a}\phi' + \left(\frac{a^*}{a}\right)^3 V_{,\phi} = 0. \quad (7.120b)$$

These are the equations that must be solved, starting from the South Pole and up to a suitably large final point. If the end point σ_1 of the no-boundary solution remains below the maximal curve, the solution is allowed, and otherwise not. Experience shows that a high level of numerical precision is required, due to the exponential growth of the scale factor (one must retain about $3\mathcal{N}$ decimal places for $\mathcal{N}$ e-folds of expansion).

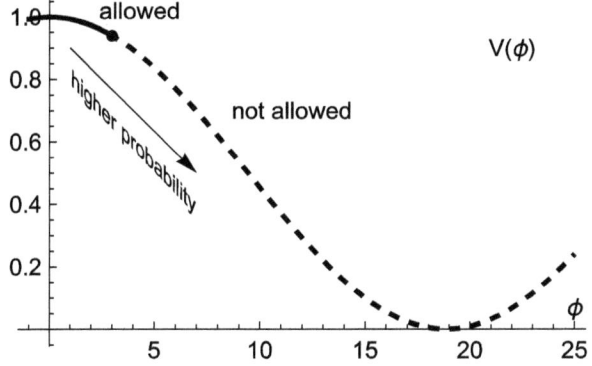

Figure 7.20 An example of a natural inflation potential $V(\phi) = V_0(1 + \cos(\phi/f))$. The only no-boundary instantons that are allowed are those that start high enough on the potential. This leads to the requirement of a minimum number of inflationary e-folds, as shown in Fig. 7.21. Since the no-boundary weighting favors lower values of the potential, the lowest allowed value (indicated by a small dot) is also the most likely starting point for inflation.

As an example, we will consider potentials of the "natural inflation" form

$$V(\phi) = V_0\left(1 + \cos(\phi/f)\right), \quad (7.121)$$

where f is a constant – see Fig. 7.20. Such potentials are of plateau type, which, as we saw in Chapter 3, is the shape preferred by current CMB observations. What one finds is that, in order for a no-boundary solution to be allowable, the scalar field must start in a sufficiently flat region of the potential. This is because a steeper potential also renders the solution more complex, leading to a violation of (7.109). But the flat part of the potential is near the top, hence requiring a solution to be allowable translates into a minimum number of inflationary e-folds.[8] The corresponding numerical results, for different values of the constant f, are shown in Fig. 7.21. What is interesting is that the minimal required number comes out as being a few dozen, which is roughly in the range required to resolve the flatness puzzle. This is quite remarkable, given that a priori it could have been much smaller or much larger. Note that the no-boundary wave function assigns a higher weighting to lower values of the potential, cf. (7.57), so that the minimum allowable number of e-folds is also the most likely number predicted by the no-boundary wave function.

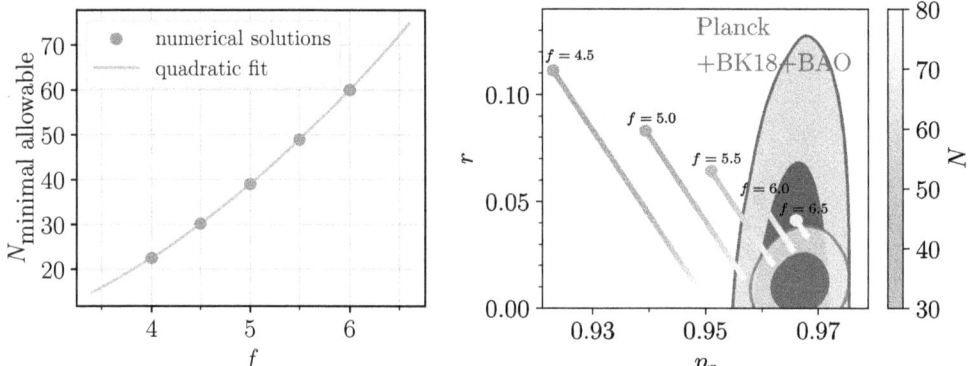

Figure 7.21 *Left panel:* minimal number of e-folds required in order for no-boundary solutions to satisfy the allowability bound (7.109), for potentials of the form $V = V_0(1 + \cos(\phi/f))$ and various f values. *Right panel:* comparison of the predictions of natural inflation models with CMB observations. A good agreement is found for sufficiently flat potentials, with $f \gtrsim 6$, and for such potentials the allowability bound implies a number of e-folds in excess of ≈ 60. Reproduced from Lehners and Quintin (2024). Published by Elsevier B.V. Funded by $SCOAP^3$ (CC-BY).

What is even more interesting is to compare these values with CMB observations. The right panel in Fig. 7.21 provides such a comparison. It shows the predictions of the spectral index n_s and the tensor-to-scalar ratio r for models of natural inflation with different f and different numbers of e-folds $\mathcal{N}$, and compared to the 68% and 95% confidence limits of the observations of the Planck satellite alone (the larger blobs in the figure), and of the observations of Planck combined with observations of the BICEP-Keck experiments as well as observations of baryon acoustic oscillations (smaller blobs). What one can see is that a value of $f \gtrsim 6$ is required in order

---

[8] For completeness: There is a tiny allowable region close to the potential minimum, which yields only $\mathcal{O}(1)$ e-folds of inflation, and does not lead to a realistic universe.

7.3 Further developments

to match observations. But for $f \geq 6$ we also find the requirement that $\mathcal{N} \gtrsim 60$. Hence for models that are in rough agreement with observations, the no-boundary wave function implies a minimum length of inflation that is sufficient to explain the flatness of the current universe, and that is sufficiently long in order to amplify primordial perturbations on CMB scales. Analogous results are found to hold for other plateau models of inflation.

Since the no-boundary wave function favors short inflationary phases, it also implies that the universe is expected to be rather small, relatively speaking. It implies that inflation did not last thousands or millions of e-folds, but significantly less. In that case, the size of the universe could be not much larger than the currently observable part. How much larger depends on the details of the potential, and thus it is unclear whether one could hope to observe the positive spatial curvature implied by the no-boundary wave function, or if we should expect that it has been diluted to the point where it becomes unobservable. Such detailed predictions will depend on a better understanding of the scalar potential.

Before proceeding, let us point out that the allowability bound rules out the pathological "necklace" no-boundary instantons that are obtained by adding additional spheres "below" the South Pole of the solutions – see Fig. 7.4 and the surrounding discussion. We had already discarded such saddles on the basis that, when adding perturbations, they develop curvature singularities. However, the allowability bound already rules them out at the background level. One can imagine regularizing such solutions by passing around the zeros of the scale factor at $\sigma = (2n+1)\pi/(2\mathsf{H})$. But whenever the time contour passes by such a zero, the scale factor necessarily becomes imaginary (on the vertical/Lorentzian line emanating from the zero), and thus, just as for the complex wormhole, we obtain $\Sigma \geq (D-1)\pi$ and a violation of (7.109).

So far, we have discussed the allowability of saddle-point solutions. However, with regard to the definition of path integrals, it is also of interest to see how off-shell configurations fare in light of the bound (7.109). We cannot say much about general metrics, but the symmetry-restricted minisuperspace examples are more tractable. Specifically, when we look at metrics where the scale factors already solve their equations of motion, and only the lapse integral remains to be carried out, we can analyze the allowability of the geometries. This can be done by similar numerical techniques to those described above.

The results for no-boundary integrals, both with a Neumann initial condition and for a Dirichlet initial condition, are shown in Fig. 7.22. Both

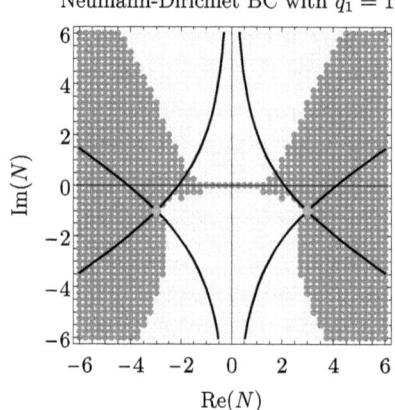 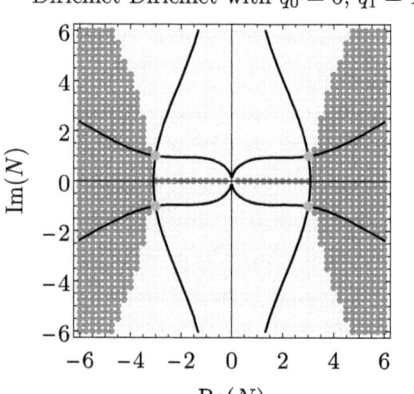

Figure 7.22 Allowable metrics are shown in light gray in the plane of the complex lapse function; disallowed metrics are shown in dark gray. *Left panel:* with an initial Neumann condition – compare with Fig. 7.12. *Right panel:* with an initial Dirichlet condition – compare with Fig. 5.5. The saddle points and steepest descent contours are also shown. Here $\Lambda = 3, q_1 = 10$. Figure reproduced from Jonas et al. (2022) *(CC-BY 4.0)*.

cases share the characteristic that the real lapse line constitutes a boundary that cannot be traversed. And in both cases, the steepest descent contours run into regions that are not allowed. This means that, if the allowability criterion is strictly enforced, then we can no longer define the sums over metrics along full thimbles. This might indicate that minisuperspace is too restrictive a framework for properly defining gravitational path integrals. Or it might simply mean that when matter is actually added, the integration contours need to be changed.

One interesting feature in the Neumann case is that, as seen in the plane of the lapse function, the saddle points reside right at the edge of the allowable domain. This can be understood analytically: For the path integral with Neumann initial conditions, the initial size of the geometries is not fixed. In fact, this initial size can become complex off-shell, and by itself it can cause the allowability bound to be violated. To see this, perturb around the saddle point by writing $\tilde{N} = \tilde{N}_\pm + \Delta$, and work to linear order in Δ. Then from (7.71) we obtain

$$q(0) \approx \pm 2\Delta \left(\frac{\Lambda}{3}q_1 - 1\right)^{1/2}, \qquad (7.122)$$

where we assume that $q_1 > \frac{3}{\Lambda}$. This implies that for $\tilde{N}_-$ we get the condition $3|\text{Arg}[\Delta]| < \pi$. Consequently, starting from the saddle point, the allowed directions are limited to $-\frac{\pi}{3} < \text{Arg}[\Delta] < \frac{\pi}{3}$. For $\tilde{N}_+$ one analogously finds $\frac{2\pi}{3} < \text{Arg}[\Delta] < \frac{4\pi}{3}$. Hence the saddle points, even though they are fully complex, are at the edge of the allowed domain, and thus the steepest descent contours are cut in "half" by the allowability criterion, with only the lower portion remaining.

For the case with Dirichlet initial conditions, the fact that the upper half plane gets separated from the lower half plane of the lapse function may be interesting. As we saw in Section 7.1.2, the unstable saddle points are in the upper half plane, while the stable ones are in the lower half plane. Therefore, if one is not allowed to cross the real lapse line, then a definition of the integral in terms of exclusively stable metrics becomes a possibility, if one can find suitable completions of the integration contours.

7.4 Postdictions, predictions, and discussion

The Hartle–Hawking no-boundary proposal is a prescription for describing and calculating the creation of space and time out of nothing, i.e., for explaining the creation of the universe. This is a grand goal, and thus we should critically assess its status. What are its main features? What can it explain/postdict? What does it predict? What are its shortcomings and the associated open questions? Let us make a list:

- The no-boundary proposal specifies a single quantum state for the entire universe, and as such provides a theory of (probabilistic) initial conditions.
- In fact, it provides a *unification of quantum gravity dynamics and initial conditions*, best expressed in the path integral approach to quantum theory, heuristically

$$\Psi(a_1, \phi_1) = \int_\mathcal{C}^{a_1, \phi_1} e^{\frac{i}{\hbar} S}. \tag{7.123}$$

The dynamics is encoded in the action S, and the state in the boundary conditions and integration ranges $\mathcal{C}$ of the integral. Thus a single mathematical object combines the dynamics and the specification of the state. This is why it is so important to understand the proper definition of gravitational path integrals. Moreover, the wave function depends on the final field values a_1, ϕ_1, and this dependence of the wave function on spatial hypersurfaces in some sense makes the integral also an intrinsically holographic object.

- The no-boundary proposal incorporates the idea of containment, namely that the universe is fully enclosed in itself, both in space and in time, without requiring any outside conditions.

- The no-boundary wave function explains how space and time became classical at the early stages of evolution of the universe. This classicalization moreover requires the presence of a dynamical attractor, such as inflation. The attractor drives spacetime to classicality, and allows for the definition of probabilities for different classical histories of the universe. This also provides a raison d'être for a primordial dynamical attractor.
- Once spacetime has become classical, one naturally obtains the framework of quantum field theory in curved spacetime. More details about how this happens will be discussed in Chapter 8.
- The big bang singularity is resolved. No-boundary instantons are finite and regular, and simply do not contain a big bang singularity. We can only say things with confidence about the universe in the regime where spacetime has already become effectively classical. Going back in time, there is no operational way of saying something about the phase when the radius of the universe was smaller than the primordial Hubble scale. The latter phase is better thought of as a quantum tunneling/nucleation event, which was required to produce space and time in the first place.
- Two immediate implications of the no-boundary idea are that the universe is finite in spatial extent, and that the overall average spatial curvature is positive. However, the spatial curvature might have been diluted to such an extent as not to be measurable today.
- The no-boundary proposal implies that quantum fields were in their vacua initially. This explains their initial conditions, and in particular offers an explanation for the Bunch–Davies vacuum of primordial cosmological perturbations. The dynamics of the universe may then excite certain fields, and a process like reheating may fill the universe with radiation and other matter.
- Homogeneous and isotropic initial conditions are favored over less symmetric ones. The initial state of the universe is thus one of relative simplicity. Combined with the previous item stating that quantum fields start out in their vacua, this provides appropriate low-entropy initial conditions for the universe. This also provides an appropriate starting point for the second law of thermodynamics.
- The no-boundary proposal appears to be consistent with the higher curvature correction terms that are expected to arise due to the nonrenormalizability of gravity, as long as the curvature of the saddle points remains sufficiently far below the Planck scale.
- Assuming the presence of an inflationary scalar field with a potential, the detailed predictions for observations of the CMB depend on the shape of the potential. This unfortunately makes it difficult to test the no-boundary

proposal. If all complex metrics are included, then the no-boundary proposal predicts that the inflationary phase should be as short as possible, in conflict with observations. However, if the allowability bound (7.109) or even the milder version (7.107) are correct, then an inflationary phase is required to start in a very flat region of the potential. Many large-field models of inflation are then ruled out straight off, since they are nowhere flat enough. For plateau models, this imposes a minimal number of e-folds. At least for phenomenologically well-motivated models, such as natural inflation or the closely related Starobinsky and Higgs inflation models, this leads to lower bounds on the number of e-folds that are sufficient for explaining the flatness puzzle, and for producing primordial fluctuations on the required scales. The no-boundary weighting implies that the lower bound on the number of e-folds is also the most likely one to have been actualized in the universe.

- The no-boundary prescription appears naturally in noncosmological settings too, and it is for example the appropriate wave function to consider in AdS spaces, as we saw in Section 6.5.

It is important to stress that, according to our current understanding, the no-boundary proposal is thus consistent with observations, and that it provides explanations for many aspects that were unexplained before, in particular the emergence of a classical spacetime from a quantum gravitational state.

That said, many open questions remain. Here are a few:

- How could we convince ourselves that the universe we observe arose in accordance with the no-boundary proposal, and not differently? In other words, is there a specific signature that would unambiguously point towards the no-boundary proposal? A detection of positive spatial curvature would certainly be strong corroborating evidence but, as we discussed, it may well be unobservably small. The problem really is that consistency with current observations is not enough, even though it is extremely nontrivial. It will be important to derive a true smoking gun observational signature.
- Can the allowability bound (7.109) be justified by a fundamental theory of quantum gravity, such as string theory? Or, if not, what replaces it? As we discussed, a restriction on complex metrics is certainly necessary, and plays an important role in assessing the observational consequences of the no-boundary proposal.
- Related to this point is the question of whether the integration contours, especially over the lapse function, can be determined from first principles. What kinds of off-shell geometries, with what kinds of singularities, should

be summed over? This question becomes especially pertinent when trying to go beyond the minisuperspace calculations we focused on here.
- When applying the no-boundary proposal to string theory, does one obtain new predictions for, or correlations between, particle physics and cosmology?
- The arguments of the wave function are field values on a 3-dimensional hypersurface. Can the no-boundary wave function be directly defined, holographically, on that hypersurface? Attempts are underway to achieve just that, though it seems fair to say that they are not fully developed yet. A basic idea is that cosmological evolution might then be mapped to renormalization group flow (i.e., a change in energy scale) in the dual, holographic theory (which would presumably not contain time at all).
- What is the meaning of the fact that the no-boundary wave function does not vanish when one sets the size of the universe to zero?
- Saddle points typically come in pairs, and may be thought of as time reverses of each other. Can there be interesting interference effects between these saddles when the universe is still very small? Could these lead to a specific no-boundary signature?
- For cosmological perturbations, an operational meaning of the probabilities we derived can readily be found. But what is the meaning of probability concerning the background, when we only get to observe a single universe? If there is a nonzero probability of creating a universe, then can/does such creation occur repeatedly, perhaps an infinite number of times, with all such universes evolving in their own, separate, spacetimes? What is the meaning of such a statement, if it remains unverifiable? Put differently, how can we verify the probabilistic nature of the no-boundary proposal, when we observe only one universe?

These points provide ample opportunities for future research.

Exercises

7.1 Estimate the South Pole value of the scalar field for no-boundary solutions in slow-roll models of inflation.

7.2 Reproduce the numerical instantons for constant equation of state, presented in Section 7.1.3. Verify the results shown for the optimized South Pole values ϕ_{SP} and for the approach to classicality. Hint: One can approximate the derivatives of the wave function by using finite differences – i.e., calculate the instantons also for slight changes in the final field values. To this end, it is useful to change the scalar field by a shift ($\phi_1 \to \phi_1 + \delta\phi_1$), and the scale factor by a fractional shift ($a_1 \to a_1(1 + \delta a_1)$), due to its significant growth.

7.3 In all minisuperspace implementations of the no-boundary proposal, we fixed the size of the universe on the final hypersurface. This is a useful, simple boundary condition – however, in astronomy we typically measure the Hubble rate rather than the size of the universe. Work out the no-boundary wave function, for gravity plus a cosmological constant, assuming spatial isotropy and homogeneity, with an initial Neumann condition and a final condition fixing the Hubble rate. Also analyze the thimbles and lapse integration contour. You may make use of Exercise 5.1.

7.4 For the potential of natural inflation (7.121), try to reproduce the numerical results for the minimum number of e-folds required for a no-boundary solution to be allowable, as shown in Fig. 7.21, using the techniques described in this chapter.

Another class of potentials that you may want to play with is of the form

$$V(\phi) = V_0 \left(1 - e^{-\sqrt{\frac{2}{3\alpha}}\phi}\right)^2, \qquad (7.124)$$

where α is a parameter. This class includes the effective potential of Higgs and Starobinsky inflation, (2.61), when $\alpha = 1$. You should find that for this class of potentials the minimum required number of e-folds for allowability is quite low, when α is near 1. Only for $\alpha > 93.9$ are more than 60 e-folds required for allowability. Beware: A high level of numerical precision is required for these calculations.

8
Interpreting the wave function

The wave function describes the entire universe, including the things one wants to observe, the apparatus used to observe these things (this could be your eye, or a measuring device, or in fact just a collection of atoms), and the rest of the universe. In this chapter we will discuss how, by subdividing the universe into various parts, the interactions (or noninteractions) between the different parts determine whether something ends up looking approximately classical or not. This property goes under the general name of decoherence. We will explain the concept with a simple example first, and then consider the cosmological setting.

8.1 Basic idea of decoherence

Let us start with an iconic quantum mechanics experiment, the double slit. Particles are emitted by a generator; there is a wall with two slits, and one measures the impact locations of the particles on a screen at some distance behind the wall – see also Fig. 5.1. Famously, the impact locations on the screen show interference fringes, even when the particles traverse the setup one at a time. This is interpreted as being due to the superposition of two particle paths, one through the left slit in the wall, and one through the right slit, so the system is described by the wave function

$$|\Psi_S\rangle = \frac{1}{\sqrt{2}} (|L\rangle + |R\rangle) . \tag{8.1}$$

If we form the corresponding density matrix,

$$\rho_S = |\Psi_S\rangle\langle\Psi_S| = \frac{1}{2} (|L\rangle\langle L| + |R\rangle\langle R| + |L\rangle\langle R| + |R\rangle\langle L|) , \tag{8.2}$$

then we can see that the off-diagonal terms are just as large as the diagonal ones, and this reflects the interference between the two particle paths.

8.1 Basic idea of decoherence

Now we imagine adding a detector that can measure whether the particle passed through the left or right slit. This detector will be in the state $|D_L\rangle$ if it measures the particle to have gone through left slit, and similarly for the right. The interaction between the particle and the detector turns the state into an entangled one,

$$|\Psi_S\rangle|D\rangle \quad \to \quad |\Psi_{SE}\rangle = \frac{1}{\sqrt{2}}\left(|L\rangle|D_L\rangle + |R\rangle|D_R\rangle\right). \tag{8.3}$$

If we now look at the density matrix again, but trace out the detector (labeled E for environment), then we find

$$\begin{aligned} \rho_S &= \text{Tr}_E\, \rho_{SE} \\ &= \text{Tr}_E\, |\Psi_{SE}\rangle\langle\Psi_{SE}| \\ &= \frac{1}{2}\left(|L\rangle\langle L| + |R\rangle\langle R| + \langle D_R|D_L\rangle\, |L\rangle\langle R| + \langle D_L|D_R\rangle\, |R\rangle\langle L|\right). \end{aligned} \tag{8.4}$$

If the measurement can detect the slit with high fidelity, then the detector states D_L and D_R are completely orthogonal, so $\langle D_R|D_L\rangle = 0$ and the off-diagonal terms disappear. Then we get a statistical mixture of two outcomes without any interference (50% probability for the particle to have gone through the left slit, and 50% to have gone through the right slit). The interaction of the particle with the environment (in the form of the detector) has decohered the state and eliminated the interference pattern. This is also exactly what is seen in actual experiments of this setup.

One may now envisage intermediate situations: If the measurement is not so effective, $0 < \langle D_R|D_L\rangle < 1$, then some interference remains, and if there is no measurement at all (so that left and right are effectively equivalent, $\langle D_R|D_L\rangle = 1$) then we recover perfect interference as before. Again, all of this agrees with actual experiments that were done on similar systems.

The main mathematical point is that when the off-diagonal terms in the density matrix become small, the state becomes a mixture of separate outcomes with a classical probability distribution.

Maybe we should clarify our aims somewhat: Normally, when learning about quantum mechanics, one tries to look precisely at situations where interference is manifest, since one wants to highlight quantum effects. But here the issue is almost the opposite: Under what circumstances do we *not* see quantum effects, such as interference? When do we see the world as behaving classically, and why? The answer is that the interactions between whatever we consider to be the system and the environment (which can be the region spatially surrounding the system, or physics occurring in the same region but mainly at different length/energy scales) reduce quantum correlations and imply that one sees definite outcomes. Another way to say it is that, due to the interactions, the environment collects records of the system and these records lend "objective" reality to the system (in the example above, it is the interaction with the measuring device that forces the particle to commit to the left or right slit, and at the same time eliminates the interference pattern, while in the absence of an interaction the interference pattern remains and we cannot tell which slit the particle passed through). We will try to make this a little more precise with equations below. The question that interests us here is: How did this happen in the early universe? How and when did the universe start behaving classically, even though we believe that it must have been described by a fully quantum state? And why do we see the primordial fluctuations, which we conjecture arose as quantum fluctuations, as classical temperature fluctuations?

Note that in quantum mechanics one may always perform a change of basis. This modifies the off-diagonal terms in the density matrix, hence an additional question is: In which basis do the off-diagonal terms become small, i.e., in which basis does the state decohere? To find an answer here, we should note that measurements represent interactions that are characterized by an interaction Hamiltonian. Now if the interaction depends on the position of a particle, say (rather than its momentum), then position states will commute with the interaction Hamiltonian and thus remain unchanged when the measurement is made. Such states are known as *pointer states*, and they constitute the preferred basis. Thus the interaction itself determines the basis in which the system is decohered. The fact that fundamental interactions depend on separation explains why we see objects by their positions in space, rather than their momenta, in everyday life.

8.2 Decoherence in the early universe

We will now apply these ideas to the wave function of the universe.

8.2.1 Structure of the wave function

As we saw, we may write the wave function in the early universe as a sum over saddle points, with products of perturbations associated with each saddle. The general structure of the wave function is thus

$$|\Psi_{\text{universal}}(a, \phi_n)\rangle = \sum_i \left(|B_i(a)\rangle \prod_n |F_n^i(a, \phi_n)\rangle \right), \quad (8.5)$$

with B denoting the background fields and F the fluctuations.

8.2 Decoherence in the early universe

We can write out a more explicit example of the no-boundary wave function in the presence of a massive scalar, in the saddle-point approximation and for an integration contour yielding a real wave function,

$$\Psi_{\rm HH} \approx e^{\frac{12\pi^2}{\hbar V(\phi_1)} - i\frac{2\pi^2}{\hbar}\sqrt{\frac{2}{3}}m\phi_1 a_1^3} \prod_{klm} e^{-\frac{\pi^2}{\hbar}a_1^2 \mathcal{R}_{1,k}^2 \frac{\epsilon(k^2-1)}{k-i\sqrt{\frac{V}{3}a_1^2-1}}}$$
$$+ e^{\frac{12\pi^2}{\hbar V(\phi_1)} + i\frac{2\pi^2}{\hbar}\sqrt{\frac{2}{3}}m\phi_1 a_1^3} \prod_{klm} e^{-\frac{\pi^2}{\hbar}a_1^2 \mathcal{R}_{1,k}^2 \frac{\epsilon(k^2-1)}{k+i\sqrt{\frac{V}{3}a_1^2-1}}}, \quad (8.6)$$

with background fields taking the final values a_1, ϕ_1, plus curvature fluctuations $\mathcal{R}_{1,k}$ with wave number k. One could easily include a further product of tensor fluctuation wave functions.

As we saw, the background wave function is driven to JWKB classicality just by the dynamics of accelerated expansion, but is this enough for the universe to appear classical? In order to assess this question, one may for instance look at the Wigner function, which is a standard tool in quantum mechanics. Because of the noncommutativity of fields and their momenta, a quantum state cannot be described by a point in phase space. The Wigner function is a quasi-probability distribution function that can be interpreted as an approximate classical probability function when it is positive. It is defined via

$$W(q,p) = \int_{-\infty}^{+\infty} \Psi^\star(q+s)\Psi(q-s)e^{\frac{i}{\hbar}2ps}ds. \quad (8.7)$$

Loosely speaking, it is the Fourier transform of the off-diagonal elements of the density matrix. The definition shows that significant off-diagonal correlations are required in order to render $W(q,p)$ negative.

Let us focus on a simple background, for gravity and a cosmological constant only. As we saw previously, for an FLRW ansatz of the form

$$ds^2 = -\frac{N^2}{q}dt^2 + q\,d\Omega_3^2, \quad (8.8)$$

the action (up to possible surface terms) is given by

$$S = 2\pi^2 \int dt \left[-\frac{3}{4N}\dot{q}^2 + 3N - N\Lambda q \right]. \quad (8.9)$$

With the momentum $p = \frac{\partial L}{\partial \dot{q}} = -\frac{3\pi^2}{N}\dot{q}$, the Hamiltonian is given by

$$H = \dot{q}p - L = -\frac{N}{6\pi^2}\left[p^2 + 36\pi^4\left(1 - \frac{\Lambda}{3}q\right)\right] = N\hat{H}. \quad (8.10)$$

For no-boundary conditions, the WDW equation $\hat{H}\Psi = 0$ is solved by

$$\Psi(q) = e^{\frac{12\pi^2}{\hbar\Lambda}} \text{Ai}\left[\left(\frac{18\pi^2}{\hbar\Lambda}\right)^{2/3}\left(1 - \frac{\Lambda}{3}q\right)\right]. \tag{8.11}$$

We may now evaluate the corresponding Wigner function, defined in Eq. (8.7). Rather surprisingly, this integral can be done exactly, with the result that

$$W(q,p) = \frac{1}{(24\pi^4\hbar\Lambda)^{1/3}} e^{\frac{24\pi^2}{\hbar\Lambda}} \text{Ai}\left[\left(\frac{36\pi^2}{\hbar\Lambda}\right)^{2/3}\left(1 - \frac{\Lambda}{3}q + \frac{p^2}{36\pi^4}\right)\right]. \tag{8.12}$$

A plot is provided in Fig. 8.1. Note that the argument of the Airy function is proportional to the Hamiltonian (8.10). The graph shows that the Wigner function has support on various crests, each of which may be interpreted as the Hamiltonian taking on different values. (Alternatively, one could think of the lower crests as corresponding to different values of the spatial curvature.) The largest crest has the smallest value of the Hamiltonian.

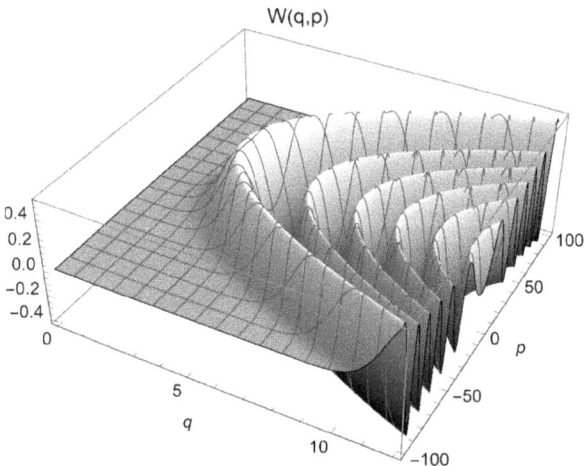

Figure 8.1 The Wigner function for a homogeneous, isotropic universe with positive cosmological constant Λ and no-boundary conditions. The crests correspond to different values of the Hamiltonian, with the largest crest yielding the smallest value of the Hamiltonian. Here $\Lambda = 1$ and $\hbar = 36$.

There are many crests, interspersed with regions in which the Wigner function goes negative. Thus it is not evident how a classical evolution follows.[1] The JWKB nature of individual saddle-point backgrounds is important, but

---

[1] Recalling that $\lim_{\beta\to\infty} \beta\text{Ai}(\beta x) = \delta(x)$, we can see that when $\hbar\Lambda$ goes to zero, the Wigner function is supported exactly on the Hamiltonian constraint, so in that limit classical physics is indeed recovered. This is important as a consistency check.

not sufficient by itself to explain why the saddles come to evolve independently. For this we will need to include more details. The fluctuations provide precisely the missing ingredient.

8.2.2 Classicalization of the background

When the wave function is a sum over saddle points, intuitively this is very much like the double-slit experiment, except that each path corresponds to the evolution of an entire universe. The question is then: How come we happen to see a single universe? We will explore the idea that it is the fluctuations that decohere the background, i.e., that interactions between background and fluctuations make one see a single background universe. Put differently, different saddle points end up effectively evolving as separate universes (whether one can say that they actually exist is however dependent on whether one could make a measurement, at least in principle, that would tell us of their existence).

Once more, we start by looking at the density matrix,

$$\rho = |\Psi\rangle\langle\Psi| = \Psi^\star \Psi. \tag{8.13}$$

We should warn the reader that when gravity is included, it is not clear that this is the correct object to consider, because the conserved current is the Klein–Gordon current rather than $\Psi^\star\Psi$. However, we have seen in Section 4.2 that in the JWKB limit the two definitions come to agree, and hence we will work in analogy with ordinary quantum mechanics here.

We can allow for some of the scalar fluctuation modes to be part of the background. We will call these Φ_B, so that for the case of two complex conjugate saddle points we have

$$\Psi = B(a, \Phi_B)F(a, \phi) + B^\star(a, \Phi_B)F^\star(a, \phi). \tag{8.14}$$

The reduced density matrix that we are interested in is the one where we take the fluctuations to be the environment, which is traced out,

$$\begin{aligned}\rho_{\text{red}}(a, a', \Phi_B, \Phi'_B) &= B(a, \Phi)B^\star(a', \Phi'_B)D_{11}(a, a') \\ &+ B(a, \Phi)B(a', \Phi'_B)D_{12}(a, a') \\ &+ B^\star(a, \Phi)B^\star(a', \Phi'_B)D_{21}(a, a') \\ &+ B^\star(a, \Phi)B(a', \Phi'_B)D_{22}(a, a'). \end{aligned} \tag{8.15}$$

We simplified the notation by writing the product over scalar fluctuation modes as a single function F, and defined the *decoherence functionals*

$$D_{11}(a,a') = D_{22}^\star(a,a') \equiv \prod_{klm} \int \mathcal{DR} F^\star(a,\mathcal{R}_k) F(a',\mathcal{R}_k), \qquad (8.16)$$

$$D_{12}(a,a') = D_{21}^\star(a,a') \equiv \prod_{klm} \int \mathcal{DR} F(a,\mathcal{R}_k) F(a',\mathcal{R}_k). \qquad (8.17)$$

There is an obvious generalization to the case of a sum over multiple saddle points, with decoherence functionals $D_{ij}(a, a', \Phi, \Phi')$. Note that a, a' simply denote two different, unrelated, values of the scale factor, and similarly for Φ_B, Φ'_B. When $i = j$ the decoherence functionals quantify the "non-classicality" of a single saddle point, and when $i \neq j$ they characterize the interference between separate saddles. Only when all components of the decoherence functionals are strongly peaked around $a = a'$ (e.g., if they are close to a delta function $\delta(a - a')$) can one say that the scale factor has decohered. Only then does the reduced density matrix represent a classical statistical mixture of definite scale factor values.

Above we left the range of k modes over which the product runs unspecified. Where exactly the split between background and fluctuations, or better between system and environment, is made depends on the physical situation. In cosmology, there is in fact a completely natural split, if we are in the presence of a horizon. Namely, we will consider modes of wavelength larger than the horizon as environment, while the shorter modes can be included in the system.

For specificity, we will continue working with the no-boundary example discussed above, although this kind of calculation will work very similarly for other models in which several saddle points are relevant. We will first focus on the interference between two saddles, called saddles 1 and 2 (we discard the homogeneous and nonnormalizable fluctuation mode with $k = 1$),

$$D_{12}(a,a') = \prod_{k=2}^{k=\bar{a}\sqrt{\frac{\Lambda}{3}}} \prod_{lm} \int \mathcal{DR} F(a,\mathcal{R}_k) F(a',\mathcal{R}_k)$$

$$= \prod_{k=2}^{k=\bar{a}\sqrt{\frac{\Lambda}{3}}} \prod_{lm} \left(\frac{[A_k(a) + A_k^\star(a)][A_k(a') + A_k^\star(a')]}{[A_k(a) + A_k(a')]^2} \right)^{\frac{1}{4}}$$

$$= \prod_{k=2}^{k=\bar{a}\sqrt{\frac{\Lambda}{3}}} \left(1 + \frac{(a^2 - a'^2)^2}{4a^2 a'^2} + \frac{\left[a^2\sqrt{\frac{\Lambda}{3}a'^2 - 1} + a'^2\sqrt{\frac{\Lambda}{3}a^2 - 1}\right]^2}{4k^2 a^2 a'^2}\right)^{-\frac{k^2}{4}},$$

(8.18)

where we used (3.112) adapted to the curvature perturbation $\mathcal{R}$ – see (7.13) – and ignored an unimportant overall phase function in the last line. We also made use of the multiplicity of spherical harmonics, which is such that the total number of l and m values is k^2 – see Appendix C. The upper limit in the product is taken to be the average horizon value, with $\bar{a} \equiv (a + a')/2$ and where we will write the difference as $\Delta \equiv (a - a')/2$. Note furthermore that the interaction between background and fluctuation indeed involves the scale factor a, and not the momentum, so that we may expect to see decoherence in the field basis.

Each term in the product (8.18) is smaller than 1, and thus we can already see that the perturbation modes will lead to decoherence between the two saddles points, even when $a = a'$. In fact, when Δ is small, we have that (writing $\mathsf{H} \equiv \sqrt{\Lambda/3}$)

$$D_{12}(a, a') \approx \prod_{k=2}^{k=\bar{a}\mathsf{H}} \left(1 + \frac{\mathsf{H}^2 \bar{a}^2}{k^2}\right)^{-\frac{k^2}{4}}. \tag{8.19}$$

This is a function that drops extremely fast once $\bar{a}$ is just a few times the size of the horizon. This implies that the two saddles decohere rapidly once the universe grows to a few times the Hubble radius associated with the vacuum energy in the early universe.

We can also look at the decoherence functional for a single saddle point, say D_{11}. An analogous calculation to that above then leads to

$$D_{11}(a, a')$$

$$= \prod_{k=2}^{k=\bar{a}\sqrt{\frac{\Lambda}{3}}} \left(1 + \frac{(a^2 - a'^2)^2}{4a^2 a'^2} + \frac{\left[a^2\sqrt{\frac{\Lambda}{3}a'^2 - 1} - a'^2\sqrt{\frac{\Lambda}{3}a^2 - 1}\right]^2}{4k^2 a^2 a'^2}\right)^{-\frac{k^2}{4}},$$

(8.20)

which differs from (8.18) merely by a sign. This time, when $a = a'$ each factor is equal to unity and thus there is no suppression, which is what we want in this case. When Δ is small, we may approximate the decoherence functional by

$$D_{11}(a,a') \approx \prod_{k=2}^{k=\bar{a}\mathsf{H}} \left(1 + \frac{4\Delta^2}{\bar{a}^2} + \frac{\mathsf{H}^2\Delta^2}{k^2}\right)^{-\frac{k^2}{4}}, \qquad (8.21)$$

which again decreases rapidly with increasing Δ, though not quite as rapidly as (8.19). Since the wave function is of JWKB form, there is also a strong correlation between the scale factor a and its conjugate momentum p_a, and hence once a takes on sharp, classical values, the expansion rate of the universe is also behaving classically.

Hence we arrive at the picture that separate saddle points decohere very fast once the universe is a little larger than the Hubble radius, while individual saddle points start displaying a classical background once the universe grows just a little further. In the case of the no-boundary wave function investigated here, this justifies treating the universe classically from $a \gtrsim$ (a few) $\cdot \frac{1}{\mathsf{H}}$ on, but also shows that the universe cannot be given a classical spacetime interpretation at smaller radii.

8.2.3 Decoherence of the fluctuations

In the previous section, we looked at the interactions between background and fluctuations. But the fluctuations also have interactions among themselves, and, although sometimes weak, these can lead to the classicalization of the fluctuations themselves. The present analysis may be seen as a refinement of the discussion above.

First we should clarify that standard, short-wavelength vacuum modes do not lead to decoherence. If they did, then we could never perform any typical quantum experiments, such as the double slit, since vacuum modes are constantly present. This implies that in cosmological models short-wavelength fluctuations, which are almost indistinguishable from Minkowski space modes, do not lead to decoherence. However, during inflation (or ekpyrosis, for that matter) fluctuations are amplified and this changes their properties, but only once they start reaching the size of the horizon. Thus we may already conclude that, during the amplification process, we do not need to worry about decoherence taking place too soon. However, at horizon crossing we may expect new effects, which we will model here.

Decoherence requires the presence of an interaction. When gravity is present, interactions are always present too. So we could for example consider the interactions of scalar fluctuations with tensor perturbations. But there is an even more direct interaction that is present: Due to the nonlinearity of general relativity, the curvature perturbation interacts with itself. We investigated an example of the observational consequences of such non-Gaussian corrections in Section 3.2.8. Here we will see that the same kind of interaction also helps in rendering the fluctuations classical.

As we discussed in Chapter 3, the curvature perturbation is conserved on large scales – i.e., $\dot{\mathcal{R}}$ approaches zero on large scales. This means that interactions involving the time derivative are suppressed compared to interactions involving the field (or its spatial derivatives). In turn, this observation implies that it is the field basis, rather than the momentum basis, that will provide the pointer states.

Even though we have applications to the curvature perturbation in mind, we will consider a generic (massless) scalar field φ here, to emphasize the generality of the decoherence process. At second order, we will assume its action (in conformal time) to be given by

$$S = \frac{1}{2} \int d\tau d^3x \, \epsilon a^2 [\varphi'^2 - (\nabla \phi^2)], \qquad (8.22)$$

and we will assume that this field evolves in a quasi–de Sitter background with $a \approx -1/(H\tau)$. We included a factor of ϵ in the action in analogy with the action of the comoving curvature perturbation. In the usual manner, the quadratic action (8.22) leads to the Hamiltonian

$$H_{\text{free}} = \frac{1}{2} \int_{\mathbf{k}} \left(\frac{H^2 \tau^2}{\epsilon} \hat{\pi}_{\mathbf{k}} \hat{\pi}_{-\mathbf{k}} + \frac{\epsilon k^2}{H^2 \tau^2} \varphi_{\mathbf{k}} \varphi_{-\mathbf{k}'} \right). \qquad (8.23)$$

Since the background has reached JWKB form, we can write down a Schrödinger equation specifying the evolution of the wave function,

$$i \frac{d}{dt} \psi_\varphi = -i H \tau \frac{d}{d\tau} \psi_\varphi = H_{\text{free}} \psi_\varphi, \qquad (8.24)$$

whose solution, for a single Fourier mode, is given by

$$\psi_\varphi = \left(\frac{A + A^\star}{\pi} \right)^{1/4} e^{-A|\varphi_{\mathbf{k}}|^2}, \qquad A = \frac{\epsilon k^3}{H^2} \frac{1 - \frac{i}{k\tau}}{1 + k^2 \tau^2}, \qquad (8.25)$$

where A denotes the correlator, cf. also Sections 3.2.6 and 6.6.1.

We will now consider an environment as well as interactions between this field and the environment (below we will specialize to the case where the environment consists of modes of the same field, but for now we keep the discussion more general). The total wave function thus becomes

$$\Psi = \psi_\varphi \psi_E \psi_{\varphi E}, \qquad (8.26)$$

with the free parts of the field/environment denoted by ψ_φ and ψ_E, respectively, while $\psi_{\varphi E}$ denotes a non-Gaussian interaction term. We will define the non-Gaussian part by using the leading cubic interaction

$$\psi_{\varphi E} = e^{\int_{\mathbf{k},\mathbf{k}',\mathbf{q}} E_{\mathbf{k}} E_{\mathbf{k}'} \varphi_{\mathbf{q}} I(\tau)}, \qquad (8.27)$$

with the short-hand notation $\int_{\mathbf{k},\mathbf{k}',\mathbf{q}} = \int \frac{d^3\mathbf{k}}{(2\pi)^3} \frac{d^3\mathbf{k}'}{(2\pi)^3} \frac{d^3\mathbf{q}}{(2\pi)^3} (2\pi)^3 \delta(\mathbf{k} + \mathbf{k}' + \mathbf{q})$. Above, I characterizes the interaction, which due to the spatial isotropy of the background only depends on the magnitudes k, k', q of the momenta. Note that the integral incorporates momentum conservation due to the delta function, implying that if $\mathbf{q}$ is a small vector (long-wavelength mode), then the large vectors $\mathbf{k}, \mathbf{k}'$ must be roughly parallel and opposite. This is the situation we will consider, where there is one long-wavelength mode (with momentum $q \ll k, k'$) that we treat as the system, with two short-wavelength modes (k, k') acting as the environment.

To address decoherence, we can now integrate out the environment, in effect averaging out the interactions with the environment E. The reduced density matrix is then given by

$$\rho_{\text{red}}[\varphi, \tilde{\varphi}] = \int \mathcal{D}E\, \Psi[\varphi, E] \Psi^\star[\tilde{\varphi}, E], \qquad (8.28)$$

where φ, φ' denote two different values of the fluctuation. When ρ_{red} becomes diagonal, i.e., peaked around $\varphi = \tilde{\varphi}$, then we can say that the fluctuation mode φ has become effectively classical. To quantify this one usually defines a *decoherence factor*

$$D[\varphi, \tilde{\varphi}] \equiv \frac{\rho_{\text{red}}[\varphi, \tilde{\varphi}]}{\sqrt{\rho_{\text{red}}[\varphi, \varphi] \rho_{\text{red}}[\tilde{\varphi}, \tilde{\varphi}]}} \qquad (8.29)$$

$$= \int \mathcal{D}E\, |\psi_E|^2\, e^{\int_{\mathbf{k},\mathbf{k}',\mathbf{q}} E_{\mathbf{k}} E_{\mathbf{k}'} (\varphi_{\mathbf{q}} I + \tilde{\varphi}_{\mathbf{q}} I^\star)} \qquad (8.30)$$

$$= \left\langle e^{i \int_{\mathbf{k},\mathbf{k}',\mathbf{q}} E_{\mathbf{k}} E_{\mathbf{k}'} (\Delta\varphi_{\mathbf{q}}) \text{Im}[I]} \right\rangle, \qquad (8.31)$$

where the last expression denotes the expectation value in the environment. Here we defined the difference $\Delta\varphi = \varphi - \tilde{\varphi}$ and ignored the real part of I, which is important for assessing the size of non-Gaussian correlation functions – see Section 3.2.8. For studying decoherence, it is rather the imaginary part (which is far larger) that is more important.

In order to calculate the expectation value, we can use the cumulant formula

$$\langle e^X \rangle = e^{\frac{1}{2}\langle X^2 \rangle + \cdots}, \qquad (8.32)$$

which as written is valid for a random variable X with zero mean $\langle X \rangle = 0$. For a single mode, this allows us to write the decoherence factor more explicitly, as a function of conformal time τ, as

$$D[\varphi, \tilde{\varphi}; \tau] \approx e^{-\frac{4\pi}{q^3}|\Delta\bar{\varphi}_\mathbf{q}|^2 \int_{\mathbf{k}+\mathbf{k}'=-\mathbf{q}} P_E(k,\tau) P_E(k',\tau)(\text{Im}[I])^2} . \qquad (8.33)$$

Here $P_E = 1/\text{Re}[A]$ denotes the power spectrum of the environment variable E, defined in the usual way as $\langle E_\mathbf{k} E_{\mathbf{k}'} \rangle = (2\pi)^3 \delta(\mathbf{k}+\mathbf{k}') P_E$. We have also rescaled the mode $\varphi_\mathbf{q} = \frac{\sqrt{2}\pi}{q^{3/2}} \bar{\varphi}_\mathbf{q}$, so that the momentum dependence is removed, and that its expectation value is equal to the variance,

$$\langle |\bar{\varphi}_\mathbf{q}|^2 \rangle = \Delta_\varphi^2 = \frac{\mathsf{H}^2}{8\pi^2 \epsilon} . \qquad (8.34)$$

Let us be a little more specific about the interaction we have in mind. Due to the nonlinearity of gravity, the comoving curvature perturbation admits a number of self-interaction terms, including

$$\mathcal{L} \supset \mathcal{R}\dot{\mathcal{R}}^2, \mathcal{R}(\partial \mathcal{R})^2 . \qquad (8.35)$$

As we discussed earlier, terms containing time derivatives will not play a big role, since the curvature perturbation is conserved on large scales. But the term involving spatial derivatives can couple modes, in particular one long-wavelength fluctuation with two short-wavelength fluctuations, $\mathcal{R}_L(\partial \mathcal{R}_S)^2$, other combinations being subdominant. The physical picture one should have in mind is that the long-wavelength perturbation changes the background in which the two short fluctuations evolve. These short fluctuations thus obtain information about the long one, and the long-wavelength mode is being rendered classical due to this interaction, in analogy with what happened for the background in the analysis of the previous section.

Given this, we will consider an interaction of the form $g\varphi(\partial\varphi)^2$ as an example, where g is a coupling constant (for the comoving curvature perturbations, it is of second order in slow-roll), though one could of course extend the present calculation to different interactions. This is the calculation that we already did when calculating an example of non-Gaussian corrections in Section 3.2.8. For convenience, let us repeat the result regarding the interaction function,

$$\text{Im}[I] = -\frac{g}{3\mathsf{H}^2 \tau}(k^2 + k'^2 + q^2) . \qquad (8.36)$$

Recall from Section 3.2.8 that the main contribution to the interaction function comes from fluctuations modes that are roughly of horizon size. Using this expression, we can straightforwardly evaluate the decoherence factor

(8.33), finding

$$D[\varphi, \tilde{\varphi}] \approx e^{-\frac{g^2}{288\epsilon^2}\left(\frac{aH}{q}\right)^3 |\Delta\bar{\varphi}_\mathbf{q}|^2}. \tag{8.37}$$

This is the result we were working towards. It shows that the fluctuation modes are indeed being decohered as the universe grows. Decoherence is effective once the exponent reaches a magnitude of order unity,

$$\frac{a}{q} \gtrsim \frac{30\epsilon}{(gH)^{2/3}} \frac{1}{H}. \tag{8.38}$$

For the example of the curvature perturbation with $g = \mathcal{O}(\epsilon^2)$, one obtains a size one to several orders of magnitude larger than the Hubble radius. Hence we learn that modes of about horizon size can decohere fluctuation modes that have left the horizon a little before, and are already somewhat larger. We obtain the physical picture that modes are being decohered in succession, as they evolve to superhorizon sizes. The decohering modes become decohered themselves after being stretched a little further. In the no-boundary case, where the universe is nucleated at about horizon size, we would then obtain the picture that the scale factor would decohere first, as the universe grows to a few horizon sizes in radius, followed by consecutive fluctuation modes. In this way, the early universe becomes classical, with fluctuation modes playing the role of observers and observed.

Note that a stronger interaction, due perhaps to couplings between different scalar fields, can reduce the characteristic decoherence size and bring it closer to the size of the horizon. In a sense, what we have calculated here is the minimal amount of decoherence we may expect to occur.

8.3 Discussion

What quantum cosmology suggests is that the wave function is a deeper concept than spacetime. Classical spacetime should be seen as an emergent property, arising only if certain circumstances (such as those discussed in Section 4.2) are met. Since everyone has their own wave function description, everyone also has their own spacetime. In practice, for inhabitants of Earth these spacetimes will be essentially indistinguishable due to the manyfold decoherence processes occurring around us, but in principle they are distinct.

The calculations of decoherence that we have performed in the last two sections indicate that, when the universe is exactly at Hubble radius size, this is not enough for decoherence yet, and the universe cannot be considered to be classical yet. Decoherence happens as the universe grows beyond the Hubble radius, which is a result also inferred by several other calculations in this book. It is precisely when the universe becomes larger than the Hubble radius that it splits up into mutually separated/inaccessible regions due to the presence of a horizon. Thus the dynamics itself leads to regions that must be considered to be an environment, inducing decoherence. But note that the information is not actually lost – the superhorizon modes later re-enter the horizon and start evolving once more, though by then in a universe filled with radiation and matter.

The process of decoherence also leads to the establishment of many records of the long-wavelength fluctuations in the universe. The splitting up of the universe into many mutually inaccessible regions facilitates this process, and combines decoherence with the notion of *quantum Darwinism*, which postulates that a classical state is achieved by generating many records of its existence in the environment. A cosmological phase in which the comoving horizon shrinks (such as inflation or ekpyrosis) therefore seems particularly well suited in this sense. It would be interesting to investigate if the ability to create so many mutually inaccessible regions also makes the universe fitter in a Darwinian sense. This would favor an inflationary phase that operates higher up on the potential (so that the horizon is smaller), and that lasts longer. This might provide an interesting balancing pressure when combined with the no-boundary proposal's predilection for short inflationary phases, low on the potential.

A further question for future research is whether the early interference of different saddle points may lead to observable effects. These would affect the very largest scales of the universe, which may present an obvious experimental difficulty if the size of the universe is nonminimal.

It seems appropriate to also comment on the interpretation of the wave function: In quantum cosmology we are led almost automatically to the relative state interpretation of quantum theory. This is because we consider the wave function of literally everything, so there is no way we can artificially split the universe up into system and classical apparatus. As we have discussed above, this in no way prevents the universe from appearing classical to us, because even small interactions between different parts of the universe can lead to decoherence. However, saying that one is led to the relative state interpretation does not necessarily imply the many-worlds interpretation – i.e., it does not imply the existence of an infinity of universes.[2]

The wave function is our description of the universe, which necessarily

---

[2] In quantum mechanics, no series of experiments can fully determine an unknown wave function, even in principle. This seems to be a counterargument to taking the actual existence of the various branches of the wave function literally, on top of the usual paradoxes arising whenever a physical infinity is invoked.

contains (at least in principle) a superposition of all possible outcomes for all possible processes. What is more, each observer has their own such description – to put it clearly: Everyone has their own wave function. The linearity of quantum theory ensures that all such descriptions are compatible and consistent with each other. The many branches and their entangled structure are precisely what is required such that objective existence is consistent for all observers (think about the example where two observers leave each other's horizon, and at a later stage re-enter it). Each branch corresponds to a potential observational outcome, and it is entangled with the rest of what can be observed in principle.

This does not resolve the question of which precise outcome will be the result of a specific experiment, i.e., which specific branch is actualized. But clearly this cannot be predetermined a priori, otherwise the probabilistic structure of quantum mechanics (which stems from the noncommutativity of certain pairs of observables) would be lost. To reinforce this point: The fact that experimental outcomes are not predetermined is what allows the future to remain free.

In classical general relativity, spacetime is a 4-dimensional construct, and no particular time slice plays a preferred role. In a sense one sees the entire evolution from past to future infinity all at once. In quantum cosmology, this is clearly different: The arguments of the wave function are only 3-dimensional (and perhaps ought to be reduced further to the information that can be accessed by a single observer at a single moment). When an observation is made, the outcome updates the wave function. This "collapse" of the wave function remains ill-understood, but let us just point out the analogy with our subjective perception of the here and now, the elusive cut between past and future that turns possibilities into actualities. One suspects that in a classical world such a perception would not exist, and that the analogy just mentioned might in fact be more than just an analogy.

9
Transitioning to different spacetimes

In Chapter 7, we saw how one could describe the nucleation of space and time from nothing. Here, we will consider the possibility that bubbles of spacetime might arise via quantum tunneling in a pre-existing universe. Even a sort of reverse process, by which spacetime can disappear into nothingness, seems possible. And finally, we will describe wormhole solutions, tunnels in space or spacetime that can connect distant regions.

We should stress from the outset that to date, these processes remain purely theoretical, as it has not been possible so far to find observational evidence for them (let alone create the required conditions in a laboratory). That said, one should keep in mind that they may have played, or may yet come to play, an important role in the history of the universe.

9.1 Vacuum decay via Coleman–DeLuccia instantons

9.1.1 Complex time tunneling in quantum mechanics

In describing tunneling across a potential barrier in quantum mechanics, the usual procedure is to consider an incoming wave packet, and to calculate how much of this wave packet extends across the potential barrier. From this a tunneling probability can be calculated. There exists a different formulation, based on Feynman's path integral formulation. This formulation leads to a simple semiclassical estimate for the tunneling rate (though in principle an exact calculation is possible), which has the advantage of being generalizable to quantum field theory and to include gravity.

As we have seen many times, in Feynman's formalism the transition amplitude from an initial state where $x(t_0) = x_0$ to a final state where $x(t_1) = x_1$

is given by the integral over all paths that link the two events,

$$\langle x_1, t_1 \mid x_0, t_0 \rangle = \mathcal{N}_{\text{norm}} \int_{x_0,t_0}^{x_1,t_1} D[x(t)] \, e^{iS}, \tag{9.1}$$

where $\mathcal{N}_{\text{norm}}$ is a normalization factor, and the action S of a unit mass particle moving in a potential $V(x)$ is given by

$$S = \int \mathrm{d}t \left(\frac{1}{2} \dot{x}^2 - V(x) \right). \tag{9.2}$$

When the final configuration can be reached from the initial configuration via a classical path, then the action S will be real valued, and the transition amplitude is simply a phase. But in a tunneling situation, there is no classical path that links the initial and final configurations. Nevertheless, the transition can be possible, as it can be mediated by a complex solution of the equations of motion. In such a case the action also becomes complex valued in general, and the transition will be suppressed, i.e., less likely than classical evolution. We will make this more precise below.

There are many situations of interest where, instead of looking for fully complex solutions, one just Wick rotates the time coordinate to Euclidean time, $t = -i\sigma$, and then looks for Euclidean solutions of the equations of motion. This is often simpler, and sufficient, as the initial and final field values must be real (although we saw in Chapter 7 that for no-boundary solutions a purely Euclidean approach is not general enough). In Euclidean time, the action becomes

$$S_E = -iS = \int \mathrm{d}\sigma \left(\frac{1}{2} (x_{,\sigma})^2 + V(x) \right), \tag{9.3}$$

which indicates that, effectively, motion in Euclidean time corresponds to motion in an *inverted potential* $-V(x)$. This largely explains why the Euclidean approach is so useful: A potential barrier turns into a potential well, thereby enabling the end points to be joinable.

As an example, take the inverted harmonic oscillator potential,

$$V(x) = -\frac{1}{2} \Omega^2 x^2, \tag{9.4}$$

where Ω is a constant. We do not care here that the potential is unbounded from below – one could imagine that it turns around at large $|x|$. The equation of motion reads

$$\ddot{x} = \Omega^2 x. \tag{9.5}$$

9.1 Vacuum decay via Coleman–DeLuccia instantons

A solution with classical energy $E = -\frac{1}{2}\Omega^2 c_{\text{turn}}^2$ is given by

$$x(t) = -c_{\text{turn}} \cosh(\Omega t) , \tag{9.6}$$

where $x(0) = -c_{\text{turn}}$ is the particle's location at the classical turnaround at $t = 0$. Classically, a particle with this energy cannot surpass the potential barrier and reach $x = +c_{\text{turn}}$ on the other side. However, in Euclidean time the potential is a simple harmonic oscillator ($x_{,\sigma\sigma} = -\Omega^2 x$) and a corresponding solution exists,

$$x(t) = -c_{\text{turn}} \cos(\Omega \sigma) = -c_{\text{turn}} \cosh(-i\Omega\sigma) , \quad 0 \leq \Omega\sigma \leq \pi . \tag{9.7}$$

This solution mediates the tunneling of the particle across the barrier. Once on the other side, the particle can continue along a classical trajectory, falling off the potential to large x values,

$$x(t) = c_{\text{turn}} \cosh \Omega t = -c_{\text{turn}} \cosh(-i\pi + \Omega t) . \tag{9.8}$$

The whole evolution can be seen as a zigzag path through the complexified time plane, along the negative real t axis first, then along the Euclidean time direction, and finally again in the positive Lorentzian direction, but shifted by $-i\pi/\Omega$.

The action along the Euclidean segment is easily evaluated, and gives rise to the suppression factor

$$e^{-\frac{S_E}{\hbar}} = e^{-\frac{1}{2\hbar}c_{\text{turn}}^2 \Omega \pi} = e^{-\frac{\pi|E|}{\hbar\Omega}} , \tag{9.9}$$

which provides an estimate for the tunneling amplitude in the limit of small $\hbar$. It depends both on the height $|E|$ of the potential barrier, and on its width $\Omega = \sqrt{|V_{,xx}|}$.

Note that the action, and thus the amplitude, remains unchanged if we deform the path taken in the complex time plane, as long as the end points remain fixed. Hence one may regard the transition as being mediated by a single, complex time solution.

9.1.2 Vacuum decay without gravity

We just saw how evolution in complexified time, in particular via a Euclidean solution, can describe quantum tunneling across a barrier. In field theory, this leads to the interesting effect of vacuum decay. Imagine a potential with two local minima, such as the one shown in Fig. 9.1, and let us assume that the universe finds itself in a state such that the scalar field ϕ sits in the higher, "false" vacuum ϕ_{fv}. Classically the field would just stay there for all times. However, in a quantum theory, there is the possibility that the field might tunnel across the barrier and end up in the lower, true vacuum $\phi_{\rm tv}$. This would happen locally, and could occur at any time – hence the quantity we would like to calculate is the rate per unit 4-volume for such a tunneling effect to happen. In addition, we would like to know what happens after the tunneling event. Achieving such an understanding will be the goal of the next two sections, where we will first analyze the case without gravity, and in the following section we will include gravity.

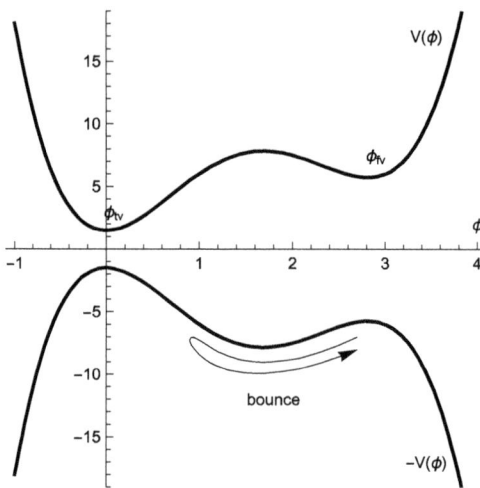

Figure 9.1 A scalar potential with two local minima. If the field sits in the false vacuum $\phi_{\rm fv}$, then it can decay via quantum tunneling by forming a bubble inside of which the field is in the lower, true vacuum at $\phi_{\rm tv}$. Here we are using the explicit form $V(\phi) = \phi^2(\phi - 3)^2 + \frac{1}{2}\phi^2 + \frac{3}{2}$.

Motivated by the discussion in the last section, we will be primarily interested in the Euclidean theory for a scalar field $\phi(x^\mu)$ evolving in a potential $V(\phi)$ with at least two local minima. The action is given by

$$S_E = -iS = \int d\sigma d^3x \left(\frac{1}{2}\delta^{\mu\nu}\partial_\mu\phi\partial_\nu\phi + V(\phi)\right). \qquad (9.10)$$

Then our quantity of interest, the decay rate per unit volume and time, can again be approximated using the path integral formalism and the associated saddle-point approximation, with[1]

$$\frac{\Gamma}{V} = \int_{\phi_{\rm fv}}^{\phi_{\rm other\ side}} \mathcal{D}\phi\, e^{\frac{i}{\hbar}S} \approx e^{-\frac{1}{\hbar}[S_E(\phi_{\rm bounce}) - S_E(\phi_{\rm fv})]} \equiv e^{-\frac{1}{\hbar}B}, \qquad (9.11)$$

---

[1] The path integral calculates the transition amplitude for tunneling to occur at a particular point, at a particular time. But since any location is as good as any other, and any time is as good as any other, this also provides us with the rate per unit 4-volume.

where it is conventional to denote the difference in Euclidean actions by the letter B. Here our boundary conditions are that the scalar field starts in the false vacuum and ends up on the other side of the potential barrier, so that it can roll down to the true vacuum. Note that the action of the false vacuum enters the formula above as a background subtraction, which is because we would like to calculate the rate of tunneling given that we start in the false vacuum. Here ϕ_{bounce} is the relevant and dominant solution describing the tunneling. It is reasonable, and can be shown rigorously to be the case, to assume that the dominant solution will have spherical $O(4)$ symmetry. In this case the solution depends only on $\rho \equiv (\sigma^2 + |\mathbf{x}|^2)^{1/2}$, and the action simplifies to

$$S_E = 2\pi^2 \int \mathrm{d}\rho\, \rho^3 \left(\frac{1}{2}(\partial_\rho \phi)^2 + V(\phi) \right). \tag{9.12}$$

The resulting field equation is

$$\phi_{,\rho\rho} + \frac{3}{\rho}\phi_{,\rho} - V_{,\phi} = 0. \tag{9.13}$$

Note that once again, in the Euclidean theory, the effective potential is flipped and the evolution takes place in an upside-down potential – see Fig. 9.1.

In the upside-down potential the barrier becomes a valley, and one can see by inspection that a solution will exist that starts near the false vacuum and rolls across the valley to the other side of the potential. If one were to continue the evolution, after coming to a momentary halt, the field would simply roll back and return to the false vacuum. For this reason a solution of this type is called a "bounce," though in the end we will only be interested in half of this solution.

The boundary conditions for a bounce solution are given by

$$\frac{\mathrm{d}\phi}{\mathrm{d}\rho}(\rho = 0) = 0, \qquad \phi(\rho \to \infty) = \phi_{\text{fv}}. \tag{9.14}$$

A numerical example is provided in the left panel of Fig. 9.2. The solution can be seen to asymptotically reach the false vacuum, and to interpolate to the true vacuum near the origin. For describing tunneling, one would only keep the $\rho \geq 0$ part. This then describes a localized region of true vacuum, surrounded by a sea of false vacuum. The region in which the field interpolates between the two vacua is called the "wall" of the instanton solution. It is a region of large gradient energy. The central region may be described as a bubble of true vacuum – see also the right panel in Fig. 9.2.

We will momentarily investigate what happens to this bubble of true

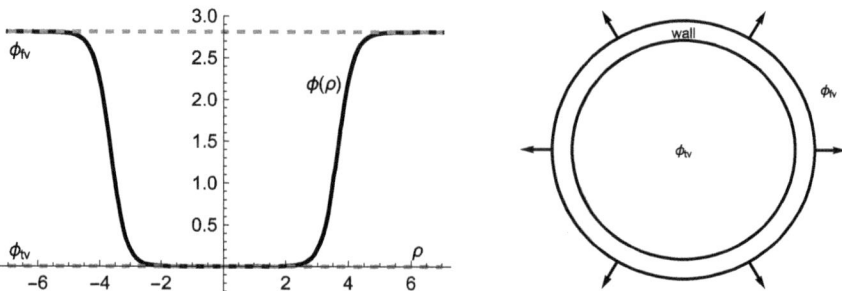

Figure 9.2 *Left:* an example of a bounce solution in the absence of gravity, and for the potential shown in Fig. 9.1. For a description, see the main text. *Right:* Via tunneling a bubble of true vacuum materializes in a sea of false vacuum, and then expands.

vacuum after it has formed. But first we can gain a little more quantitative understanding of bounce solutions by considering the special case in which the heights of the two minima are very similar. To this end, assume now that the potential is of the form

$$V(\phi) = U(\phi) + \frac{\Delta U}{2a}(\phi - a), \quad U(a) = U(-a), \quad U_{,\phi}(a) = U_{,\phi}(-a) = 0, \tag{9.15}$$

where U is a symmetric function, $U(-\phi) = U(\phi)$, with minima at $\phi = \pm a$, and ΔU is a small positive number. The equation of motion (9.13) can now be approximated by

$$\phi_{,\rho\rho} = U_{,\phi}, \tag{9.16}$$

where we have neglected a term proportional to ΔU as well as the friction term proportional to $\phi_{,\rho}$. This can be justified in retrospect. The approximate equation of motion can be integrated to yield

$$\frac{1}{2}\phi_{,\rho}^2 - U(\phi) = -U(a), \tag{9.17}$$

where the integration constant was fixed by considering the asymptotic false vacuum region. The scalar ϕ interpolates between the two extrema $\pm a$ as we go into the core of the solution. When integrated, Eq. (9.17) determines ϕ up to an integration constant $\bar{\rho}$, which we can take to be the value of ρ at which ϕ reaches its average value,

$$\rho - \bar{\rho} \equiv \int d\phi \frac{1}{\sqrt{2[U(\phi) - U(a)]}}. \tag{9.18}$$

9.1 Vacuum decay via Coleman–DeLuccia instantons

This implies that $\bar\rho$ specifies the location of the wall. For example, if we take

$$U_{\text{ex}} = \frac{1}{8}(\phi^2 - a^2)^2, \tag{9.19}$$

then the profile of ϕ is simply

$$\phi = a\tanh\left(\frac{a}{2}(\rho - \bar\rho)\right). \tag{9.20}$$

This shows that when the two minima are far apart compared to the energy difference of the minima, the interpolating region is of rather small extent, and consequently the approximation that we are investigating presently is called the *thin wall approximation*.

We still need to calculate the location $\bar\rho$ and the damping coefficient B that was defined in (9.11). It is natural to subdivide the integration required for determining B into three regions, outside the wall, around the wall, and inside the wall. Outside of the wall, the field is in the false vacuum, thus

$$B_{\text{outside}} = 0. \tag{9.21}$$

Inside the wall, the field is in the true vacuum and thus

$$B_{\text{inside}} = 2\pi^2 \int_0^{\bar\rho} d\rho\, \rho^3 [U(-a) - U(a)] = -\frac{\pi^2}{2}\bar\rho^4 \Delta U. \tag{9.22}$$

Meanwhile, just around the wall we have

$$B_{\text{wall}} = 2\pi^2 \bar\rho^3 \int d\rho [\tfrac{1}{2}\phi_{,\rho}^2 + U(\phi) - U(a)], \tag{9.23}$$

$$\equiv 2\pi^2 \bar\rho^3 S_1, \tag{9.24}$$

where S_1 denotes the wall tension. Using (9.17) one can rewrite the integral for the tension in useful ways,

$$S_1 = 2\int d\rho[U(\phi) - U(a)] = \int_{\phi_{\text{tv}}}^{\phi_{\text{fv}}} d\phi\sqrt{2[U(\phi) - U(a)]}. \tag{9.25}$$

For our example potential (9.19), we have $S_1 = 2a^3/3$.

Putting it all together, we have $B = -\frac{\pi^2}{2}\bar\rho^4 \Delta U + 2\pi^2 \bar\rho^3 S_1$. Since we are considering a solution to the Euclidean equations of motion, the action must be stationary, which it is at

$$\bar\rho = \frac{3S_1}{\Delta U} \quad\Rightarrow\quad B = \frac{27\pi^2}{2}\frac{S_1^4}{(\Delta U)^3}. \tag{9.26}$$

Note that $\bar\rho$ is large when ΔU is small, which justifies the thin wall approximation. In realistic early-universe applications the radius of the bubble, when it comes into existence, would typically be of microscopic size.

It is rather straightforward to describe the evolution of the bubble of true vacuum after it has nucleated. The Euclidean solution in fact not only provides initial conditions for the subsequent evolution, but via analytic continuation also provides a solution describing the growth of the bubble. Thus, one would take the bubble profile $\phi(\rho)$ and analytically continue from Euclidean to Lorentzian time,

$$\rho = (|\mathbf{x}|^2 + \sigma^2)^{1/2} \to (|\mathbf{x}|^2 - t^2)^{1/2}. \tag{9.27}$$

This changes the $O(4)$ symmetry of the solution into an $O(1,3)$ symmetry. The wall location then evolves according to the simple relation

$$\rho = \bar{\rho}, \tag{9.28}$$

which, since $\bar{\rho}$ is expected to be microscopic, implies that the bubble wall quickly approaches the speed of light ($\rho \approx 0$) and then continues expanding at the speed of light. In this way, bubbles of the new, true vacuum phase grow in all directions and "convert" their surrounding regions to true vacuum at the speed of light.

This description without gravity already highlights many of the important features of phase transitions, but a deeper understanding can be gained by including gravity. This is the topic to which we will turn next.

9.1.3 Vacuum decay with gravity

The inclusion of gravity does not fundamentally change the picture of vacuum decay developed in the previous section. In fact, as seen from the outside, the evolution of bubbles of true vacuum remains almost identical, except that the outside spacetime is now dynamical itself. The description of the inside of bubbles is however somewhat more complicated, as we will see.

The most important difference that gravity entails is that now it is not only the energy difference of potential minima that counts, but their absolute magnitude as well. When a scalar field sits at a local minimum of the potential, it acts like a cosmological constant. And here there is a vast difference: A positive cosmological constant leads to exponential expansion of space. By contrast, a negative cosmological constant leads to an unstable universe, which is prone to catastrophic collapse. For this reason, we will focus only on the case where both the false as well as the true vacuum reside at positive values of the potential.

In the absence of gravity, a proof exists that shows that the most likely instantons have $O(4)$ symmetry. With gravity, such a proof is not available. However, there is no reason to think that asymmetric configurations

9.1 Vacuum decay via Coleman–DeLuccia instantons

should dominate. In fact, our experience with no-boundary instantons has also shown that the most symmetric configurations have the highest likelihood. For this reason we will keep the $O(4)$ symmetry, and choose a corresponding metric, which is the Euclidean version of a closed FLRW universe,

$$ds^2 = d\sigma^2 + \rho^2(\sigma)d\Omega_3^2, \qquad (9.29)$$

where ρ specifies the radius of curvature of the 3-spheres (in FLRW one would denote this by a, but we will use the more conventional letter ρ in the present context). The equations of motion are then given by

$$0 = \phi'' + 3\frac{\rho'}{\rho}\phi' - V_{,\phi}, \qquad (9.30)$$

$$\rho'^2 = 1 + \frac{\kappa^2 \rho^2}{3}\left(\frac{1}{2}\phi'^2 - V\right), \qquad (9.31)$$

where $' \equiv \frac{d}{d\sigma}$. We will explicitly keep factors of $\kappa^2 \equiv 8\pi G$, as these will be useful later on. The action is

$$S_E = 2\pi^2 \int d\sigma \left[\frac{1}{2}\rho^3 \phi'^2 + \rho^3 V + \frac{3}{\kappa^2}(\rho^2 \rho'' + \rho \rho'^2 - \rho)\right]. \qquad (9.32)$$

We are looking for a solution that interpolates between the two potential minima – see again Fig. 9.1. The solution of Einstein's equations with constant positive vacuum energy is de Sitter spacetime, whose Euclidean version is a sphere. This means that we are looking for a solution that looks like a sphere at both ends, with a wall region in between. In other words, we expect the gravitational tunneling solution to be compact (which is a useful property, as it ensures that the action will be finite). The equations of motion (9.30)–(9.31) show that, if we were to start or end exactly at the potential minimum, the scalar field would simply stay there, and there would be no interpolation. Thus we must actually look for a solution that starts away from the minimum, and also ends away from the second local minimum. Denoting the initial and final scalar field values by ϕ_0 and ϕ_m, respectively, we are looking for a solution obeying the boundary conditions

$$\phi(0) = \phi_0, \quad \phi'(0) = 0, \quad \rho(0) = 0, \quad \rho'(0) = 1, \qquad (9.33)$$

at $\sigma = 0$, and

$$\phi(\sigma_{\max}) = \phi_m, \quad \phi'(\sigma_{\max}) = 0, \quad \rho(\sigma_{\max}) = 0, \quad \rho'(\sigma_{\max}) = -1, \qquad (9.34)$$

at some to-be-determined end point $\sigma_{\max}$. We will refer to such solutions as CDL instantons, after their discoverers S. Coleman and F. De Luccia.

A numerical example, for the explicit potential shown in Fig. 9.1, is shown

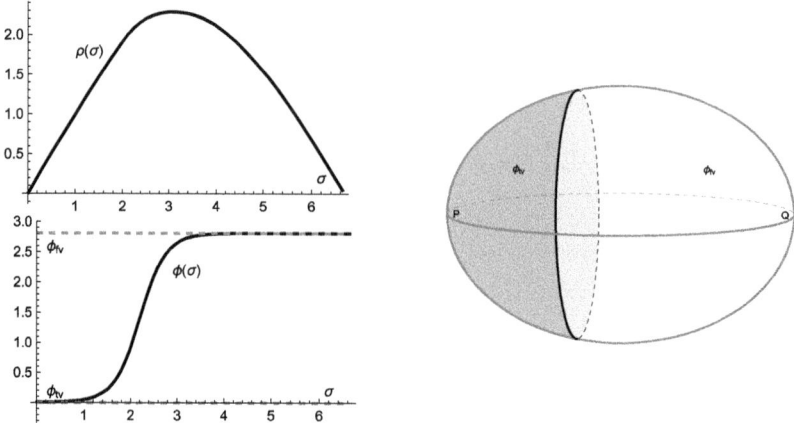

Figure 9.3 *Left:* a numerical solution of a CDL instanton, for the potential shown in Fig. 9.1 and with $\kappa^2 = 0.1$. The starting value of the scalar field is $\phi \approx 0.00794896$. *Right:* a sketch of the shape of a CDL instanton. The center of the bubble ($\sigma = 0$) is located at the point P, while Q corresponds to $\sigma_{\max}$. The wall separating false from true vacuum is indicated by the vertical black circle. Initial conditions for subsequent evolution are provided by the data on the gray (horizontal) meridian that includes P and Q. Note that in this picture σ is an angular coordinate.

in the left panels in Fig. 9.3. There one can see that the solution is indeed compact, and represents a somewhat deformed ball. The profile of the scalar field is very similar to the one without gravity. Note that the start and end points of the scalar field are in fact very close to the local minima of the potential, although there exist examples in which the start/end points can be significantly further away from the end points. It is useful to realize that the CDL instanton in fact only depends on the region of the potential between the start and end points – hence one could arbitrarily deform the potential outside of this region (and thus, e.g., move the local minima further away) and still find the same CDL solution.

Unfortunately, there exists no good way of approximating CDL instantons analytically. The action, and thus the decay rate, typically also has to be calculated numerically. In the exercises, we will provide a few more details regarding the numerical procedure used to find such instantons. However, there is a criterion that helps in assessing whether a CDL solution is likely to exist at all.

It provides only an indication, though, as it is built on approximating the background by pure de Sitter with scale factor $\rho = \frac{1}{\mathsf{H}} \sin(\mathsf{H}\sigma)$ with

$\mathsf{H} = \sqrt{V(\phi_\text{fv})/3}$, and where $V(\phi_\text{fv})$ is the potential value at a pole of the putative bubble. Then one can look at the behavior of the scalar field near the barrier, by defining $\delta\phi = \phi - \phi_\text{top}$, where ϕ_top is the field value where the potential barrier is located. The equation of motion (9.30) leads to

$$\delta\phi'' + 3\mathsf{H}\cot(\mathsf{H}\sigma)\delta\phi' = V_{,\phi\phi}(\phi_\text{top})\delta\phi. \tag{9.35}$$

Solutions that come to rest at either end of the bubble (i.e., at $\sigma = 0$ and $\sigma = \pi/\mathsf{H}$) are of the form

$$\delta\phi = \sum a_n \cos(n\mathsf{H}\sigma), \tag{9.36}$$

for some integration constants a_n. By direct substitution one sees that these solutions only exist when

$$-V_{,\phi\phi}(\phi_\text{top}) = n(n+3)\mathsf{H}^2. \tag{9.37}$$

The CDL solution corresponds to $n = 1$. Solutions with more transitions across the potential barrier are thus also seen to exist, for $n \geq 2$. We will discuss these *oscillating instantons* a little further in the exercises, where we will see that they do not provide the dominant decay channel.

We should note an important point: CDL solutions to (9.37) with $n = 1$ (or perhaps rather its nonlinear extension) can be found by adjusting where exactly $\sigma = 0$ is located, i.e., where on the potential the Euclidean solution starts. This does not have to be at or even near a local minimum of the potential. For tunneling one just needs the solution to start at one side of the potential barrier and interpolate to the other side. The existence of a CDL solution may then be inferred via an *undershoot/overshoot* argument. If the scalar starts too close to ϕ_top it will not make it far enough and roll back before $\sigma = \pi/\mathsf{H}$ can be reached. This is an undershoot solution. If, on the contrary, ϕ is released too far from ϕ_top then the "friction" term $\cot \mathsf{H}\sigma$ changes sign early on and becomes an anti-friction term, pushing the scalar out too far. This is an overshoot solution. By tuning the starting value, one can thus find a solution that comes to rest precisely at the opposite pole of the bubble. This tuning of the initial value must be done numerically in general – see again Fig. 9.3.

There is one more loose end we must tie up. Namely, we must answer the question of how the scalar field, which by assumption sits in the false vacuum initially, reaches the starting point for the tunneling solution, if that starting point is not necessarily at the local minimum of the potential but can be somewhere on the slope of the potential barrier. For this we have to recall the results of Section 6.6, where we showed that de Sitter spacetime has a temperature. This temperature implies that the scalar field undergoes thermal fluctuations, and that it can fluctuate to the starting point of the tunneling solution. Thus, in some sense, one could say that gravitational false vacuum decay is activated via thermal fluctuations. A corollary is that one should regard false vacuum decay as occurring within a single Hubble volume.

We are now ready to discuss the evolution of the bubble after tunneling. As we saw in the flat case, an $O(4)$ dependence in the Euclidean solution transforms into an $O(1,3)$ dependence in the Lorentzian solution. The inclusion of gravity does not change this. Hence it is useful to start by looking at the $O(1,3)$ slicing of de Sitter spacetime.

We will choose coordinates centered on the point P, which in the embedding of de Sitter spacetime as a hyperboloid (see Eq. (2.11) and Fig. 2.2) would be located at[2]

$$z = 0, \quad x_1 = 1, \quad x_i = 0 \quad (i = 2,3,4). \tag{9.38}$$

The light cone is defined by the intersection of de Sitter spacetime with the curve

$$-z^2 + \delta^{ij} x_i x_j = 0. \tag{9.39}$$

One can cover the region outside the light cone $-z^2 + \delta^{ij} x_i x_j > 0$ with the coordinates and metric

$$z = \sin\lambda \sinh\theta, \tag{9.40}$$

$$x_1 = \cos\lambda, \tag{9.41}$$

$$x_i = \sin\lambda \cosh\theta \, n^i, \tag{9.42}$$

$$ds^2 = d\lambda^2 + \sin^2\lambda(-d\theta^2 + \cosh^2\theta \, d\Omega_2^2), \tag{9.43}$$

with $\theta \in \mathbb{R}$, $\lambda \in [0,\pi)$, $n^i \in S^2$. Meanwhile, inside the light cone $-z^2 + \delta^{ij} x_i x_j < 0$, one can use the coordinates and metric

$$z = \sinh\gamma \cosh\beta, \tag{9.44}$$

$$x_1 = \cosh\gamma, \tag{9.45}$$

$$x_i = \sinh\gamma \sinh\beta \, n^i, \tag{9.46}$$

$$ds^2 = -d\gamma^2 + \sinh^2\gamma(d\beta^2 + \sinh^2\beta \, d\Omega_2^2), \tag{9.47}$$

with $\gamma \in \mathbb{R}$, $\beta \in \mathbb{R}_+$, $n^i \in S^2$. The only region not covered so far is the inside light cone of the point Q, located diametrically opposite P at $z = 0, x_1 = -1, x_i = 0$. This can be done just as above, by flipping the sign of

---

[2] We assume $H = 1$ here, as this calculation is messy enough with all the new coordinate names that one has to introduce.

x_1. The left panel in Fig. 9.4 shows a few representative lines of constant γ (inside the light cone) and constant λ (outside the light cone) on the conformal (Penrose) diagram of de Sitter spacetime. One may verify that these lines are expressed as the curves $\sin(\tau)/\cos(\chi) = constant$ in the conformal coordinates (6.138)–(6.139).

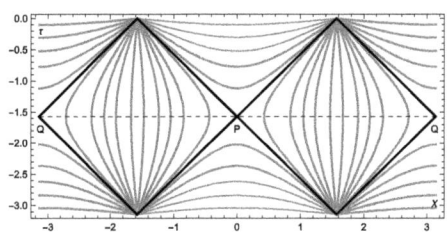 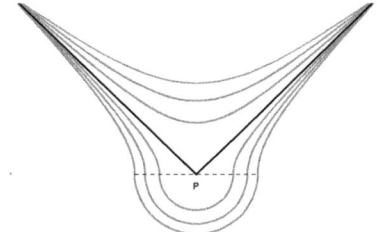

Figure 9.4 *Left:* the $O(1,3)$ "open universe" slicing of de Sitter spacetime, shown here in 2 dimensions. The vertical coordinate is conformal time τ, and the horizontal axis shows an angular coordinate χ, cf. the conformal de Sitter metric (6.139) and Fig. 6.10. Here the spatial coordinate has range $(-\pi, \pi]$ because the graph shows dS^2. *Right:* The meridian (dashed line) provides initial data for future evolution. One can see that the bubble is analytically continued to the outside light cone of the point P. Note that in this figure the bubble is rotated by a quarter turn compared to the right panel in Fig. 9.3, such that P is now seen head-on.

Let us now go back to the bubble metric (9.29), which we will write slightly more explicitly as

$$ds^2 = d\sigma^2 + \rho^2(\sigma) \left(d\psi^2 + \cos^2 \psi \, d\Omega_2^2 \right), \quad \psi \in \left[-\frac{\pi}{2}, \frac{\pi}{2} \right], \tag{9.48}$$

where the meridian $\psi = 0$ includes both poles P and Q, and is indicated by the horizontal gray line in the right panel of Fig. 9.3, and by the dashed line in Fig. 9.4. If we now analytically continue the solution via $\psi = i\theta$, we find

$$ds^2 = d\sigma^2 + \rho^2(\sigma) \left(-d\theta^2 + \cosh^2 \theta \, d\Omega_2^2 \right), \tag{9.49}$$

which one should compare with Eq. (9.43). The meridian $\psi = 0$ thus provides initial data for Lorentzian evolution, and outside of the light cone this can be obtained directly via analytic continuation, cf. also the right panel of Fig. 9.4.

The evolution inside the light cone is not directly determined by analytic continuation, although it also respects the $O(1,3)$ symmetry. A such, it proceeds according to the open $K = -1$ Friedmann equations (1.36)–(1.37), with metric

$$ds^2 = -dt^2 + a^2(t) \left(d\beta^2 + \sinh^2 \beta \, d\Omega_2^2 \right). \tag{9.50}$$

The future evolution inside the light cone in fact depends on the shape of the potential outside of the region explored during tunneling.

Now that we have described CDL tunneling mathematically, let us make a few comments on its potential relevance. CDL describes how a bubble of a new vacuum can form inside of an existing universe, and how that bubble can expand, thereby effectively becoming its own universe.

One interesting effect that could occur is that, inside bubbles, the constants of nature might be different. This would be the case if the values of certain constants are linked to the value of the scalar field (in string theory, this situation is in fact ubiquitous). Physical constants could then evolve via successive tunneling events.

A question is also whether our current vacuum could decay in the future. There is in fact some evidence that the potential of the Higgs boson is only metastable (though this statement depends on a number of assumptions, and could be changed by the presence of new particles at high energies). If that happens to be confirmed, then sooner or later (hopefully later) our region of the universe will tunnel to a region of the potential that is negative. In such an event, it is likely that our region of the universe will collapse, due to the negative vacuum energy.

A pertinent question is: Could we detect that we live in the aftermath of a bubble nucleation? Circumstantial evidence for bubble nucleations could come from observing the universe to be open, since this is a generic prediction. So far observations are however consistent with a flat spatial geometry of the universe. But there is another possibility, stemming from the realization that a bubble is not completely separated from its parent universe. In fact, it could occur that a second bubble forms very close to an already expanding bubble. In such a case the two bubbles might collide. The constant time surfaces inside of bubbles are such that the bubble wall is always to the past, regardless of where one lives inside the bubble – see Fig. 9.4. Thus, if we are currently in such a bubble, then we might be able to detect whether it collided with another bubble by looking into our past. In such a case one would expect to see ring-like disturbances in the CMB. None have been detected to date, but this is not entirely conclusive, as a long inflationary phase (occurring inside the bubble) could wash away the traces of a bubble collision. Evidently, cosmologists will keep looking for such traces.

Before deepening our study of CDL bubbles, let us point out a couple of differences with the no-boundary proposal. The first is that CDL describes transitions out of a pre-existing universe, while no-boundary solutions describe the nucleation of space and time from nothing. A mathematical difference lies also in the way that a Lorentzian solution arises: The no-boundary instantons are oriented such that the South Pole is at the "bottom," with a succession of constant time slices describing Lorentzian evolution. By contrast, it is a constant-angle slice through the CDL bounce that provides initial conditions for the Lorentzian growth of the bubble. Related to this is also the fact that no-boundary instantons lead to closed spatial slices, while CDL bounces lead to open spatial universes. Finally, we can remark that CDL bounces are always allowable, as the metric is purely Euclidean, while we saw that no-boundary instantons necessarily develop complex field values and can sometimes be excluded based on the allowability criterion.

9.1.4 Negative modes

In this section, we will tie up a couple of loose ends. In particular, we will justify that we calculated the right quantity, namely that the path integral indeed gives the decay rate, and we will also justify that the Euclidean CDL solution was indeed the relevant one to consider.

For ease of notation, we will briefly revert back to quantum mechanics, and assume that the metastable potential of the form shown in Fig. 9.1 is that of a particle with position states $|x\rangle$. Then the evolution from an initial to a final state proceeds via the Hamiltonian $\mathcal{H}$, with a complete set of eigenstates $\mathcal{H}|n\rangle = E_n|n\rangle$, according to

$$\langle x_1|e^{i\mathcal{H}t}|x_0\rangle = \sum_n e^{iE_n t}\langle x_1|n\rangle\langle n|x_0\rangle . \tag{9.51}$$

If we now assume that the relevant energy eigenvalue picks up an imaginary part, $E_0 = E_{\text{real}} - i\frac{\Gamma}{2}$, then the probability density to stay in the false vacuum $|x_{\text{fv}}\rangle$ scales as

$$|\langle x_{\text{fv}}|x_{\text{fv}}\rangle|^2 \sim e^{-\Gamma t} . \tag{9.52}$$

Thus the imaginary part of the energy provides the decay rate out of the false vacuum. And the ground state energy can be calculated in the Euclidean approach to quantum theory, as the leading contribution in the large time T limit – see Eq. (7.2).

Now for the functional integral over histories: For a single "bounce" solution x_B describing the particle tunneling over the potential barrier and back, we have

$$\int_{\text{single bounce}} \mathcal{D}x\, e^{iS} \approx F\, e^{-S_E(x_B)} . \tag{9.53}$$

But the path integral instructs us to sum over all possible paths the particle could take. Thus we must sum over bounces, which can be located at different times $t_1, t_2, \cdots, t_m$ on the interval $[-\frac{T}{2}, \frac{T}{2}]$, and occur in sequence. We will assume the dilute gas approximation, which means that we assume that these bounces are well separated from each other. Summing over all such locations gives a term

$$\frac{1}{m!} \int_{-T/2}^{T/2} dt_1 dt_2 \cdots dt_m = \frac{T^m}{m!}, \qquad (9.54)$$

where the division by $m!$ arises because the bounces are indistinguishable. The contribution to the integrand for such a multibounce solution is simply $F^m e^{-m S_E(x_B)}$. Then if we sum over all contributions, we find that the sum exponentiates, and we obtain

$$\int_{\text{all bounces}} \mathcal{D}x \, e^{iS} \approx \sum_m \frac{(F e^{-S_E(x_B)} T)^m}{m!} \qquad (9.55)$$

$$\approx e^{F e^{-S_E(x_B)}}. \qquad (9.56)$$

The dilute gas approximation is justified, as the largest terms in the sum $\sum_m x^m/m!$ arise when $x \sim m$, which in our context means that the bounce density $m/T \sim F E^{-S_E}$ is indeed very small, and thus the bounces are typically widely separated. Now we can see that if F contains an imaginary part, this will provide us with the decay rate

$$\Gamma = 2 \operatorname{Im}[F] \, e^{-S_E(x_B)}. \qquad (9.57)$$

It remains to see whether F contains an imaginary part. To figure this out, we can look at fluctuations around the bounce solution. Up to quadratic order we have

$$S_E = S_E(x_B) + \frac{1}{2} \int_{\delta x(\sigma_0)=0}^{\delta x(\sigma_1)=0} d\sigma \left((\delta x_{,\sigma})^2 + V''(x_B)(\delta x)^2 \right) + \cdots, \qquad (9.58)$$

where $V'' = V_{,xx}$. There is no linear term as we are expanding around a solution of the equation of motion. The fluctuations must vanish at the end point in order for the full solution to still satisfy the required boundary conditions. It is useful to expand the fluctuations into a complete set of eigenfunctions of the fluctuation operator,

$$\delta x = \sum_n c_n \delta x_n, \qquad (9.59)$$

where $\int d\sigma \, \delta x_n \, \delta x_m = \delta_{nm}$, and

$$\left[-\frac{d^2}{d\sigma^2} + V''(x_B) \right] \delta x_n = \omega_n \delta x_n, \qquad (9.60)$$

9.1 Vacuum decay via Coleman–DeLuccia instantons

where ω_n are the (real) eigenvalues. The integral over fluctuations then turns into a product of simple Gaussian integrals, which result in a prefactor

$$\int [\mathcal{D}x]\, e^{iS} \sim \frac{1}{\sqrt{\Pi_n \omega_n}}\, e^{-S_E(x_B)}. \tag{9.61}$$

Because of the square root, we can see that if one of the eigenvalues is negative and all others positive, then the prefactor will indeed be imaginary, and we will have a consistent picture of vacuum decay. But what about additional negative modes? Could one have three, or five negative modes? Here we must briefly think about the meaning of the fluctuations: If there is a negative mode, then it indicates a field direction along which one can lower the action. Hence additional negative modes indicate field directions that lead to more dominant solutions. As a result, the bounce solution with a single negative mode will be the relevant solution describing the decay of a false vacuum.

The preceding discussion carries over almost unchanged to the case of field theory. When gravity is included, it also remains conceptually the same, but there is the usual difficulty that arises from diffeomorphism invariance. When calculating fluctuations around a curved spacetime tunneling solution, we must be careful to eliminate gauge degrees of freedom, and retain only the physical fluctuation mode. For the remainder of this section, we will outline the associated calculation. It is entirely similar in spirit to the calculation of cosmological fluctuations in Chapter 3, but with a couple of additional technical complications: The first is that the spatial sections of the tunneling solutions are spheres, and the second is that the comoving curvature perturbation turns out not to be the best variable to use. We will see momentarily how to fix these issues.

We start by writing out a general scalar perturbation of the bubble metric (9.29), but in Euclidean conformal time $\rho\, d\eta \equiv d\sigma$,

$$ds^2 = \rho^2 \left[(1+2A)d\eta^2 + 2B_{,i} dx^i d\eta + ((1+2\psi)\gamma_{ij} + 2E_{;ij})\, dx^i dx^j \right], \tag{9.62}$$

$$\phi_{\text{total}} = \phi + \delta\phi, \tag{9.63}$$

where γ_{ij} is the metric on the 3-sphere and ; denotes its covariant derivative. Here we denoted the full scalar field by ϕ_{total}. In principle one could now expand the perturbations in terms of spherical harmonics, as we have done before. But it turns out that all higher harmonics increase the energy. In other words, they have purely positive eigenvalues. The only sector that can harbor negative modes is the homogeneous sector, where the perturbations depend solely on the time variable η and have no spatial dependence. Hence we will only look at that case.

The (quadratic) action for the perturbations becomes

$$S^{(2)} = \frac{\pi^2}{\kappa^2} \int d\eta \rho^2 \left[-6\psi'^2 + 12\mathcal{H} A \psi' - 2(\mathcal{H}' + 2\mathcal{H}^2 + 1)A^2 + 6\psi^2 + 12\psi A \right.$$
$$\left. + \kappa^2 \left(\delta\phi'^2 + \rho^2 V_{,\phi\phi} \delta\phi^2 - 6\phi' \psi' \delta\phi - 2\phi' \delta\phi' A - 2\rho^2 V_{,\phi} \delta\phi A \right) \right], \tag{9.64}$$

where $\mathcal{H} \equiv \rho'/\rho$ and here $' \equiv d/d\eta$. Interestingly, B and E have disappeared from the action for homogeneous perturbations. Note that A is a nondynamical variable, leaving us with the dynamical variables ψ and $\delta\phi$. Their canonically conjugate momenta are

$$\Pi_\psi = \frac{12\pi^2 \rho^2}{\kappa^2} \left(-\psi' - \frac{\kappa^2}{2} \phi' \delta\phi + \mathcal{H} A \right), \quad \Pi_{\delta\phi} = 2\pi^2 \rho^2 (\delta\phi' - \phi' A). \tag{9.65}$$

Under a gauge transformation, which takes the form of a time reparametrization $\eta \to \eta + \lambda$ here, all variables change,

$$\delta\psi = -\mathcal{H}\lambda, \quad \delta\Pi_\psi = \frac{-12\pi^2 \rho^2}{\kappa^2} \lambda,$$
$$\delta(\delta\phi) = \phi'\lambda, \quad \delta\Pi_{\delta\phi} = 2\pi^2 \rho^2 (\phi'' - \mathcal{H}\phi')\lambda, \quad \delta A = \lambda' + \mathcal{H}\lambda. \tag{9.66}$$

Hence we can use this gauge freedom to set one of the variables to zero, and thus fix the gauge. It is tempting to set the scalar perturbation $\delta\phi$ to zero, and thus to recover comoving gauge, which proved very useful in calculating inflationary perturbations. However, this is a problematic choice, as we know that bounce solutions continue to exist when gravity is turned off, i.e., in the limit $\kappa^2 \to 0$. This suggests that it makes more sense to eliminate one of the gravitational degrees of freedom. In this regard, a very useful choice is

$$\Pi_\psi = 0. \tag{9.67}$$

It allows us to replace A, so that the action depends solely on ψ and $\delta\phi$,

$$S^{(2)} = \frac{6\pi^2}{\kappa^2} \int d\eta \rho^2 \left[-\frac{2}{\phi'} \psi \delta\phi' - \frac{2(\mathcal{H}\phi' - \phi'')}{\phi'^2} \psi \delta\phi + \frac{\kappa^2}{2} \delta\phi^2 + \frac{6Q}{\kappa^4 \phi'^2} \psi^2 \right], \tag{9.68}$$

where we defined

$$Q \equiv 1 - \frac{\kappa^2}{6} \phi'^2 = 1 - \frac{\kappa^2}{6} \rho^2 \phi_{,\sigma}^2. \tag{9.69}$$

Note that $\delta\phi$ and ψ are canonically conjugate variables. We can now choose a single linear combination of these as our physical variable – the preceding discussion points towards using $\delta\phi$ itself. The orthogonal combination, in this case ψ, can then be integrated out (its path integral reduces to a simple Gaussian integral), with the end result

$$S^{(2)} = \pi^2 \int d\sigma \, \delta\phi \left[-\frac{d}{d\sigma}\left(\frac{\rho^3}{Q}\frac{d}{d\sigma}\right) + \rho^3 U[\phi,\rho] \right] \delta\phi, \qquad (9.70)$$

$$U[\phi,\rho] \equiv \frac{V_{,\phi\phi}}{Q} + \frac{2\kappa^2 \phi_{,\sigma}^2}{Q} + \frac{\kappa^2}{3Q^2}\left(6\rho_{,\sigma}^2 \phi_{,\sigma}^2 + \rho^2 V_{,\phi}^2 - 5\rho\rho_{,\sigma}\phi_{,\sigma} V_{,\phi}\right), \quad (9.71)$$

where we have reverted to standard (nonconformal) Euclidean time σ.

Now we need to find the eigenvalues of the fluctuation operator. In defining the eigenvalues, there is an a priori arbitrary overall (weight) function that one can choose. The most sensible way to fix this is by using the standard norm provided by general relativity,

$$\|\delta\phi\|^2 \equiv \int d^4x \sqrt{g}\,(\delta\phi)^2 = 2\pi^2 \int d\sigma\, \rho^3\,(\delta\phi)^2. \qquad (9.72)$$

With this choice, it follows that the eigenfunctions Φ_n with eigenvalues λ_n satisfy

$$-\frac{1}{\rho^3}\frac{d}{d\sigma}\left(\frac{\rho^3}{Q}\frac{d\Phi_n}{d\sigma}\right) + U\Phi_n = \lambda_n \Phi_n, \qquad (9.73)$$

with the boundary condition that they vanish at the end points (i.e., at $\sigma = 0$ and at $\sigma = \sigma_{\max}$).[3]

The fluctuation equation must typically be solved numerically. For this purpose, one needs the expansion of the effective potential near the end points,

$$U = V_{,\phi\phi} + \left(\frac{1}{6}\kappa^2 V_{,\phi}^2 + \frac{1}{8}V_{,\phi}V_{,\phi\phi\phi}\right)\sigma^2 + \mathcal{O}(\sigma^4), \qquad (9.74)$$

where the background fields are evaluated at the end point. A regular solution, close to $\sigma = 0$, then behaves as

$$\Phi = \Phi_{\text{norm}}\{1 + \frac{1}{8}(V_{,\phi\phi} - \lambda)\sigma^2 + \mathcal{O}(\sigma^4)\}, \qquad (9.75)$$

with Φ_{norm} being a normalization constant. An analogous expansion in $\xi = \sigma_{\max} - \sigma$ is valid near $\sigma_{\max}$. In general there also exists a singular solution $\Phi^{\text{sing}} \propto \frac{f(\lambda - \lambda_n)}{\xi^2}$ for some function f with the property that $f(0) = 0$. The idea is that one adjusts the eigenvalue λ in such a way that this singular

---

[3] The flat space limit of this equation is obtained by letting $\kappa^2 \to 0$ and $\rho \to \sigma$, which correctly yields $\left(-\frac{d^2}{d\sigma^2} - \frac{3}{\sigma}\frac{d}{d\sigma} + V_{,\phi\phi}\right)\Phi_n = \lambda_n \Phi_n$.

branch disappears. In this way one can determine the quantized energy eigenvalues of the eigenfunctions.

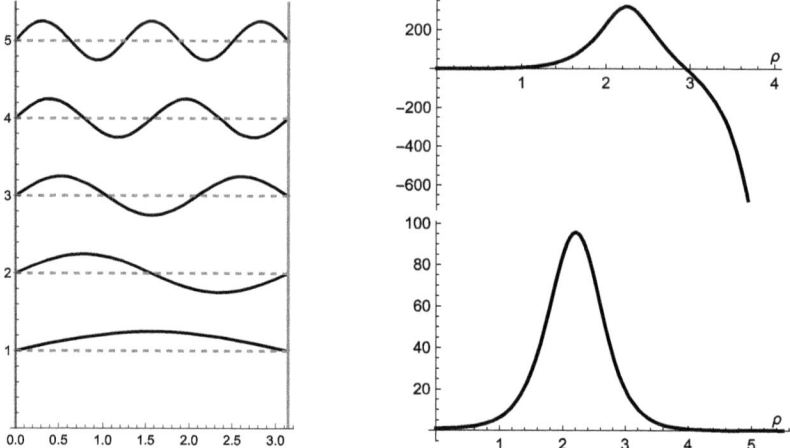

Figure 9.5 *Left:* the wave functions of a particle in a box illustrate the nodal theorem, according to which the energy eigenvalues grow with the number of nodes of the wave function. *Right:* On the top, a test solution with zero eigenvalue $\lambda_n = 0$ displays a single node, and thus suggests the existence of a negative mode. This negative mode is shown explicitly in the bottom graph, for the CDL solution described in Fig. 9.3. Here we have $\lambda_0 \approx -0.72058$.

We are chiefly interested in the question of how many negative modes exist around a given solution – that is to say, how many eigenfunctions have a negative eigenvalue λ_n. There is a useful trick that gives a clear indication as to how many negative modes one can expect, and it is based on the nodal theorem. Instead of trying to prove a general theorem, it is sufficient to take inspiration from the quantum mechanics of a particle in a box – see the left panel in Fig. 9.5. In this case, the wave function must also satisfy the boundary condition that it vanishes at the enclosing walls. The energy then grows according to the number of nodes of the wave function. Put differently, if we look at a wave function with m nodes, then we know that there exist m additional solutions with fewer nodes and lower eigenvalues.

The idea then is to simply solve the fluctuation equation (9.73) with $\lambda_n = 0$. This will not yield a proper eigenfunction in general. However, the number of nodes of the zero eigenvalue test function tells us how many lower, i.e., negative, eigenvalues there should exist. Such a solution is shown in the top-right panel of Fig. 9.5, for the CDL bounce that we had discussed earlier in connection with Fig. 9.3. As one can clearly see, it crosses zero once, and thus one expects there to exist a single negative mode. By using the optimization method described above, one can find this mode explicitly, and it is shown in the bottom-right graph in the figure. This is the only negative mode of the solution.

Note that the negative mode is concentrated near the wall of the bubble. Physically, this mode would have the effect of shrinking or enlarging the bubble in the wall region.

This calculation confirms the interpretation of the CDL bubble as describing the decay of the false vacuum, via nucleation of a bubble of true vacuum.

9.2 Disappearing act: bubbles of nothing

We just saw how one can describe the emergence of a bubble of a new vacuum inside of an existing universe, while in Chapter 7 we saw how one can describe the emergence of spacetime from nothing. Perhaps surprisingly, there is also the possibility, discovered by E. Witten, that spacetime may disappear – i.e., that spacetime may disappear into nothingness, again via tunneling. We will describe this process now.

To this end, we have to consider not the standard 4-dimensional Minkowski vacuum but rather a topologically nontrivial vacuum, known as the Kaluza–Klein vacuum, in which a fifth (circular) dimension is added. The metric is thus

$$\mathrm{d}s^2 = -\mathrm{d}t^2 + \mathrm{d}x^2 + \mathrm{d}y^2 + \mathrm{d}z^2 + R^2 \mathrm{d}\phi^2\,, \quad \phi \sim \phi + 2\pi\,, \qquad (9.76)$$

where the angular coordinate ϕ has period 2π and the corresponding circle direction has radius R. This metric, which can be trivially continued to Euclidean time $\sigma = it$, is a solution of the 5-dimensional Einstein equations in the absence of matter. We will discuss the potential relevance of such solutions with extra dimensions in more detail in Chapter 10 – the main idea is that the metric degrees of freedom of small, compact extra dimensions can look like gauge fields and scalar fields from the point of view of 4 dimensions. Thus additional dimensions can allow for a geometrical unification of gravity and matter.

For our present purposes, we will simply accept (9.76) as our starting spacetime. If we are investigating the possibility of tunneling out of this spacetime, then we must search for a Euclidean solution with same asymptotic structure as the Euclidean version of (9.76), but which can be analytically continued to a different Lorentzian evolution. Such a solution could then mediate the decay of the Kaluza–Klein vacuum – i.e., it could provide the saddle point for a path integral having as initial boundary a space-like slice of (9.76), and as final boundary a space-like slice of the new vacuum.

Of particular interest in this respect are the 5-dimensional equivalents of the Schwarzschild black hole solution. With time rotated into the Euclidean, they are of the form

$$ds^2 = \frac{dr^2}{1 - \frac{\alpha}{r^2}} + r^2 d\Omega_3^2 + \left(1 - \frac{\alpha}{r^2}\right) R^2 d\phi^2, \qquad (9.77)$$

where α is a parameter analogous to the mass M of the ordinary Schwarzschild solution. By following the same arguments as those given when analyzing the Euclidean Schwarzschild solution in Section 6.4, it is straightforward to see that the solution above is singular unless $\alpha = R^2$ and $r \geq R$,

$$ds^2 = \frac{dr^2}{1 - \frac{R^2}{r^2}} + r^2 \left(d\theta^2 + \cos^2\theta \, d\Omega_2^2\right) + \left(1 - \frac{R^2}{r^2}\right) R^2 d\phi^2, \quad r \geq R. \tag{9.78}$$

Now the solution caps off smoothly at $r = R$. More precisely, if we look at the circle direction ϕ, we can see that at large r the circle has radius R, just as in the Kaluza–Klein vacuum. However, as r becomes smaller the circle radius shrinks, and at $r = R$ it rounds off smoothly – see also Fig. 9.6.

Figure 9.6 *Left:* The Kaluza–Klein vacuum solution is a product of 4-dimensional Minkowski spacetime with a small, circular, fifth direction. *Right:* The bubble of nothing solution can mediate the nucleation of a hole in this spacetime. After "materialization," the hole expands outwards and converts more and more space to nothing.

In a similar manner as for CDL bubbles, one can now find an analytic continuation to Lorentzian spacetime by rotating the angle θ to $t = i\theta$, with the result that

$$ds^2 = \frac{dr^2}{1 - \frac{R^2}{r^2}} + r^2 \left(-dt^2 + \cosh^2 t \, d\Omega_2^2\right) + \left(1 - \frac{R^2}{r^2}\right) R^2 d\phi^2, \quad r \geq R. \tag{9.79}$$

For CDL bubbles, the analytic continuation only covered the outside of the light cone of the nucleation point, and one had to solve the equations of motion to obtain the subsequent evolution inside the light cone. Here, there simply is no spacetime for $r \leq R$. There is only a hole in the spacetime, which starts at rest and then expands, with its area given by $4\pi R^2 \cosh^2 t$, as implied by a constant t slice of (9.79). Thus the hole expands with a speed quickly approaching that of light, and converts more and more space to nothing. Other such holes may form elsewhere, and even collide with each other. Note that the bubble of nothing mediates a topological instability of the Kaluza–Klein vacuum.

To figure out how likely this process might be, we have to calculate the tunneling rate, defined in Eq. (9.11). In order to calculate the difference in action between the bubble of nothing and the Kaluza–Klein vacuum, we should first notice that both are solutions to the vacuum Einstein equations – hence their Ricci scalars (and thus their Einstein–Hilbert actions) both vanish. However, because the metric is being held constant asymptotically, one needs to add GHY terms at a fixed value of the radius r,

$$B = \frac{1}{8\pi G_5} \int_{\partial \mathcal{M}} \sqrt{h} \left(K_{\text{bubble}} - K_{\text{Kaluza–Klein}} \right). \tag{9.80}$$

At the end of the calculation, the boundary will be sent to infinity. In order to calculate the extrinsic curvatures, we may use the formula that $K = \nabla_\mu n^\mu = \frac{1}{\sqrt{h}} \partial_r (\sqrt{h} n^r)$, where n^μ is the unit normal to the surface $\partial \mathcal{M}$. For the bubble of nothing, we have $\sqrt{h} = r^3 R \sqrt{1 - \frac{R^2}{r^2}} \sin^2 \phi \sin \psi$ and $n^r = \sqrt{1 - \frac{R^2}{r^2}}$. This leads to

$$K_{\text{bubble}} = \frac{3r^2 - R^2}{r^3 \sqrt{1 - \frac{R^2}{r^2}}}, \quad K_{\text{Kaluza–Klein}} = \frac{3}{r}. \tag{9.81}$$

Putting it all together, we find

$$B = \frac{1}{8\pi G_5} (2\pi^2)(2\pi R) r^3 \sqrt{1 - \frac{R^2}{r^2}} \left[\frac{3r^2 - R^2}{r^3 \sqrt{1 - \frac{R^2}{r^2}}} - \frac{3}{r} \right]$$

$$= \frac{\pi^2 R^3}{4 G_5} + \mathcal{O}\left(\frac{1}{r^2}\right), \tag{9.82}$$

where we can drop subdominant terms in the limit that the boundary is moved to infinity. If we relate the 4- and 5-dimensional Newton constants as $G_5 = 2\pi R G_4 = R/4$, where the last relation uses units where $8\pi G_4 = 1$, then we obtain a nucleation rate per unit volume per unit time of

$$\frac{\Gamma}{V} \approx e^{-B} \approx e^{-\pi^2 R^2}. \tag{9.83}$$

For an asymptotic circle radius that is large in Planck units, this rate is (perhaps fortunately) highly suppressed.

A few comments: 1. Similar bubble of nothing solutions may be found in theories with different shapes and different numbers of extra dimensions. 2. Arguments have been put forward that the bubble of nothing also admits a single negative mode in its fluctuation spectrum, which would be essential to justify the interpretation of a semiclassical decay process. However, this is a topic that has not been studied very thoroughly thus far. 3. It remains unclear to what extent we should worry about such instabilities. Even disregarding the fact that there exists no evidence to date for additional dimensions, it is known that in certain theories additional fields can stabilize the Kaluza–Klein vacuum – in particular, the presence of fermionic fields has such a stabilizing effect. Here, also, a complete overview is still lacking.

But perhaps the main point to be taken from this discussion is the ease with which one can use a Euclidean/complex time formalism to describe semiclassical processes that would otherwise be difficult to handle, let alone to discover.

9.3 Wormholes

Wormholes – tunnels in space or in spacetime – have long figured prominently in science fiction, mostly as a quick way of getting from one part of the universe to a distant part. In this section, we will discuss to what extent such descriptions are realistic according to presently understood physics, and also what other effects wormholes may cause. We will treat Lorentzian wormholes first, and Euclidean ones next.

9.3.1 Lorentzian wormholes

A wormhole is a solution to the Einstein-plus-matter equations of motion that connects two distant regions of spacetime via a tunnel, reaching a minimum size b, say, at the throat. For a sketch see Fig. 9.7. Since we are analyzing the existence of classical solutions here (rather than transition amplitudes), we will require the fields to take real values. The very simplest metric describing such a wormhole is the Morris–Thorne metric

$$\mathrm{d}s^2 = -\mathrm{d}t^2 + \mathrm{d}l^2 + (b^2 + l^2)(\mathrm{d}\theta^2 + \sin^2\theta \mathrm{d}\phi^2), \tag{9.84}$$

where t is a time-like coordinate, l a radial one, and θ, ϕ angular coordinates (for simplicity we have assumed the tunnel to have spherical symmetry). At large $l \to \pm\infty$, the metric reduces to that of flat Minkowski spacetime.

Meanwhile, at $l = 0$ the tunnel reaches a minimum radius of size b, implying that the wormhole has minimum area $4\pi b^2$ at $l = 0$. Such a wormhole can connect two separate universes, or connect arbitrarily distant (almost flat) regions of a single universe. This last possibility is the one usually considered for fast interstellar or intergalactic travel in science fiction works. Note that this metric does not contain a horizon (which would have corresponded to the location where $g_{tt} = 0$). This is helpful, as a horizon is classically a one-way membrane, which would have made it impossible to re-emerge from the wormhole.

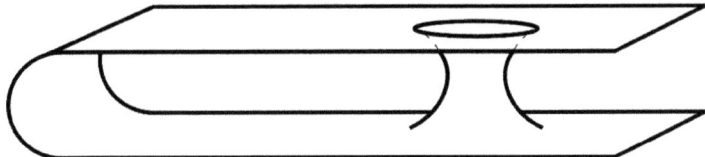

Figure 9.7 A sketch of a wormhole connecting arbitrarily distant parts of our universe (depicted as a 2-dimensional sheet here), in order to provide a shortcut for intergalactic travel.

Writing down a metric is of course not sufficient to be in possession of a genuine solution of the gravity–matter system. One must also determine what kind of matter is required to support the solution. To this end, we may calculate the required stress–energy tensor. Using $R_{l\theta l\theta} = R_{l\phi l\phi} = -R_{\theta\phi\theta\phi} = -\frac{b^2}{b^2+l^2}$, we may deduce that the Morris–Thorne wormhole would require matter with the property that

$$T_{tt} \equiv \rho = -\frac{b^2}{(b^2 + l^2)^2}, \tag{9.85}$$

$$T_{ll} \equiv p\, g_{ll} = -\frac{b^2}{(b^2 + l^2)^2}, \tag{9.86}$$

$$T_{\theta\theta} \equiv p_\theta\, g_{\theta\theta} = \frac{b^2}{b^2 + l^2}, \tag{9.87}$$

$$T_{\phi\phi} \equiv p_\phi\, g_{\phi\phi} = \frac{b^2 \sin^2\theta}{b^2 + l^2}. \tag{9.88}$$

The components of the stress–energy tensor can be interpreted as representing the energy density ρ, the radial pressure p, and the angular pressures $p_{\theta,\phi}$ of the matter. Thus, as we can see, the energy density ρ is required to be negative in order for this wormhole solution to exist, as is the radial pressure p. This implies that this geometry would require a matter source violating the null energy condition (which requires $\rho + p \geq 0$). No stable, physically reasonable matter with this feature is currently known (with a small proviso, to which we will return below: Quantum fluctuations can violate the null energy condition for short time spans, though not on average).

One can investigate much more elaborate metric ansätze to try to build more realistic wormholes, but in fact there exists a general argument that shows that all traversable Lorentzian wormholes must violate the null energy condition. By traversable we mean that at the very least a light signal can pass through the wormhole (should the solution pass this test, then one would try to see if a human could also travel through the tunnel). Consider then a family (congruence) of null geodesics $x^\mu(\lambda)$, parametrized by an affine parameter λ and with tangent vector k^μ. We are interested in the behavior of the cross-sectional submanifold at constant λ. To define a metric on this subspace, we need an additional, auxiliary, null vector n^μ, which we take to satisfy $k^\mu n_\mu = -1$. Then a transverse metric is given by

$$h_{\mu\nu} = g_{\mu\nu} + k_\mu n_\nu + k_\nu n_\mu. \tag{9.89}$$

The rate of change of a deviation vector v^μ is given by

$$\nabla^{(\lambda)} v^\mu \equiv k^n \nabla_n v^\mu = D_{\mu\nu} v^\nu, \tag{9.90}$$

where the index n refers to the normal direction here and $D_{\mu\nu}$ is the deviation tensor. Only the projection $P_{\mu\nu} = h_\mu^\rho h_\nu^\sigma D_{\rho\sigma}$ actually contributes to this equation. One can then decompose the deviation tensor into expansion θ (trace), shear $\sigma_{(\mu\nu)}$ (trace-free symmetric part), and vorticity $\omega_{[\mu\nu]}$ (antisymmetric part),

$$P_{\mu\nu} = \frac{1}{2}\theta h_{\mu\nu} + \sigma_{\mu\nu} + \omega_{\mu\nu}. \tag{9.91}$$

The expansion θ can be thought of as specifying how a family of null geodesics grows apart or comes closer together. Its evolution is governed by the Raychaudhuri equation

$$\frac{d\theta}{d\lambda} = -\frac{1}{2}\theta^2 - \sigma^{\mu\nu}\sigma_{\mu\nu} + \omega^{\mu\nu}\omega_{\mu\nu} - R_{\mu\nu}k^\mu k^\nu. \tag{9.92}$$

If we are considering a congruence that is hypersurface orthogonal (such as radial null geodesics), then these are specified by some scalar function f according to $k^\mu \propto \nabla^\mu f$, and in that case the vorticity ω vanishes. This is the case of interest to us.

In a wormhole geometry, by definition the null geodesics must at first come closer together (focus) and then, after reaching the throat, they must move apart again (defocus). This means that in the first part of the geometry we need $\frac{d\theta}{d\lambda} < 0$, which (9.92) shows is easy to achieve. In the second part of the geometry, the wormhole must expand again, and here we need defocusing, $\frac{d\theta}{d\lambda} > 0$. This is where the problem lies: On the right-hand side of (9.92), the expansion and shear terms are negative. Moreover, the Einstein equations imply that for a null vector k^μ we have $R_{\mu\nu} k^\mu k^\nu = T_{\mu\nu} k^\mu k^\nu$. Thus, in order to obtain defocusing, we would need to have

$$T_{\mu\nu} k^\mu k^\nu < 0, \qquad (9.93)$$

which corresponds to a violation of the null energy condition. Thus, for matter satisfying the null energy condition, traversable Lorentzian wormholes are impossible.

We should mention a notable, and illuminating, special case. Quantum fluctuations can lead to momentary, localized, violations of the null energy condition (they do however satisfy an averaged null energy condition). This can lead to a small amount of defocusing. It has in fact been possible to construct examples of traversable wormholes making use of this effect (the details can be found in the references – see Appendix E). However, because the violation of the null energy condition is necessarily tiny, it turns out that the wormhole must be very long, longer in fact than the ambient distance between the two mouths of the wormhole. Thus wormholes of this kind represent a detour, rather than a shortcut, in terms of intergalactic travel. This is another instance where quantum theory and relativity work remarkably well together – the quantum fluctuations are precisely such that the ensuing wormhole respects causality, which is one of the basic principles that relativity is built upon.

The no-go result that we just discussed applies specifically to traversable wormholes, as it was built on analyzing light rays passing through the tunnel. If we relax the assumption of traversability, then it turns out that it is in fact quite straightforward to find Lorentzian wormholes, with reasonable matter support. We will look at the best-known example in Exercise 9.5. Such wormholes cannot be used for travel, but they have started stimulating more interest in recent years, as they appear to be related to entangled quantum states, which also have the property of exhibiting space-like separated connections. This new interplay between gravity and quantum theory is a research area in rapid development.

9.3.2 Euclidean wormholes

We will now look at Euclidean wormholes, which may rather be thought of in the context of transition amplitudes (much like the Coleman–DeLuccia solutions we described earlier), either connecting different parts of the universe (if the two mouths of the wormhole are located in our universe) or having the interpretation of mediating the nucleation of a "baby universe" (if only one mouth is located in our universe, and the second one corresponds to an initial configuration for the evolution of a separate universe) – see Fig. 9.8.

We will explicitly describe the Euclidean wormhole solutions first found by S. Giddings and A. Strominger (GS), and which have subsequently been generalized in many contexts, and then discuss their potential implications as well as some associated puzzles.

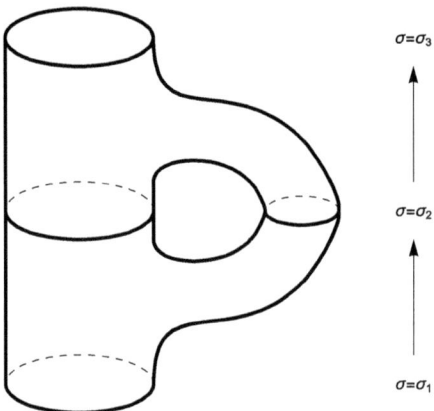

Figure 9.8 A Euclidean space with a baby universe pinching off and then reattaching itself. Different 3-dimensional sections through this space can have different topologies, as sketched here with the slices at $\sigma = \sigma_{1,2,3}$. The entire "handle" may also be regarded as a wormhole connecting distant parts of the space.

These solutions make use of a 2-form gauge field $B_{\mu\nu}$, which may be regarded as a generalization of the electromagnetic vector potential, except that it now contains an extra index. This field, which appears naturally in string theory, is antisymmetric in its indices and has a field strength $H_{\mu\nu\rho} = 3\partial_{[\mu} B_{\nu\rho]}$, also antisymmetric in its indices. In 4 dimensions, the 2-form could be dualized to a scalar A according to

$$H_{\mu\nu\rho} = \epsilon_{\mu\nu\rho}{}^{\sigma} \partial_{\sigma} A, \qquad (9.94)$$

where $\epsilon_{\mu\nu\rho\sigma}$ is a (Levi-Civita) tensor antisymmetric in all of its indices, and that takes the numerical values $\{0, \pm 1\}$ in a local frame. The pseudoscalar A is often called an *axion* field, and accordingly the wormholes we will describe are sometimes called axionic wormholes. Here we will work directly with the 2−form.

We will consider this 2-form field coupled not only to gravity, but also to an additional scalar field ϕ, with Euclidean action

$$S_E = \int d^4x \sqrt{g} \left(-\frac{1}{2}R + \frac{1}{2}\nabla_\mu \phi \nabla^\mu \phi + \frac{1}{12f^2} e^{-\beta\phi} H_{\mu\nu\rho} H^{\mu\nu\rho} \right), \quad (9.95)$$

where β and f are coupling constants. In a string theory context, ϕ is called the *dilaton* field, and the associated exponential coupling is typical.

The equations of motions are

$$R_{\mu\nu} = \partial_\mu \phi \partial_\nu \phi + \frac{1}{2f^2} e^{-\beta\phi} H_{\mu\rho\sigma} H_\nu^{\rho\sigma} - \frac{1}{6f^2} e^{-\beta\phi} H_{\gamma\rho\sigma} H^{\gamma\rho\sigma} g_{\mu\nu}, \quad (9.96)$$

$$\frac{1}{\sqrt{g}} \partial_\mu (\sqrt{g} g^{\mu\nu} \partial_\nu \phi) = -\frac{\beta}{12f^2} e^{-\beta\phi} H_{\gamma\rho\sigma} H^{\gamma\rho\sigma}, \quad (9.97)$$

$$\partial_\mu (\sqrt{g} e^{-\beta\phi} H^{\mu\rho\sigma}) = 0. \quad (9.98)$$

We will focus on a homogeneous, spherically symmetric ansatz

$$ds^2 = N_E(\sigma)^2 d\sigma^2 + a(\sigma)^2 d\Omega_3^2, \quad (9.99)$$

$$\phi = \phi(\sigma), \quad (9.100)$$

$$H_{0ij} = 0, \quad H_{ijk} = q\epsilon_{ijk}, \quad (9.101)$$

where ϵ_{ijk} is the volume form on the 3-sphere, and q determines the amount of form-field *flux* threading through the sphere (more explicitly, $\int_{S^3} H = 2\pi^2 q$). The Euclidean lapse function is denoted N_E above. With this ansatz, the equations of motion reduce to

$$\frac{\dot{a}^2}{N_E^2} - 1 = \frac{1}{6N_E^2} a^2 \dot{\phi}^2 - \frac{q^2}{6f^2 a^4} e^{-\beta\phi}, \quad (9.102)$$

$$\ddot{\phi} + \left(\frac{3\dot{a}}{a} - \frac{\dot{N}_E}{N_E} \right) \dot{\phi} + \frac{\beta q^2 N_E^2}{2f^2 a^6} e^{-\beta\phi} = 0. \quad (9.103)$$

The GS wormhole solution is given by

$$a(\sigma) = \sigma, \quad (9.104)$$

$$N_E(\sigma) = \left(1 - \frac{a_0^4}{\sigma^4} \right)^{-1/2}, \quad (9.105)$$

$$e^{\beta\phi(\sigma)} = \frac{q^2}{6f^2 a_0^4} \cos^2 \left[\frac{\beta}{\beta_c} \arccos(\frac{a_0^2}{\sigma^2}) \right], \quad (9.106)$$

where β_c denotes the maximal value of the dilaton coupling above which no solution exists, with

$$a_0^4 = \frac{q^2}{6f^2} \cos^2 \left(\frac{\pi}{2} \frac{\beta}{\beta_c} \right), \quad 0 \leq \beta < \beta_c = \frac{2\sqrt{2}}{\sqrt{3}}. \quad (9.107)$$

The geometry above asymptotes to flat space as $\sigma \to \infty$, and the range of the Euclidean "time" coordinate is restricted to $\sigma \in [a_0, \infty)$, with $\sigma = a_0$ corresponding to a mere coordinate singularity. In fact, the solution above represents a semiwormhole, which one would use to describe the nucleation of a baby universe. A full (smooth) wormhole, connecting distant parts of the universe, is obtained by gluing two such solutions together at the throat $\sigma = a_0$. The action of the semiwormhole is calculated to be

$$S_{\text{E,GS}} = \frac{4\pi^2 q}{\beta f} \sin\left(\frac{\pi}{2} \frac{\beta}{\beta_c}\right). \qquad (9.108)$$

For the full wormhole, one would multiply the result by two.

Two comments on the dilatonic coupling: First, one can also obtain a dilaton-independent wormhole by taking the limit $\beta \to 0$ in the above solution. And second, in string theory solutions there can be several axionic fields, each with a different dilatonic coupling. Then β represents the coupling to an effective axionic direction (in the space of all axions), and this coupling can differ from the couplings of the individual axions. It has typically been in this way that actual stringy wormholes have been constructed.

There exist many generalizations of these wormhole solutions, not only to multiple axions, but also for instance including a dilaton mass or a more complicated dilaton potential (these solutions are usually "only" known numerically). Multiwormhole solutions, in which the size of the tunnel grows and shrinks several times, can also be found. Given that a wide range of such Euclidean solutions exist, we now turn our attention to their possible physical effects.

Using techniques entirely analogous to those used in Section 9.1.4, one can show that the GS wormholes discussed explicitly above contain no negative modes. By contrast, the more elaborate solutions just mentioned (especially those in which the scale factor oscillates) contain negative modes. Typically the number of negative modes increases with the complexity of the solution. However, the main point is that in certain theories there exist wormhole solutions without negative modes. One expects those to contribute directly to the gravitational path integral. Put differently, any gravitational transition amplitude will be affected by the presence of wormholes, whose two ends can appear at arbitrary locations in all geometries interpolating between initial and final conditions. Can we estimate the effect of such wormholes?

One should note first of all that the presence of wormholes introduces a kind of nonlocality into the "ambient" spacetime, as any two arbitrarily distant points can be connected. That said, especially if the wormholes have a characteristic thickness that is of order the Planck length, one may hope that an effective description of all these wormholes may emerge. A further helpful insight is that the properties of the wormholes do not depend on the separation between their mouths in the ambient spacetime, since the wormholes have no support in the ambient space. In the following we will show how this kind of situation can be modeled.

A single wormhole introduces factors in the action at the locations x_1, x_2 of its mouths. This provides a contribution of the form

$$\frac{1}{2}\Delta_{ab}\mathcal{O}_a(x_1)\mathcal{O}_b(x_2), \qquad \Delta_{ab} \sim e^{-S_E^{(\text{wormhole})}}, \qquad (9.109)$$

where we leave the explicit form of the insertions implicit, but where we expect the magnitude of such a contribution to be suppressed by the wormhole action. We also assume an implicit sum over the discrete a, b labels. Since the wormholes can attach anywhere, one has to integrate this factor over all possible x_1, x_2 locations, so that the effect of a single wormhole may be expressed as

$$C \equiv \frac{1}{2} \int d^4x_1 \sqrt{g_1} \int d^4x_2 \sqrt{g_2} \Delta_{ab}\mathcal{O}_a(x_1)\mathcal{O}_b(x_2). \qquad (9.110)$$

But of course there could be multiple wormholes. Thus we need to sum over all numbers n of wormholes, each contributing a factor of C, though we must also divide each contribution by $n!$ to avoid overcounting, since the wormholes are indistinguishable. If the wormholes have a small thickness, they may certainly be treated as being independent. The total contribution is thus of the form

$$\sum_n \frac{C^n}{n!} = e^C = e^{\frac{1}{2} \int d^4x_1 \sqrt{g_1} \int d^4x_2 \sqrt{g_2} \Delta_{ab}\mathcal{O}_a(x_1)\mathcal{O}_b(x_2)}, \qquad (9.111)$$

where we recognize the Taylor expansion of the exponential function, and thus realize that the total contribution exponentiates.

Now there is an interesting manipulation that one can do, by reverse engineering a Gaussian integral over some parameters α_a and rewriting

$$e^C = \prod_a \int d\alpha_a \, e^{-\frac{1}{2}\alpha_a \Delta_{ab}^{-1} \alpha_b} \, e^{\alpha_a \int d^4x \sqrt{g} \mathcal{O}_a(x)}. \qquad (9.112)$$

Quite remarkably, this rewriting has managed to replace the nonlocal expression (9.111) with a local one, although it required the introduction of a new set of parameters α_a. Note that these parameters appear as coupling constants, changing the already existing couplings of the theory. More explicitly, if a certain operator $\mathcal{O}_a(x)$ was already present in the theory with coupling λ_a (for the case where $\mathcal{O}$ is the identity, the coupling corresponds to the cosmological constant), this changes the coupling to $\lambda_a - \alpha_a$. Thus we see that the effect of the wormholes is to change the coupling constants of the theory! Depending on one's point of view, this may introduce a new level of quantum gravitational uncertainty into physics or, more optimistically, in a specific theory it may allow us to calculate the expected values of the coupling constants, at least in a statistical sense since (9.112) contains a Gaussian measure.

We can develop the interpretation of the alpha parameters a little further. As we discussed previously, the sprouting of a wormhole may be seen as the creation of a baby universe. If this reconnects to the original universe, then it may be seen as the annihilation of a baby universe. We may thus introduce baby universe creation and annihilation operators $\hat{a}_a^\dagger$, $\hat{a}_a$, satisfying the usual commutation relations $[\hat{a}_a, \hat{a}_b^\dagger] = \delta_{ab}$, and with a, b now labeling different kinds of baby universes. If we denote by $|0\rangle$ the state where there are no baby universes, then one can straightforwardly define states containing, e.g., n baby universes of a type according to

$$|n_a\rangle = \frac{(\hat{a}_a^\dagger)^n}{n!}|0\rangle. \tag{9.113}$$

We can then also define the operators

$$\hat{A}_a = \hat{a}_{a^*}^\dagger + \hat{a}_a, \qquad \hat{A}_a|\alpha\rangle = \alpha_a|\alpha\rangle, \tag{9.114}$$

where a^* labels the CPT conjugate (meaning charge conjugated, parity transformed and time reversed) baby universe of a, and where we have introduced *alpha vacua* $|\alpha\rangle$ that are eigenstates of $\hat{A}_a$ (note that the $\hat{A}_a$ commute amongst themselves and with the Hamiltonian). The presence of wormholes may then be seen as correcting the Euclidean action according to

$$S_\mathrm{E} \to S_\mathrm{E} - \sum_a \hat{A}_a \int \mathcal{L}_a = S_\mathrm{E} - \sum_a \alpha_a \int \mathcal{L}_a. \tag{9.115}$$

Several scenarios are now possible. One can for instance consider the possibility that we live in a certain alpha vacuum. This would be a kind of superselection sector of the theory. In that case all values α_a would be fixed, and there would be no interference with other alpha vacua – the baby universes would remain in the same alpha vacua for all times. A second possibility is that we would live in a superposition of alpha vacua. In that case, the coupling constants would only be determined statistically.

A third possibility that has been considered is to assume that the total Hilbert space of baby universes is only 1-dimensional. In such a case, there would be no information loss into baby universes, because they would not have any information storage capacity. Such a drastic assumption would require highly nontrivial cancellations between the contributions of different baby universes. In other words, in the gravitational path integral there would have to be many nontrivial cancellations between contributions of different topologies. Toy models have been constructed that actually lend support to this hypothesis, though this is still a research area in development.

It should be stressed that it has been impossible to perform precise calculations of the effects of wormholes in realistic theories so far. However, the arguments that we reviewed here should make it clear that their effects may well end up being truly profound, and that they are likely to reveal important quantum gravitational properties of the microphysical makeup of spacetime.

Exercises

9.1 Extend the thin wall approximation to the case where gravity is included. Derive the new bubble nucleation rate to leading order in gravity corrections, assuming that the potential value of the true vacuum is negligibly small.

9.2 Numerically verify the example of a CDL solution shown in Fig. 9.3. For this, you may devise an optimization algorithm making use of overshoot/undershoot until the solution is reached to the desired level of precision. The origin $\sigma = 0$ is a regular singular point. Hence you must also derive the appropriate series expansions for the fields.

9.3 Extending the work of the previous exercise, try to find examples of oscillating instantons, in which the scalar field crosses the potential barrier several times before the scale factor returns to zero. You should find that these solutions have one extra negative mode for each additional crossing of the potential barrier, indicating that there are field directions in which the action can be lowered. Such solutions thus do not provide the dominant decay channel, and only the CDL solution with a single interpolation describes the decay of the false vacuum reliably.

9.4 One of the greatest mysteries in cosmology is the fact that the dark energy is currently tiny, but nonzero. In this exercise we will explore the possibility that CDL bubble nucleations could help in explaining its small value. For this purpose, consider the "washboard" potential, which is an inclined sinusoidal potential,

$$V(\phi) = M^4 \cos\left(\frac{\phi}{f}\right) - \epsilon \frac{\phi}{2\pi f} + V_0, \tag{9.116}$$

where M is a mass scale, f a parameter, and ϵ a small constant, while V_0 characterizes all other contributions to the vacuum energy. What are the requirements on the parameters if this model is to explain the current value of dark energy? Is there a reason to think that smaller values of the potential are more likely than larger ones? Is such a model viable within the standard inflationary scenario?

Also consider this model in the context of a cyclic universe. Disregarding the details of cosmic bounces, might the washboard model offer a possible explanation of the current value of dark energy?

9.5 Consider again the Schwarzschild solution (6.77). Rather surprisingly, the fully extended solution contains a wormhole as part of its geometry. This is known as the *Einstein–Rosen* bridge. It will be useful to take another look at the conformal diagram, presented in Fig. 6.3. Start by looking at the $t = 0$ space-like slice, and change coordinates to $w^2 = r - 2M$. Describe the resulting space. Is the wormhole traversable? Discuss the meaning of the horizon. If you feel like exploring this further, look at other space-like slices (note that these are not given by constant t surfaces in general, as the t coordinate becomes space-like inside of the horizon).

10
Going deeper yet: stringy cosmology

So far, we have focused on the quantization, either via canonical or path integral methods, of 4-dimensional general relativity coupled to various matter fields. At the semiclassical level, this approach can yield valuable insights into quantum effects in cosmology. However, ultimately we will want to know what a full theory of quantum gravity implies for the creation, development, and future evolution of our universe. Here we are limited by the fact that, to date, no full theory of quantum gravity exists. The most developed candidate theory is string theory, and for this reason we will look at several of its new features in the present chapter.

We should clarify at the outset that most of what is known with confidence about string theory pertains to situations that are not directly relevant to cosmology: The only solutions for which quantum corrections are under control are certain time-independent solutions, and the most comprehensive understanding of string theory is in situations where the space asymptotically approaches Anti-de Sitter space. Thus, it should be understood that string cosmology is still in its infancy. That said, it is clear that string cosmology is a topic of great interest and promise.

In this chapter, we will simply introduce a few of the conceptually distinct features that string theory implies and whose existence would revolutionize our geometrical view of the universe. More specifically, we will discuss extra dimensions and some of their phenomenological consequences. Moreover, we will discuss branes and present two examples of their dynamics: a model of brane inflation, and a model of the big bang as a brane collision. This chapter will (necessarily) be less pedagogical than the rest of the book, and can in no way replace a course on the fundamentals of string theory. But hopefully it will provide the reader with a glimpse of the new ideas involved, and an entry point into the literature.

10.1 Strings require extra dimensions

10.1.1 Lightning review of string theory

We will start with a lightning review of how string theory is constructed. As the name says, the theory starts with the quantization of strings, which have a 2-dimensional worldsheet consisting of one space (σ) and one time (τ) dimensions (the letters σ, τ are conventional in this context and unrelated to earlier uses in this book). The quantization is usually performed in Euclidean time, and then the two dimensions are conventionally denoted σ^α with $\alpha \in (0,1)$.

In analogy with the action of a relativistic point particle, whose action is the spacetime length it traces out, the action for a string is given by its area, called the *worldsheet*. We can write it in terms of the metric induced on the worldsheet, which in an ambient D-dimensional Minkowski space is the pullback of the Minkowski metric $\gamma_{\alpha\beta} = \frac{\partial X^\mu}{\partial \sigma^\alpha} \frac{\partial X^\nu}{\partial \sigma^\beta} \eta_{\mu\nu}$ with $\mu, \nu, \ldots = 0, \ldots, D-1$ and X^μ being the spacetime coordinates of the string. The *Nambu–Goto* action is thus

$$S_{\rm NG} = -T \int d\mathcal{A} = -T \int d^2\sigma \sqrt{-\det(\gamma)}, \qquad (10.1)$$

where T denotes the tension (mass per unit length) of the string. The tension is conventionally replaced by $\alpha' = \frac{1}{2\pi T} = l_s^2$, which is also related to the string length scale l_s.

The Nambu–Goto action is difficult to quantize, because of the square root. A better suited action is the *Polyakov* action

$$S_{\rm P}^{\rm Mink} = \frac{1}{4\pi\alpha'} \int d^2\sigma \sqrt{g} g^{\alpha\beta} \partial_\alpha X^\mu \partial_\beta X^\nu \eta_{\mu\nu}, \qquad (10.2)$$

which you will show to be equivalent to the Nambu–Goto action in Exercise 10.1. It involves D scalars X^μ living on the worldsheet, and which again have the interpretation of being the coordinates of the string in the spacetime it moves in (called the *target space*). In addition to Poincaré and diffeomorphism invariance, the Polyakov action has the important property of being invariant under rescalings of the metric $g_{\alpha\beta}(\sigma) \to \Omega^2(\sigma) g_{\alpha\beta}(\sigma)$. This property is called *Weyl invariance*.

The Polyakov action can be generalized such that the string moves in the presence of more general background fields, the action becoming

$$S_{\rm P} = \frac{1}{2\pi\alpha'} \int d^2\sigma \sqrt{g} \left(g^{\alpha\beta} \partial_\alpha X^\mu \partial_\beta X^\nu G_{\mu\nu} + i\epsilon^{\alpha\beta} \partial_\alpha X^\mu \partial_\beta X^\nu B_{\mu\nu} + \alpha' \Phi R \right). \qquad (10.3)$$

Here $G_{\mu\nu}$ is now a general target space metric, $B_{\mu\nu}$ is an antisymmetric field (the factor of i being due to working in Euclidean time), while Φ is a scalar field called the *dilaton* and which couples to the 2-dimensional Ricci scalar (more on this below). Furthermore, one can add fermionic fields and construct a supersymmetric worldsheet theory (supersymmetry is a symmetry that relates bosons and fermions, whose details we will not require here). We will not write out the fermions explicitly.

When quantizing the theory, by which one means quantizing the fields $X^\mu(\sigma)$ and their fermionic counterparts, it is important to verify that the classical symmetries remain preserved, i.e., that no anomalies are present. The Weyl symmetry ensures that classically the trace of the stress–energy tensor vanishes. Requiring this to also be true in the quantum theory leads to the condition that the following functions (related to the renormalization group β-functions) must vanish:

$$\beta_{\mu\nu}(G) = \alpha' \left(R_{\mu\nu} + 2\nabla_\mu \nabla_\nu \Phi - \frac{1}{4} H_{\mu\lambda\rho} H_\nu{}^{\lambda\rho} \right) + \mathcal{O}(\alpha'^2), \tag{10.4}$$

$$\beta_{\mu\nu}(B) = \alpha' \left(-\frac{1}{2} \nabla^\lambda H_{\lambda\mu\nu} + \nabla^\lambda \Phi H_{\lambda\mu\nu} \right) + \mathcal{O}(\alpha'^2), \tag{10.5}$$

$$\beta(\Phi) = \frac{1}{4}(D-10) + \alpha' \left(\nabla^\mu \Phi \nabla)\mu \Phi - \frac{1}{2} \Box \Phi - \frac{1}{24} H^{\mu\nu\lambda} H_{\mu\nu\lambda} \right) + \mathcal{O}(\alpha'^2), \tag{10.6}$$

where $H_{\mu\nu\rho} = 3\partial_{[\mu} B_{\nu\rho]}$ is the field strength associated with $B_{\mu\nu}$ (see also Section 9.8) and where we have kept the fermionic fields implicit. Several utterly remarkable things have happened:

- If the string length $\sqrt{\alpha'}$ is small, then we can solve these equations perturbatively. To leading order, they require $D = 10$ – i.e., they *fix the dimension of spacetime*.
- At the next order in α' the background fields have to satisfy equations of motion that *correspond to the Einstein equations* coupled to $B_{\mu\nu}$ and Φ. So even though we started simply by quantizing a string, we find the equations of general relativity coupled to matter (or of supergravity, when adding fermions). This shows that string theory is indeed a theory of quantum gravity. It is clear that corrections to the Einstein equations can be found by working to higher orders in α'.
- There are *no arbitrary constants* that appear in the quantization process, everything is fixed.

Let us elaborate slightly on the last point. The statement that everything is fixed is was a slight exaggeration: One can add additional fields (some of which are related to the existence of branes – see below), and this leads to the existence of five superficially distinct string theories. However, these are all related to each other (and to the unique supergravity theory in 11 dimensions – see Section 10.2.3) when considering spacetimes with compact spatial sections.

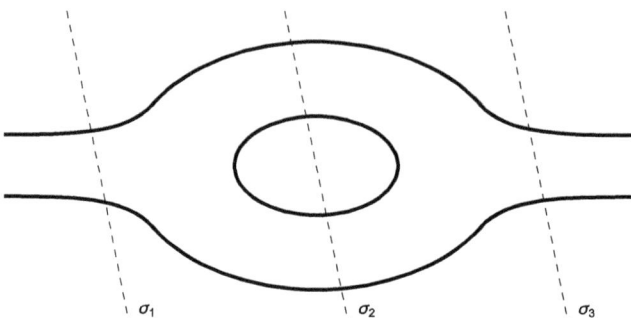

Figure 10.1 A worldsheet with a nontrivial topology automatically encodes string interactions. In this case successive slices at $\sigma_{1,2,3}$ show a string splitting in two and rejoining later. The strength of such processes, i.e., the string coupling, is determined by the dilaton Φ.

The more basic point is that all coupling constants are determined by the expectation values of fields. To this end, it is useful to recognize that the equations of motion can be reproduced by an effective action, whose bosonic part is

$$S_{10} = \frac{1}{2\kappa_0^2} \int d^{10}X \sqrt{-G} e^{-2\Phi} \left(R - \frac{1}{12} H^{\mu\nu\lambda} H_{\mu\nu\lambda} + 4\nabla^\mu \Phi \nabla_\mu \Phi \right), \quad (10.7)$$

where corrections depending on powers of α' could in principle be added. In this action, there are no explicit coupling constants, except for an overall factor of $1/\kappa_0^2$, which can be changed by a shift in Φ (meanwhile $\sqrt{\alpha'}$ is dimensionful and just provides a unit of length). The expectation value of Φ determines the string coupling,

$$e^{\langle\Phi\rangle} \equiv g_s. \quad (10.8)$$

This follows from the Φ coupling in the action (10.3): When Φ is constant, then this term is proportional to the Euler character of the worldsheet. For illustration, consider a worldsheet with a hole in it – see Fig. 10.1. If one considers consecutive slices through the worldsheet, one realizes that such a configuration describes a string that splits into two strings, which rejoin at a later time. In this manner, the string coupling determines the strength of string interactions. Moreover, as the figure shows, the interactions are smooth, and not localized at a point – in this manner string theory avoids the infinities that plague quantum field theories.

Hence we arrive at the picture that different values of the background fields $G_{\mu\nu}, B_{\mu\nu}, \Phi, \ldots$ do not describe different theories, as one might have guessed from (10.3), but rather describe different solutions to a unique string theory. It is time now to discuss in more detail how one tries to describe our universe in this framework.

10.1.2 Compactifications and cosmological implications

The most obvious consequence of the preceding discussion is that the universe is expected to have a total of 10 (or, perhaps even 11) dimensions. Since it does not seem viable to have additional time dimensions (see, e.g., Section 7.3.2 for one aspect of this problem), we must assume the existence of an additional 6 or 7 spatial dimensions. Physics as we know it requires 3 large spatial dimensions – for instance, if there are additional large spatial dimensions, then planetary orbits are not stable and star systems cannot exist. This means that, if additional dimensions exist, they must be hidden from view. The simplest manner in which this can be achieved is to assume that the extra dimensions are compact and of small (microscopic) volume.[1]

Think of the surface of a garden hose: It has 1 large dimension (along its length) and a small, circular dimension forming the tube. Roughly speaking, if you have a ruler that is larger than the circumference of the tube, the hose will appear 1-dimensional. If you can probe the hose on shorter scales, a 2nd dimension is revealed. This is in fact the original example of an additional dimension, considered by T. Kaluza and O. Klein, and leading to the name Kaluza–Klein (KK) compactification. It is instructive to review this mechanism in more detail.

We will focus on what happens to the gravitational action when 1 dimension is compactified on a circle. We will start with the Einstein–Hilbert action in $D+1$ dimensions, and will denote quantities in $D+1$ dimensions with hats,

$$S = \frac{1}{2} \int d^{D+1}x \sqrt{-\hat{G}} \hat{R}. \tag{10.9}$$

We now assume that the Dth spatial dimension is described by the coordinate $x^D \equiv z$ and that it takes the form of a circle, hence $0 \leq z < 2\pi$.

---

[1] A second possibility is that the spacetime curvature is very strong in the extra dimensions, effectively confining physics to a 3-dimensional subspace. This is harder to implement in concrete examples.

Furthermore, we assume that the metric depends only on the D-dimensional coordinates x^μ, $\mu = 0, ..., D-1$, and not on z (we will return to this point below). The metric then decomposes into

$$\hat{G}_{\mu\nu}(x), \quad \hat{G}_{\mu D}(x), \quad \hat{G}_{DD}(x). \tag{10.10}$$

The index structure suggests that the $(D+1)$-dimensional metric reduces to a D-dimensional one, plus a vector field A_μ (with one μ index), as well as a scalar field ϕ (with no index in the lower dimensions). It turns out to be useful to consider the following metric decomposition:

$$d\hat{s}^2 = e^{2a\phi}ds^2 + e^{-(D-2)a\phi}(dz + A_\mu dx^\mu)^2, \quad a^2 = \frac{1}{(D-1)(D-2)}, \tag{10.11}$$

with $A_\mu = A_\mu(x^\nu)$ and $\phi = \phi(x^\nu)$. The ϕ-dependent terms are chosen such that the dimensionally reduced theory has canonically normalized Einstein–Hilbert and scalar terms, without additional couplings to ϕ. Plugging the metric (10.11) into the original action (10.9), one obtains

$$S = \frac{\text{Vol}(S^1)}{2} \int d^D x \sqrt{-G} \left(R - \partial^\mu \phi \partial_\mu \phi - \frac{1}{4} e^{-2(D-1)a\phi} F^{\mu\nu} F_{\mu\nu} \right), \tag{10.12}$$

where $F_{\mu\nu} = 2\partial_{[\mu} A_{\nu]}$ is the field strength associated with A_μ. Thus pure gravity in $D+1$ dimensions has turned into gravity coupled to a Maxwell field and a scalar in D dimensions. This suggests a way of unifying forces in a purely geometrical way by adding suitable additional compact dimensions.

A few comments: The gauge invariance of the electromagnetic potential is inherited from the diffeomorphism invariance in the higher dimensions. Under a coordinate change $\delta x^\mu = \Lambda^\mu$, the metric transforms as $\delta \hat{G}_{\mu\nu} = \nabla_\mu \Lambda_\nu + \nabla_\nu \Lambda_\mu$. If we write $\Lambda_D \equiv \Lambda$, then we obtain $\delta A_\mu = \partial_\mu \Lambda$, which is indeed a $U(1)$ gauge invariance. If the compact manifold has a larger symmetry, then this will be inherited by the gauge field. Sometimes the gauge symmetry can be even further enhanced due to an interplay of diffeomorphism invariance and gauge symmetries of additional fields. Via KK reduction gauge groups containing the Standard Model of particle physics have been constructed from compactifications on 6 spatial dimensions (although typically these then contain many additional fields as well).

The scalar field ϕ can be seen to determine the radius of the Dth dimension (with coordinate z), as is evident from the metric (10.11). In the present theory, nothing fixes the value of ϕ, and thus the radius is arbitrary and could potentially be easily changed (since ϕ is massless). This is an example of the *moduli stabilization problem*. In general many scalar fields without potentials (referred to as moduli) arise from compactifications, and these moduli determine the size and shape of the extra dimensions. Such fields must acquire a potential that fixes their values, otherwise the solution will be in conflict with observations. Below we will describe a mechanism by which such potentials can be generated.

The extra dimensions must have been held pretty much fixed in size and shape (over the last 13.8 billion years, at least), because they determine the values of the lower-dimensional coupling constants, and no variations of such couplings have been measured. And, of course, one would like the couplings to take on values that are in agreement with what we observe. Let us take the example of Newton's constant. It appears as the coefficient of the Einstein–Hilbert term in 4 dimensions. When we look back at the dimensionally reduced action (10.12), we see that the integration over the extra dimensions has produced a volume factor, in this case $\text{Vol}(S^1) = 2\pi$. More generally, when compactifying down from 10 dimensions, this implies that the 4-dimensional Newton constant G_N is related to the higher-dimensional one via

$$\frac{1}{8\pi G_N} = \frac{1}{\kappa^2} = \frac{\text{Vol}(\mathcal{M}_6)}{\kappa_0^2} e^{-2\langle\Phi\rangle}. \tag{10.13}$$

In similar ways, all other coupling constants are obtained, and it is evidently a nontrivial task to find a compactification that reproduces precisely the observed characteristics of the Standard Model (an exact agreement has not been found to date even though there seems to be no obstruction in principle – it is just that the number of possibilities is vast).

We may also wonder how large extra dimensions could in principle be. The main way we would notice the presence of extra dimensions would be in a change to Newton's law on short scales. In 4 dimensions the gravitational potential falls off radially as $1/r$, and for each extra dimension one obtains an extra factor of $1/r$. But it is difficult to measure the gravitational potential on short scales, due to the weakness of gravity compared to electromagnetism. Delicate torsion balance experiments currently put an upper limit of $\mathcal{O}(10)$ microns on the size of extra dimensions, vastly larger than nuclear scales.

Up to this point, we mostly focused on the connection between extra dimensions and particle physics. But of course we should ask whether the compactified theory can naturally lead to an expanding universe, such as we observe. This turns out to be a surprisingly thorny problem, as we will now explain for the case where the evolution corresponds to dark energy or inflation.

Let us assume that the extra dimensions have their moduli fixed, so that coupling constants in the low-energy theory in fact remain constant. Then we can write the metric quite generally as follows

$$d\hat{s}^2 = e^{2A(y)} ds^2 + g_{mn}(y) dy^m dy^n, \quad (10.14)$$

where the internal coordinates are called y^m with $m, n, \ldots = 1, \ldots, 6$, the 4-dimensional metric is ds^2 (with coordinates x^μ) and $A(y)$ is known as a *warp factor*. The 10-dimensional Einstein equations then imply

$$\hat{R}_{\mu\nu} = R_{\mu\nu} - \hat{g}_{\mu\nu}(\Box^{(y)} A + 2\partial_m A \partial^m A) = \hat{T}_{\mu\nu} - \frac{1}{8}\hat{g}_{\mu\nu}\hat{T}, \quad (10.15)$$

where we have left the stress–energy tensor implicit, and $\hat{T}$ is its 10-dimensional trace. Multiplying this equation by $\hat{g}^{\mu\nu}$ and rearranging, we obtain

$$2\Box(e^{2A}) = R^{(4)} + \frac{1}{2}e^{2A}\left(-\hat{T}^\mu_\mu + \hat{T}^p_p\right). \quad (10.16)$$

The matter content of the 10-dimensional supergravity theories that are the low-energy limits of string theory consists solely of scalar fields and antisymmetric tensor fields (plus fermions, which do not significantly affect spacetime globally). The stress–energy tensor of a p-form gauge field with field strength $F_{M_1 \ldots M_{p+1}}$ is given by

$$T_{MN} = F_{MN_1 \ldots N_p} F_N{}^{N_1 \ldots N_p} - \frac{1}{2(p+1)} g_{MN} F^2. \quad (10.17)$$

If we respect the assumed symmetry, implying that for antisymmetric tensor fields either all indices must be in the internal y^p directions, or there could be four indices in the large x^μ directions with the additional indices being internal, then one may verify straightforwardly that the combination $-\hat{T}^\mu_\mu + \hat{T}^p_p$ is always positive.

Now comes the crux of the argument: We can integrate Eq. (10.16) over the internal manifold. Since the manifold is compact, and since the integrand on the left-hand side is a total derivative, the left-hand integral will give zero if no singularities are present. Together with the positivity of the stress-energy contribution, this implies that $R^{(4)}$ must be negative, and this excludes a de Sitter solution (which would have required $R^{(4)} = 4\Lambda > 0$) as well as other accelerated FLRW solutions, cf. Eq. (1.31).

As with any no-go result, one must try to understand how to break its assumptions in order to make progress. In the present case, several avenues present themselves, in particular the inclusion of higher-derivative corrections, the inclusion of singularities in the compact manifold (e.g., in the form of certain branes – see the next section), or allowing for (hopefully mild) time dependence of the compact dimensions. All of these possibilities are being actively pursued.

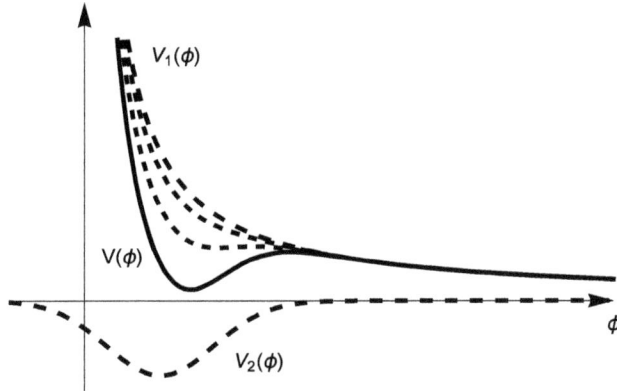

Figure 10.2 An illustration of the Dine–Seiberg problem: Two asymptotically vanishing potentials $V_{1,2}$ need to be combined with suitable coefficients (here we show the sum $V_1(\phi) + \beta V_2(\phi)$ with increasing β) until a stabilizing potential $V(\phi)$ (thick curve) with a local minimum is obtained. This minimum is necessarily away from the asymptotic region of weak coupling, and is thus under less computational control.

However, what the no-go result shows is that obtaining an accelerated universe from string theory is not simple (as C. Vafa once said: "there is no harmonic oscillator model for de Sitter"[2]). One can make the resulting challenges somewhat more precise in terms of the *Dine–Seiberg problem*, first described by M. Dine and N. Seiberg. The issue is a very general one in string theory.

If we think about the moduli (scalar fields) that are the parameters of compactifications, then one of their general properties is that there is always some field direction in which they have the effect of decompactifying the spacetime (this must be the case because 10-dimensional Minkowski space with all form fields vanishing is always a solution), and moreover this direction is associated with weak coupling (as an example, consider the $a\phi \to -\infty$ limit of the KK scalar in (10.11)). In other words, their potential (once they obtain one) falls off to zero along this field space direction. In this regime, the solutions are under good control, but they are not realistic. Obtaining a nontrivial vacuum then requires balancing at least two such potentials against each other, and this will necessarily take place in a region

---

[2] C. Vafa in a talk given at the meeting *Forefronts in Cosmology and Numerical General Relativity*, Salzburg, Austria, July 5–7 2018.

of field space where the coupling is not so weak anymore – see Fig. 10.2 for an illustration. Then one progressively loses control over the solutions, in the sense that one cannot reliably calculate all correction terms. Hence it will be intrinsically hard to find a realistic cosmological solution of string theory.[3]

It is certainly surprising that a universe like ours does not arise naturally/easily in string theory. And it is fair to say that the consequences of this fact are still ill-understood. At its most pessimistic, it could mean that string theory is simply wrong. As with any physical theory, this is a possibility one has to keep in mind. Or it could mean that an important ingredient or technique has not been discovered yet – after all, string theory is still a theory under development. But it could also mean that our universe is in some mathematical sense very rare, and ultimately the difficulty in finding it as a string theory solution might help in understanding why it seems so precisely tuned, and help in understanding how it fits into the larger context of conceivable stringy universes.

Leaving this open question aside for now, we will continue with a discussion of how cosmology might reveal clues about string theory. Above, we looked at the dimensional reduction of fields that do not depend on the extra dimensions and that are massless from the point of view of the lower dimensions. More generally, we could have considered the full Fourier expansions of all fields. In the simple case of a reduction on a circle of radius R, and assuming the presence of a scalar field Φ in the higher dimensions, we would have

$$\Phi(x^\mu, z) = \sum_{n \in \mathbb{Z}} \Phi_n(x^\mu) \, e^{in\frac{z}{R}}, \qquad (10.18)$$

where we assumed here that z has range $[0, 2\pi R)$, and where the reality of the scalar implies that $\Phi_n^* = \Phi_{-n}$. To understand the main point, we may ignore the effects of curvature and look at the kinetic term, which becomes

$$-\frac{1}{2} \int d^{D+1}x \, (\partial \Phi)^2 = -\pi R \int d^D x \sum_{n \in \mathbb{Z}} \left(\nabla^\mu \Phi_n \nabla_\mu \Phi_{-n} + \frac{n^2}{R^2} \Phi_n \Phi_{-n} \right). \qquad (10.19)$$

A single massless field in $D+1$ dimensions thus gives rise to a massless field in the lower dimensions, accompanied by an infinite tower of KK modes with masses $m^2 = n^2/R^2$. When the extra dimension is small, even the lowest

---

[3] Specific instances of these difficulties have been studied in depth over the last few years, and they are collectively known as *swampland conjectures*. The obstructions they highlight are best understood in asymptotic regions of field space – very little is known about what is possible, and what is not possible, in the interior regions of field space.

KK mode can already have a very large mass, so that it would have gone unnoticed in particle physics experiments to date. Such modes can however have cosmological implications, as we will discuss momentarily. Before doing so, notice that the KK modes are also charged, in this case under $U(1)$. This is because the higher-dimensional gauge transformation $\delta x^D = \delta z = \Lambda$ inserted in (10.18) implies the shift

$$\Phi_n \to e^{i\frac{n\Lambda}{R}} \Phi_n, \qquad (10.20)$$

which in turn implies a charge n/R. An analogous treatment implies that gravity and form fields will also lead to their own towers of massive KK modes upon dimensional reduction.

On general grounds, one expects the massive modes to remain frozen when the Hubble rate is still very high. This can be seen from their effective equation of motion, which in an FLRW background takes the familiar form

$$\ddot{\phi} + 3H\dot{\phi} + m^2\phi = 0. \qquad (10.21)$$

When $H \gg m$, the damping term is large and the field freezes, $\dot{\phi} \approx 0$. As the Hubble rate decreases, at some point $H \sim m$ and the modes can oscillate. This can occur either because they have some classical energy, or, even if classically the fields reside at the minimum of the potential, because of quantum fluctuations that are always present. These massive fields thus act as harmonic oscillators with a constant frequency, with $\phi \sim e^{\pm imt}$. In that sense these massive fields act as *standard clocks*. Now if there is a coupling between such KK modes and the field that gives rise to the primordial curvature perturbations, then the oscillations of the KK modes will lead to a (generally small) characteristic imprint on the CMB.

We can make this a little more precise. For specificity, we will assume that the background universe follows a power-law expansion/contraction evolution, with

$$a(t) = a_0 |t|^p, \quad a\tau = \frac{t}{1-p}, \qquad (10.22)$$

where p is a constant and τ is the conformal time. As we discussed in Chapter 2, many different types of cosmological evolution can be modeled this way, by choosing p and the range of t appropriately – e.g., $p > 1$ and $0 < t < \infty$ for inflation, or $0 < p \ll 1$ and $-\infty < t < 0$ for ekpyrosis. On subhorizon scales, then, the KK modes oscillate as e^{imt}. Meanwhile, the density/curvature perturbations giving rise to the CMB evolve as $e^{-ik\tau}$ (for comoving wave number k), since on such small scales they are well approximated by the Minkowski vacuum state – see Chapter 3. Now we assume that there exists a coupling S_c between the KK mode and the curvature perturbation, for instance of the form

$$S_c = \int_{\tau_i}^{\tau_f} d\tau \, g(t) \, e^{imt} \, e^{-ik\tau} \,, \tag{10.23}$$

where $g(t)$ is a coupling function that varies more slowly than the two exponential factors. In general, the full integrand above contains fast oscillations, with many cancellations, and does not lead to sizeable interactions between KK and curvature perturbation modes. But at a saddle point, also known in this context as a *resonance*, there are no cancellations and a significant interaction arises. The condition for a resonance is

$$0 = \frac{d}{d\tau}(mt - k\tau)\,|_{t=t_r} \tag{10.24}$$

$$= m\frac{dt}{d\tau} - k\,|_{t=t_r} \tag{10.25}$$

$$= m\,a(t_r) - k\,. \tag{10.26}$$

The coupling term is then well approximated by

$$S_c \approx g(t_r)\,e^{imt_r - ik\tau_r} \approx g(t_r)\,e^{-i\frac{p}{1-p}m\,t\left(\frac{k}{m}\right)}\,. \tag{10.27}$$

Here is the interesting part: In the exponent $t\left(\frac{k}{m}\right)$ is the inverse function of the scale factor function $a(t)$, cf. (10.22) and (10.26). This means that the massive KK mode leaves an imprint on the CMB (in the Fourier or k-space) that provides direct information about the background evolution. This is something that the curvature perturbations in the CMB by themselves do not provide, and this is the reason why, even though the CMB has been observed to exquisite precision, it is not known with great confidence yet whether it was caused by inflation, ekpyrosis, or some other cosmological phase. The fact that the KK modes act as standard clocks would allow precisely such an inference, and this is evidently a good reason to look for such signatures (which, to date, have not been observed yet). One brief comment: Towers of massive fields are an automatic outcome of string compactifications, but it should be clear that other massive fields, not necessarily related to string theory, could also play the role of standard clocks.

If there is a whole tower (or even several) of massive modes, then they can start affecting the evolution and expansion history of the universe in sequence as the Hubble rate decreases, especially if there is a mechanism that transfers energy into these massive modes. Here, let us just point out that when the Hubble rate becomes much smaller than their mass, $H \ll m$, such modes keep oscillating (though they also get damped over time) and act like dark matter – their energy density redshifts just like pressure-free matter, $\rho_\phi \propto a^{-3}$, as described in Section 2.1.6 in the context of reheating. Hence massive KK modes can potentially explain the origin of (at least part of) the dark matter that is observed. We will comment on another manner in which stringy dark matter can arise in the next section.

10.2 Braneworlds

10.2.1 Dp-branes

When considering the motion of open strings in spacetime, it is possible to restrict their motion to submanifolds by imposing Dirichlet boundary conditions on some of the coordinates at the end points of strings; see Fig. 10.3 (meanwhile, closed strings cannot be restricted in this way and can travel anywhere). These submanifolds are called Dp-branes, where D stands for Dirichlet, the number p specifies their number of spatial dimensions, and the name "brane" arises from "membrane."

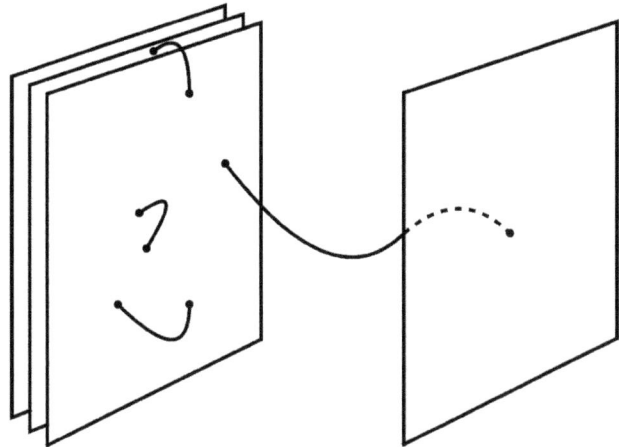

Figure 10.3 Strings can end on branes, which are themselves dynamical objects in string theory. Strings ending on the same brane, or on stacked branes, have massless modes that lead to gauge theories on the branes. Strings stretched between separated branes are massive. One influential idea of string theory is that we could be made of strings living on a brane (or on a stack) with 3 spatial and 1 time dimension. Matter on separated branes would then automatically appear as dark matter to us.

These branes also turn up as solutions to the effective supergravity theories, in particular of so-called type IIA or IIB (for completeness let us note that, in addition, there exist branes that are not related to the end points of strings). The bosonic (type II) supergravity actions in 10 dimensions contain form fields (antisymmetric tensors) in addition to the universal fields in (10.7), and these form fields provide the charges carried by branes. The form fields, which are bosonic, arise from products of string worldsheet fermions.

If we focus on the metric, the dilaton ϕ, and a form field, the action is given by[4]

$$S_{10D} = \frac{1}{2\kappa_{10}^2} \int d^{10}x \sqrt{-g} \left(R - \frac{1}{2} \nabla^M \phi \nabla_M \phi - \frac{1}{2n!} e^{\frac{5-n}{2}\phi} F_{[n]}^2 \right), \quad (10.28)$$

where we have restricted to a single $(n-1)$-form gauge field with antisymmetric field strength $F_{[n]}$ and $F_{[n]}^2 = F^{M_1 M_2 \cdots M_n} F_{M_1 M_2 \cdots M_n}$. The coupling to the dilaton arises because we are working in Einstein frame, where the gravitational term is canonical.

In general dimensions, a supergravity action may contain many scalars and form fields. In 10 dimensions, only even n appears for the type IIA theory, and only odd n for type IIB (in each case the $H_{[3]}$-field of (10.7) is also present), as well as fermionic partners for all fields. We will not require the full form of the various supergravity actions, since we simply wish to illustrate a few general properties.

The equations of motion following from (10.28) are

$$R_{MN} = \frac{1}{2} \phi_{,M} \phi_{,N}$$
$$+ \frac{1}{2(n-1)!} e^{\frac{5-n}{2}\phi} \left(F_{MN_1 \cdots N_p} F_N{}^{N_1 \cdots N_p} - \frac{n-1}{n(D-2)} g_{MN} F_{[n]}^2 \right), \quad (10.29)$$

$$0 = \nabla^{M_1} \left(e^{\frac{5-n}{2}\phi} F_{M_1 \cdots M_n} \right), \quad (10.30)$$

$$\Box \phi = \frac{5-n}{4n!} e^{\frac{5-n}{2}\phi} F_{[n]}^2. \quad (10.31)$$

Let us work by analogy: Electrons are point particles tracing out 1-dimensional worldlines and coupling to the electromagnetic potential A_μ, which is a 1-form. Branes then arise as objects of p spatial dimensions tracing out $(p+1)$-dimensional worldvolumes in spacetime, and carrying the charges C_{p+1} of antisymmetric $(p+1)$-form gauge fields with $(p+2)$-form field strengths. In the simplest case, assuming maximal symmetry, i.e., Poincaré symmetry in $p+1$ dimensions and $SO(9-p)$ symmetry in the transverse

---

[4] We are using a different normalization of the scalar kinetic term here, which is common in the supergravity/string theory literature.

$9 - p$ dimensions, their solution is given by

$$ds^2 = H^{-\frac{7-p}{8}} \eta_{\mu\nu} dx^\mu dx^\nu + H^{\frac{p+1}{8}} \delta_{mn} dy^m dy^n, \tag{10.32}$$

$$F_{m\mu_1\cdots\mu_{p+1}} = \epsilon_{m\mu_1\cdots\mu_{p+1}} \partial_m (H^{-1}), \tag{10.33}$$

$$e^\phi = H^{\frac{3-p}{4}}, \tag{10.34}$$

$$H(r) = 1 + \frac{\mathcal{M}}{r^{7-p}}, \tag{10.35}$$

where $p = n - 2$ and where $H(r)$ is a harmonic function in the transverse space, with radial coordinate $r = \sqrt{y_m y^m}$. By harmonic function, one means that H satisfies Laplace's equation in the transverse space,

$$\Box^{(y)} H = 0. \tag{10.36}$$

This equation can also be extended to include delta function sources on the right-hand side, at the location of the brane(s). In the explicit expression for the harmonic function in (10.35), $\mathcal{M}$ sets the mass scale of the solution and is related to the tension of the brane. Note that the entire solution is determined by the harmonic function. You are encouraged to verify that the expressions above indeed solve the equations of motion.

A few comments: 1. The use of the letter p to designate the number of spatial dimensions originates from the possibility of interpreting the sound of the word p-brane as "pea brain." 2. There are reasons to think that branes may be just as fundamental as strings, but to date it has not been possible to consistently quantize objects with more spatial dimensions than strings, so in our treatment we will consider branes as solutions to the effective supergravity theories of string theory. 3. One can also couple branes to the dual $(7 - p)$-forms, in which case one considers branes with $6 - p$ spatial dimensions. 4. Gauge fields can arise on the worldvolume of branes due to open strings ending there – see Fig. 10.3. If multiple branes are stacked on top of each other, the overall gauge group can be significantly enhanced. 5. The brane solution can be generalized to much more elaborate, curved backgrounds, and can include nontrivial cycles that the brane wraps around. When the cycles are small, this reduces the number of large dimensions that the brane occupies. 6. There exist many other types of branes, being solutions to different supergravity theories in different dimensions and different matter content. Some are related to each other via dimensional reduction. 7. The solutions can be generalized such that they describe multiple branes. These branes can then also intersect at various angles, and in various dimensions. Overall, this leads to a vast range of possible solutions containing higher-dimensional extended objects.

One can also write down an action for the branes themselves. In analogy with strings, for which the action (10.1) was given by the area traced out in spacetime, for Dp-branes the action is the worldvolume they trace out, plus a coupling to the dilaton and a topological term (not dependent on the metric) involving the $(p+1)$-form gauge field the branes are charged under,

$$S_{\mathrm{D}p} = -T_p \int_{W_{p+1}} \mathrm{d}^{p+1}z\, e^{\frac{p-3}{4}\phi} \sqrt{-\det(\gamma_{\mu\nu})} + \mathcal{C}_p \int_{W_{p+1}} \mathrm{d}^{p+1}z\, A_{[p+1]}, \tag{10.37}$$

where $\gamma_{\mu\nu} = \frac{\partial x^M}{\partial z^\mu}\frac{\partial x^N}{\partial z^\nu} G_{MN}$ is the pullback of the spacetime metric onto the worldvolume W_{p+1}, $A_{[p+1]} = \frac{1}{(p+1)!} \epsilon^{\mu_0 \dots \mu_p} \frac{\partial x^{M_0}}{\partial z^{\mu_0}} \dots \frac{\partial x^{M_p}}{\partial z^{\mu_p}} A_{M_0 \dots M_p}$ is the pullback of the form field, T_p is the brane tension (energy per volume), and $\mathcal{C}_p$ is its charge. The brane coordinates are named z^μ here. In the action above, we neglected the coupling to the $B_{\mu\nu}$ field as well as to any gauge fields on the worldvolume, since we will not make use of these here – they can be found in the literature recommendations in Appendix E.

The brane action is written in Einstein frame above, so that it can be coupled to (10.28). If we had considered it in string frame, the gravitational part would have contained a dilaton factor of $e^{-\Phi}$. This factor indicates that Dp-branes are nonperturbative objects arising at order $\frac{1}{g_s}$ in the string coupling constant, cf. Eq. (10.8). This means that they are heavy when the string coupling is small, and vice versa. The brane action may also be used to calculate its motion and fluctuations in a particular background in which it is embedded, and we will see an example of this usage below.

The most symmetric brane solutions typically preserve some amount of supersymmetry (which is a symmetry relating bosons and fermions), which means that the solutions remain invariant under a subset of supersymmetry transformations. There are two important consequences: The first is that in such a case the tension T_p is equal to the charge $\mathcal{C}_p$. Moreover, supersymmetric brane solutions are protected from quantum corrections, because the corrections caused by bosons and those caused by fermions cancel each other out. Thus they are trustworthy as full string theory solutions. Unfortunately, in cosmology supersymmetry is immediately broken, as it is not compatible with time dependence.[5] This is one of the main reasons why it has been difficult to find string theoretic cosmological solutions to date.

With this cautionary note out of the way, let us point out that branes offer a great many new ways of constructing stringy solutions that could po-

---

[5] For people familiar with supersymmetry: Time dependence of the metric usually prevents the existence of a nonzero Killing spinor.

tentially be relevant to our universe. The main idea is that we might live on a *braneworld* – i.e., we could live on a 3-brane in a higher-dimensional spacetime. In fact, there might be a whole stack of branes, which would facilitate obtaining large gauge groups (suitable for embedding the Standard Model of particle physics) on the brane. If, in addition, other branes exist that are not coincident but still nearby (in the extra dimensions), then matter on those branes would automatically appear dark to us, since at low energies only closed strings (which include graviton excitations) could travel between the branes, while open strings (making up ordinary matter and gauge fields, but not gravity) would remain confined to our stack of branes. This would be a very natural setting for understanding the nature of dark matter. Note also that in the transverse space, branes appear as singular objects. Thus they can help to evade the no-go theorem that we described around Eq. (10.16).

10.2.2 Brane inflation

In braneworld models, a challenge is that the brane must be able to expand or contract (and perhaps to wiggle slightly), while the additional features of the solution, such as the geometry of the compactification manifold, must remain essentially static in order to agree with experimental bounds on the time dependence of coupling constants. That said, some other branes might be dynamical and move with respect to "our" brane. This provides opportunities for building cosmological models.

Let us go through an illustrative example in more detail. This involves D3-branes and their "antiparticles", anti-D3-branes, moving in a curved static background. Type IIB supergravity contains a 5-form field strength,[6] and admits a background solution that is a warped product of AdS_5 and a 5-sphere,

$$ds^2 = \frac{r^2}{R^2}\eta_{\mu\nu}dx^\mu dx^\nu + \frac{R^2}{r^2}dr^2 + d\Omega_5^2, \qquad (10.38)$$

$$(F_{[5]})_{rtx^1x^2x^3} = \frac{4r^3}{R^4}, \qquad (A_{[4]})_{tx^1x^2x^3} = \frac{r^4}{R^4}, \qquad (10.39)$$

where we have also provided the expression for the 4-form gauge potential in an obvious coordinate basis (where μ, ν run over $0, 1, 2, 3$ and r is the fifth AdS coordinate, while R is the radius of curvature of both the AdS space and the 5-sphere). This solution may be regarded as the near-horizon limit of a D3-brane (or a stack of such branes), cf. Eqs.(10.32)–(10.35) with

---

[6] The 5-form has the additional property of being self-dual, which means that $F_{[5]} = \star F_{[5]}$, where $\star$ denotes the Hodge star, or more explicitly
$F_{M_1M_2M_3M_4M_5} = \frac{1}{5!}\epsilon_{M_1M_2M_3M_4M_5}{}^{N_1N_2N_3N_4N_5}F_{N_1N_2N_3N_4N_5}$.

$p = 3, b = 0$, and $\mathcal{M} = R^4$, and in the limit where $r \to 0$, so that the harmonic function becomes

$$H(r) = 1 + \frac{R^4}{r^4} \to \frac{R^4}{r^4} \text{ as } r \to 0. \qquad (10.40)$$

In order to perform this identification, one has to rewrite the transverse space in (10.32) in polar coordinates. We should also note, without derivation, that R^4 increases in proportion to the number of branes.

If we consider a probe D3-brane, with action (10.37) with $\mathcal{C}_3 = T_3$, moving in this background along the radial direction at position $r_1(t)$, then the explicit source action becomes

$$S_{D3} = -T_3 \int_{W_4} d^4x \frac{r_1^4}{R^4} \sqrt{1 - \frac{R^4}{r_1^4}\dot{r}_1^2} + T_3 \int_{W_4} d^4x \frac{r_1^4}{R^4} \qquad (10.41)$$

$$\approx T_3 \int_{W_4} d^4x \frac{1}{2}\dot{r}_1^2, \qquad (10.42)$$

up to terms of $\mathcal{O}(\dot{r}^4)$. There is no potential, and the motion of the brane is free. A way to interpret this result is to note that the contributions to the potential from the gravity part (which are attractive) and from the gauge field part (which are repulsive) have exactly canceled. One can show that this is a consequence of supersymmetry. Such a setup will therefore not lead to any interesting cosmological dynamics.

However, if we add an anti-D3-brane, then this will break supersymmetry and generate a potential. One can show that the warping of the geometry separates the anti-D3-brane from the stack of D3-branes and drives it to small r. The setup we would like to describe is one where we have the antibrane, well separated from a stack of D3-branes, and with a single D3-brane in between. To calculate its interaction with the antibrane, we can also change perspective and consider the anti-D3-brane as a probe in the background provided by the stack of branes plus the single brane located at r_1. The associated harmonic function is

$$H = \frac{R^4}{r^4} + \frac{R^4}{nr_1}, \qquad (10.43)$$

where the term due to the single D3-brane is suppressed by n, the number of branes in the stack.[7] As will be shown in Exercise 10.3, the resulting

---

[7] Schematically, the harmonic function satisfies the equation $\Box^{(y)} H = nC\delta r + C\delta(r - r_1)$, where C is determined by the tension of a D3-brane.

effective brane action can be approximated by

$$S_{D3} \approx \int d^4 z \left[T_3 \frac{1}{2} \dot{r}_1^2 - (T_3 - C_3) H^{-1} \right]. \tag{10.44}$$

For the anti-D3-brane, located at $r_0 \ll r_1$, we have $C_3 = -T_3$, so that we obtain an effective interbrane potential,

$$V(r_1) = 2T_3 H^{-1} \approx 2T_3 \frac{r_0^4}{R^4} \left(1 - \frac{r_0^4}{nr_1^4} \right), \tag{10.45}$$

with $r_1(t)$ playing the role of an effective scalar field (inflaton) from the 4-dimensional point of view. The potential above is very flat, and if this were the end of the story, it could drive a successful inflationary phase.

At this point we encounter an obstacle that has turned out to be the main problem with cosmological models in string theory: Once we consider the full solution, additional terms appear that tend to spoil the model one has built. In the present case, there is one term that we neglected so far, and this is a conformal coupling of the brane modulus to the spacetime curvature $R^{(4)}$. This term, which one can show to be necessarily present, takes the form

$$\mathcal{L}_c \sim \frac{T_3}{12} r_1^2 R^{(4)}. \tag{10.46}$$

If we canonically normalize r_1, setting $\sqrt{T_3} r_1 \equiv \varphi$, and add the conformal coupling term to the effective potential, we get the approximate form

$$V = V_0 + \frac{1}{12} \varphi^2 R^{(4)} \approx V_0 + \frac{1}{3} \varphi^2 V_0, \tag{10.47}$$

since during inflation $R^{(4)} \approx 4V_0$. The correction term, unfortunately, has a disastrous effect on the second slow-roll parameter:

$$\eta = \frac{V_{,\varphi\varphi}}{V} \approx \frac{2}{3}. \tag{10.48}$$

This is far too large to sustain a prolonged inflationary phase, and is commonly known as the *eta problem*. Similar dangerous terms also tend to arise when devising mechanisms to stabilize moduli, in particular the volume and shape moduli of the internal dimensions.

Let us add a couple of remarks regarding brane inflation: In the model we considered above, the inflaton has the simple interpretation of being the distance modulus between branes. This suggests an elegant end to inflation: When the brane and antibrane approach each other, they may eventually collide and annihilate each other. At that point, the inflaton field entirely disappears. Moreover, the annihilation provides a natural reheating process, filling the universe with matter and hot radiation.

Many variants of the scenario above have been studied. The current understanding suggests that the ingredients for a successful inflationary phase are abundantly present in string theory, but no fully controlled example has been constructed to date.

10.2.3 End-of-the-world branes and the big bang

There exists another prominent approach to string phenomenology, with obvious potential applications to cosmology, involving branes that are not located in the middle of spacetime, but are rather forming the edge of spacetime. The idea for such "end-of-the-world" branes stems from the connection between the heterotic string theory and 11-dimensional supergravity. Instead of reviewing the construction of the heterotic string, we will start with supergravity and see (in a simplified manner) how the low-energy limit of heterotic string theory arises. For people who are interested in the details of this construction, references are provided in Appendix E.

If one restricts to fields with maximal spin 2, then 11 dimensions constitute the upper limit for constructing a theory of supergravity, and in fact the theory in 11 dimensions is unique. There is ample evidence by now that this special theory is also related to string theory via various dualities. Neglecting fermions, the action of 11-dimensional supergravity is given by

$$S_{11} = \frac{1}{2\kappa_{11}^2} \int_{M^{11}} \sqrt{-\hat{g}} \bigg[\hat{R} - \frac{1}{24} \hat{G}_{MNPQ} \hat{G}^{MNPQ} \\ - \frac{\sqrt{2}}{1728} \epsilon^{M_1...M_{11}} \hat{A}_{M_1 M_2 M_3} \hat{G}_{M_4...M_7} \hat{G}_{M_8...M_{11}} \bigg], \tag{10.49}$$

where 11-dimensional quantities have hats. Here κ_{11} determines the strength of gravity in 11 dimensions, $\hat{G}_{MNPQ} = 4! \partial_{[M} \hat{A}_{NPQ]}$ is the 4-form field strength associated with the 3-form gauge field $\hat{A}_{MNP}$, and $\hat{A}_{(3)} \hat{G}_{(4)} \hat{G}_{(4)}$ is a Chern–Simons term that is topological in nature and that will not play a role below.

The idea now is to compactify this theory not on a circle, as we did for the simplest Kaluza–Klein example, but instead on a line segment with two end points. A useful way to construct a line segment is to take a circle, choose a diameter, and identify points located at right angles on either side of the diameter. This corresponds to an identification under a $\mathbb{Z}_2$ reflection symmetry – see Fig. 10.4. The end points are fixed points under this operation, and they correspond to 10-dimensional planes, usually called *orbifolds*, in between which an 11-dimensional bulk is sandwiched.

10.2 Braneworlds

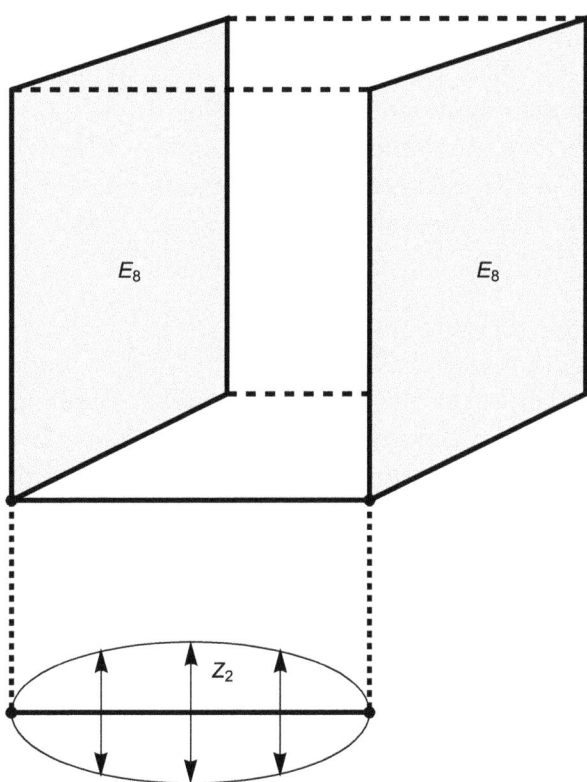

Figure 10.4 The relation between 11-dimensional supergravity and the $E_8 \times E_8$ heterotic string theory, as discovered by P. Hořava and E. Witten. When the 11th dimension is a line segment, anomaly cancellation forces one to put an E_8 gauge theory on each end-of-the-world brane. The inter-brane distance is proportional to a positive power of the string coupling, so that the 11th dimension only appears at strong coupling. At the bottom, we illustrate how a line segment may be viewed as a $\mathbb{Z}_2$ identification of a circle.

Even though 11-dimensional supergravity is a classical theory, one can calculate the form of potential gravitational and gauge anomalies, and such calculations offer a glimpse of the quantum theory. Anomalies of this type can arise in 10 dimensions, and their absence requires the addition of the following boundary action,

$$S_{\text{boundary}} = \sum_{i=1,2} -\frac{1}{8\pi\kappa_{11}^2}\left(\frac{\kappa_{11}}{4\pi}\right)^{2/3} \int_{M_{10}^{(i)}} \sqrt{-g}\left[\text{tr}(\text{F}^{(i)})^2 - \frac{1}{2}\text{trR}^2\right], \quad (10.50)$$

involving the gauge field strengths $F_{(i)}$ and a gravitational term. It turns out that anomaly cancellation requires the gauge group to be E_8 on both branes. This is rather remarkable, because it is a large group that can easily accommodate the Standard Model, and because the theories on both branes are chiral.

From the 10-dimensional point of view, the distance between the branes is a scalar field, and this is none other than the dilaton of string theory. But the dilaton is also related to the string coupling. Putting these facts together, one can show that the following relation holds,

$$R_{11} = g_s^{\frac{2}{3}}, \tag{10.51}$$

where R_{11} is the radius of the 11th dimension and g_s is the string coupling constant. We arrive at the picture that, at weak coupling, the two branes sit on top of each other and the theory one recovers is the low-energy limit of 10-dimensional heterotic $E_8 \times E_8$ string theory. Meanwhile, at strong coupling, the two E_8 orbifolds separate and an 11th dimension opens up. If one imagines the branes as being dynamical, then their motion might lead to interesting cosmological dynamics. They might even collide, and this would look very much like a big bang from the point of view of end-of-the-world inhabitants – see also the discussion further below. Note that at such an orbifold collision, the 11th dimension would momentarily disappear, but at the same time the string coupling would go to zero. Hence one may reasonably hope such a big bang to be tractable, although a detailed analysis of this scenario has not been undertaken so far.

To make contact with daily life, one has to consider a further Kaluza–Klein reduction on a 6-dimensional internal manifold, typically a so-called Calabi–Yau space[8] (that the orbifolds also wrap around). In doing so, one finds that in order to obtain realistic coupling constants, one has to take the orbifold direction to be larger than the radius of the 6-dimensional compactification manifold. In turn this implies that, going up in energy, one would first discover the orbifold dimension and an effective 5-dimensional world, and only at even higher energies would one see all 11 dimensions.

This suggests a change of perspective: We should first dimensionally reduce 11-dimensional supergravity down to 5 dimensions, and then compactify on a line segment. Let us sketch how this is done: We start by reducing the metric according to

$$\mathrm{d}s_{11}^2 = V_{\mathrm{CY}}^{-2/3} g_{mn} \mathrm{d}x^m \mathrm{d}x^n + V_{\mathrm{CY}}^{1/3} g_{ab} \mathrm{d}x^a \mathrm{d}x^b, \tag{10.52}$$

---

[8] A Calabi–Yau (CY) space has vanishing Ricci scalar curvature, and a reduced holonomy. This means that if one moves a vector around and brings it back to its starting point, then in a 6-dimensional CY space it will not in general be changed by an element of $SO(6) \simeq SU(4)$, but only by an element of $SU(3)$. This property allows a compactification on a CY space to preserve 1/4 of the supersymmetries.

where V_{CY} denotes the volume of the Calabi–Yau and g_{ab} its metric. Indices $m, n, \ldots$ run over 5 spacetime dimensions including the line segment direction, which we label 11 (and we also write $x^{11} = y$) and $a, b, \ldots$ denote CY directions. Due to the boundary terms (10.50), the 4-form satisfies a modified Bianchi identity,

$$(\mathrm{d}\hat{G})_{11MNPQ} = -\frac{1}{4\sqrt{2}\pi} \left(\frac{\kappa_{11}}{4\pi}\right)^{2/3} (\delta(y-1) - \delta(y+1)) \{\mathrm{tr} R \wedge R\}_{MNPQ}, \tag{10.53}$$

where we have used the "standard embedding," which amounts to identifying the gauge connection with the spin connection so that $\mathrm{tr}(\mathrm{F}^{(1)})^2 = \mathrm{tr} R^2$, while $\mathrm{tr}(\mathrm{F}^{(2)})^2 = 0$ (this breaks the gauge symmetry on one brane according to $E_8 \to E_6 \times SU(3)$, and then one can consider more elaborate solutions that break E_6 further into the Standard Model gauge group – in this way phenomenologically rich solutions can be constructed). This equation tells us that there are oppositely charged sources at the end points, which we have placed at $y = \pm 1$. The Bianchi identity, as well as the form field equation of motion, can be solved to leading order by setting

$$\hat{G}_{abcd} = \frac{\alpha}{\sqrt{2}} \epsilon_{abcd}{}^{ef} \omega_{ef} \theta(y), \tag{10.54}$$

where

$$\theta(y) = \begin{cases} +1 & \text{for } -1 \leq y < +1 \mod 4, \\ -1 & \text{for } -3 < y < -1 \mod 4, \end{cases} \tag{10.55}$$

and

$$\alpha = \frac{\pi}{6}\left(\frac{\kappa_{11}}{4\pi}\right)^{\frac{2}{3}} \beta, \tag{10.56}$$

where $\beta = -\frac{1}{8\pi^2}\int_{\mathcal{C}} \mathrm{tr} R \wedge R$ is the first Pontryagin class of the Calabi–Yau (the integration being done over the (2,2)-cycle $\mathcal{C}$ corresponding to the Kähler form ω_{ab}) and takes integer values – thus $\hat{G}$ is quantized. Putting all of this together, we arrive at the 5-dimensional theory specified by the action

$$S_5 = \frac{1}{2\kappa_5^2} \int_{5D} \sqrt{-g} \left[R - \frac{1}{2} V_{CY}^{-2} \partial_m V_{CY} \partial^m V_{CY} - 6\alpha^2 V_{CY}^{-2} \right]$$
$$+ \frac{1}{2\kappa_5^2} \left\{ -12\alpha \int_{4D, y=+1} \sqrt{-g}\, V_{CY}^{-1} + 12\alpha \int_{4D, y=-1} \sqrt{-g}\, V_{CY}^{-1} \right\}. \tag{10.57}$$

This theory contains a potential term $6\alpha^2/V_{CY}^2$, which means that Minkowski space will not be a solution. In fact, in line with our earlier discussion, the vacuum of this theory is given by a configuration of end-of-the-world branes, a fact already suggested by the 4-dimensional source actions at $y = \pm 1$. Explicitly, one finds the solution

$$ds^2 = h^{2/5}(y) \left[A^2 \left(-d\tau^2 + d\vec{x}^2 \right) + B^2 \, dy^2 \right],$$
$$V_{CY} = B \, h^{6/5}(y),$$
$$h(y) = 5\alpha y + C, \tag{10.58}$$

where A, B, and C are arbitrary constants. The y coordinate spans the orbifold $S^1/\mathbb{Z}_2$ with fixed points at $y = \pm 1$. If we were to "unroll" the solution, by which we mean $\mathbb{Z}_2$-reflecting it across the end points, we would obtain a downward-pointing kink at $y = -1$ and an upward-pointing kink at $y = +1$. These are the characteristics of *domain walls*, which is another name for branes with codimension 1. These kinks ensure that the singular source terms (present in the stress–energy tensor) are matched by the Einstein tensor. Note that the brane at $y = -1$ has negative tension, while that at $y = +1$ has positive tension. In fact, it is the $\mathbb{Z}_2$ symmetry that keeps the negative-tension brane stable.

The solution just presented is a minimal solution. In general, many more fields as well as bulk branes can be added in order to render the solution more realistic. Here we simply wish to illustrate some basic properties of the end-of-the-world branes. What is of special cosmological interest is their dynamics. The solution above is static, with the moduli A, B, C being constants that determine the size of the extra dimensions as well as the scale factors on the branes. In order to see how these evolve, we will promote these moduli to time-dependent fields and consider the ansatz

$$ds^2 = h^{2/5}(\tau, y) \left[A^2(\tau) \left(-d\tau^2 + d\vec{x}^2 \right) + B^2(\tau) dy^2 \right],$$
$$V_{CY} = B(\tau) \, h^{6/5}(\tau, y),$$
$$h(\tau, y) = 5\alpha y + C(\tau), \quad -1 \le y \le +1. \tag{10.59}$$

This ansatz satisfies the τy Einstein equation identically, which is important as otherwise that equation would act as a constraint. The so-called moduli space approximation then amounts to plugging this ansatz into the 5-dimensional action (10.57), which leads to an effective description. We find

$$S_{\text{mod}} = -6 \int_{4D} d\tau d^3 x \, A^2 B I_{\frac{3}{5}} \left[\left(\frac{A'}{A} \right)^2 - \frac{1}{12} \left(\frac{B'}{B} \right)^2 + \frac{A'B'}{AB} \right.$$
$$\left. - \frac{1}{25} \frac{I_{-\frac{7}{5}}}{I_{\frac{3}{5}}} C'^2 + \frac{3}{5} \frac{I_{-\frac{2}{5}} A'C'}{I_{\frac{3}{5}} A} \right], \tag{10.60}$$

where primes denote τ derivatives and

$$I_n = \int_{-1}^{1} dy \, h^n = \frac{1}{5\alpha(n+1)}[(C+5\alpha)^{(n+1)} - (C-5\alpha)^{(n+1)}]. \quad (10.61)$$

The following field redefinitions are very useful,

$$a^2 \equiv A^2 \, B \, I_{\frac{3}{5}}, \quad (10.62)$$

$$e^{\phi_1/\sqrt{2}} \equiv B \, (I_{\frac{3}{5}})^{3/4}, \quad (10.63)$$

$$5\alpha \left[\frac{(1+e^{2\sqrt{2/3}\phi_2})^{5/4} + (1-e^{2\sqrt{2/3}\phi_2})^{5/4}}{(1+e^{2\sqrt{2/3}\phi_2})^{5/4} - (1-e^{2\sqrt{2/3}\phi_2})^{5/4}} \right] \equiv C, \quad (10.64)$$

because they convert the moduli space action (10.60) into the simple form

$$S_{\mathrm{mod}} = \int_{4D} d\tau d^3x \left[-6a'^2 + a^2(\phi_1'^2 + \phi_2'^2) \right], \quad (10.65)$$

where we see that a represents the effective 4-dimensional scale factor, while $\phi_{1,2}$ are two scalar fields. We can also relate 5-dimensional physical quantities, such as the distance d between the branes, as well as the CY volume and scale factors at the locations $y = \pm 1$ of the branes, to the moduli fields, according to

$$d = \frac{1}{3(2\alpha)^{1/4}} e^{\phi_1/\sqrt{2} - \sqrt{3/2}\phi_2} [(1+e^{2\sqrt{2/3}\phi_2})^{3/2} - |1 - e^{2\sqrt{2/3}\phi_2}|^{3/2}],$$

$$(10.66)$$

$$V_{\mathrm{CY}\pm} = (2\alpha)^{3/4} e^{\phi_1/\sqrt{2}} \begin{cases} \left(\cosh\sqrt{2/3}\phi_2 \right)^{3/2}, \\ \left(-\sinh\sqrt{2/3}\phi_2 \right)^{3/2}, \end{cases} \quad (10.67)$$

$$a_\pm = (2\alpha)^{1/8} \, a \, e^{-\phi_1/2\sqrt{2}} \begin{cases} \left(\cosh\sqrt{2/3}\phi_2 \right)^{1/4}, \\ \left(-\sinh\sqrt{2/3}\phi_2 \right)^{1/4}. \end{cases} \quad (10.68)$$

From the expression for the moduli space action (10.65), we can see that the scalars have no potential, which is a consequence of them being moduli of a supersymmetric solution. Thus their motion is free, except for one significant restriction, which becomes apparent from the expression for $V_{\mathrm{CY}-}$ above. Since the physical volume is a positive quantity, we see that we must require $\phi_2 < 0$ – i.e., a knowledge of the higher-dimensional origin of the effective action informs us that there is a boundary to moduli space. At this boundary the scalar field trajectories get reflected, and this can have interesting cosmological consequences, for instance in a multifield inflationary or ekpyrotic model it can lead to the sourcing of curvature perturbations by isocurvature perturbations.

The crux of the matter is once again, however, the question of whether appropriate potentials can be generated for the scalar fields. At the perturbative level, as we saw, there is no potential. However, nonperturbative effects can generate potentials. In this concrete example, the virtual exchange of Euclidean 2-branes wrapping a cycle in the CY manifold and stretching between the two end-of-the-world branes leads to a potential, whose specific shape strongly depends on the values of additional moduli that we have not considered here. It is presently unknown whether a fully realistic potential can be generated in this way, while all other moduli get stabilized. Once again, we have a setting where many interesting ingredients are present, but where it has not been possible yet to put them all together in such a way that a realistic cosmological-plus-particle-physics model emerges. In the exercises, we will explore a simple example that highlights why it is generally difficult (in supersymmetric theories) to generate either an inflationary potential that is flat enough, or an ekpyrotic potential that is steep enough.

We should mention that the end-of-the-world model arising from heterotic M-theory was the inspiration for the ekpyrotic model, since it suggested that the brane collision might be what appears as the big bang to us. If one takes this hypothesis seriously, then it follows that there must have been a phase of evolution prior to the big bang. And since the brane collision can quite naturally incorporate reheating, the concept was investigated whether cosmological perturbations could have been formed prior to the brane collision. Over the course of a few refinements this led to the ekpyrotic model, which is now usually presented without any reference to branes and higher dimensions. However the higher-dimensional view of the brane collision is quite revealing. The particular solution envisioned in the ekpyrotic case corresponds to the solution for the moduli fields given by

$$a = |2y_0\tau|^{1/2}, \qquad e^{\phi_1} = |2y_0\tau|^{3/2\sqrt{2}}, \qquad e^{\phi_2} = |4\alpha y_0\tau|^{\sqrt{3}/2\sqrt{2}}. \qquad (10.69)$$

From (10.66) one can see that, for small τ, the 5-dimensional distance d between the branes is given by $d \approx 2y_0\tau$. Hence $2y_0$ has the interpretation of being the collision velocity. Meanwhile, both the CY volumes as well as the scale factors on the branes approach a nonzero constant at the collision (even though the effective 4-dimensional scale factor reaches zero). Thus, only the orbifold dimension shrinks to zero, and in fact the spacetime near the collision can be approximated by the Milne spacetime. Combined with the fact that the string coupling constant goes to zero at the collision – see (10.51) – one can see that such a cataclysmic event may in fact be rather mild from the higher-dimensional point of view, and can perhaps be fully resolved in quantum gravity.

10.3 Perspectives

String theory introduces fascinating new elements: strings that are so highly constrained that they determine the number of spacetime dimensions and require spacetime to follow generalizations of Einstein's equations; internal manifolds whose geometries determine the properties of particles; branes that interact, intersect, and localize matter; and dualities that sometimes exchange some of these elements with each other, yet provide an equally valid description of the system under study. What is striking is that this provides a framework that unifies and generalizes essentially all known concepts of physics. In fact, it is the only approach to quantum gravity that unifies gravity with all other interactions, and this is an obvious reason for including the present chapter in this discussion. And yet, what in some sense is perhaps the biggest surprise, it has proven extremely nontrivial to connect string theory with cosmology.

Of particular difficulty is the realization of an accelerated phase of expansion. One could of course just conclude that inflation is likely wrong, but the discovery of dark energy complicates this conclusion (as well as the fact that on a phenomenological level inflation provides the simplest explanation for the primordial perturbations seen in the CMB). It currently seems without doubt that the universe has been undergoing accelerated expansion over the last 5 billion years or so. This means that, if string theory truly is to describe nature, then it must be able to explain this behavior. An interesting hint has appeared recently in observations by the Dark Energy Spectroscopic Instrument (DESI) indicating that dark energy may be decreasing over time (DESI Collaboration; Abdul Karim, M., et al. (2025)). Such a time-dependent dark energy would seem to fit better into the framework of string theory – it will certainly be interesting to follow related developments over the coming years.

If one looks at the development of our descriptions of nature, then a common trend has been for our theories to become more general, while at the same time allowing for more and more solutions. In Newton's theory of gravity there are many places where one can live in the universe, but in some sense all locations are equivalent to all others, as spacetime is simply an immutable stage on which stars and planets can revolve. With the advent of general relativity, spacetime becomes dynamical, so that now there is a plethora of possibilities for spacetime evolution too. Something then must select the specific shape of spacetime we are experiencing. In string theory the possibilities multiply rather drastically once more, as there are now different numbers of large and small dimensions, great varieties of configurations for objects of various dimensions, leading to vast numbers of possible low-energy worlds with their specific particle physics, their own "chemistry," in various numbers of dimensions. The space of solutions is phenomenal, and again we must find a principle that determines why we see the universe as it is.

Efforts are underway (in particular via the so-called swampland program), to specify more precisely what may lead to a consistent solution and what may not. This is a welcome development, as it reduces the number of possibilities. Still, at present it seems highly unlikely that the number of consistent solutions will come down to a single one. This implies that, in the wave function of the universe, there will be very many branches representing the possible evolutions that could lead to the present state (and vastly more that could lead to completely different states). Faced with this realization, it is clear that we must find principles, which may well be independent of our search for a unified theory of all forces, that select solutions. More quantum mechanically, one should say: We must find principles that give probabilities to solutions.

In this book, we studied one such example in detail, namely the no-boundary proposal. Below, an exercise is included that shows that, in a toy model inspired by string theory, the no-boundary wave function can assign probabilities to stringy ingredients such as the values of p-form fluxes. But one should keep in mind that in truth there might be a vast sum of possibilities, ranging from a nucleation of the universe with a simple expanding phase to the present; along with an evolving universe from which a new universe like ours could tunnel; along with many tunneling events prior to "our" tunneling; as well as big bang-like transitions, perhaps caused by brane collisions, that may change the gauge group of particle physics or the configuration of braneworlds, or even the number of large dimensions. In principle a universal wave function will contain all of these possibilities. The task of quantum cosmology will be to find a bearing in this vast, and to date still rather nebulous, space.

Exercises

10.1 Work out the equation of motion (for X^μ) stemming from the Nambu–Goto action (10.1). Then derive the equations of motion implied by the Polyakov action (10.2), for X^μ and $g_{\alpha\beta}$, and show that they agree with the Nambu–Goto equation of motion.

10.2 The "universal" 10-dimensional effective action (10.7) has a surprising feature, which is the apparently wrong sign of the dilaton kinetic term. To assess whether this is problematic or not, one should transform the action to Einstein frame, defined as the frame in which the Einstein–Hilbert action takes its canonical form. This can be achieved by the following field redefinition:

$$G_{MN} = e^{\frac{1}{2}\Phi} G_{E,MN}, \qquad (10.70)$$

where $G_{E,MN}$ is the Einstein frame metric. Implement this redefinition and discuss the outcome.

10.3 Derive the action of a probe brane moving radially and in a purely time-dependent manner in an equidimensional brane background.

10.4 This exercise will illustrate the difficulty in obtaining flat positive or steep negative potentials in supergravity. In 4-dimensional models with minimal supersymmetry, the scalar fields are described by a Kähler potential $K(A, A^\star)$, which is a Hermitian function of a complex scalar field A, and a superpotential $W(A)$, which is a holomorphic function of A. The kinetic term is then given by $-K_{,AA^\star}\partial^\mu A \partial_\mu A^\star$, and the potential by the famous formula

$$V = e^K \left[K^{,AA^\star} D_A W D_{A^\star} W^\star - 3 W W^\star \right], \qquad (10.71)$$

where $D_A W = (\partial_A + K_{,A})W$. As a representative example, consider the Kähler potential $K = -\frac{1}{2}(A - A^\star)^2$ and verify that it encodes the kinetic term for two canonically normalized real scalar fields. Then consider the superpotential (which is rather typical in stringy models) of the form $W = w e^{cA}$, where w, c are real constants, and analyze the resulting scalar potential.

10.5 Do no-boundary solutions arise in string theory and, if so, what do they predict? In this exercise you will use a toy model to show that the answer to the first question is "yes," and that the no-boundary proposal results in probabilities for solutions with different amounts of flux. Start by considering the following theory in $D = 8$ dimensions,

$$S_8 = \frac{1}{2} \int d^8 x \sqrt{-\hat{g}} \left(\hat{R} + \alpha \hat{R}^4 - \frac{1}{2\cdot 4!} q^2 F_{(4)}^2 \right), \qquad (10.72)$$

which includes a higher-derivative curvature correction and a 4-form flux term with coupling constant q. Such curvature corrections are indeed expected to arise in string theory – this theory is a toy model in the sense that it is not known whether a single such term could be dominant over all others that are thought to appear as well. The first task is to perform a redefinition of the metric by a conformal factor in order to end up in Einstein frame. For this, you may make use of (D.72) with a conformal factor $e^{2\varphi}$, with $e^{-6\varphi} \equiv 1 + 4\alpha \hat{R}^3$. Then rescale φ so as to obtain a canonically normalized scalar. In a second step, compactify the theory on a 4-sphere, keeping the volume modulus of the sphere in analogy with (10.11), and with flux threading the sphere as

$$F_{[4]} = 2n_4 \text{vol}(S^4). \tag{10.73}$$

You may take the metric of the large dimensions to be of FLRW form. Derive the effective scalar potential in 4 dimensions and analyze its qualitative properties with regard to no-boundary solutions. (You can, if you feel like it, also numerically look for explicit no-boundary solutions of the reduced theory.)

Appendix A
Constants of nature and cosmological quantities

Constants of nature

A few useful constants, specified to nine significant figures (if known):

speed of light (by definition)	$c \equiv 299\,792\,458 \, \mathrm{m\,s^{-1}}$
reduced Planck constant	$\hbar = \dfrac{h}{2\pi} = 1.054\,571\,82 \cdot 10^{-34} \, \mathrm{J\,s}$
Newton's constant	$G = 6.674\,30(15) \cdot 10^{-11} \, \mathrm{m^3 kg^{-1} s^{-2}}$
Boltzmann's constant	$k_B = 1.380\,649 \cdot 10^{-23} \, \mathrm{J\,K^{-1}}$
Stefan–Boltzmann constant	$\sigma = \dfrac{\pi^2 k_B^4}{60 \hbar^3 c^2}$
	$= 5.670\,374\,42 \cdot 10^{-8} \, \mathrm{W m^{-2} K^{-4}}$
elementary charge	$e = 1.602\,176\,63 \cdot 10^{-19} \, \mathrm{C}$
electron mass	$m_e = 9.109\,383\,71 \cdot 10^{-31} \, \mathrm{kg}$
proton mass	$m_p = 1.672\,621\,92 \cdot 10^{-27} \, \mathrm{kg}$
neutron mass	$m_n = 1.674\,927\,50 \cdot 10^{-27} \, \mathrm{kg}$
Avogadro number	$N_A = 6.022\,140\,76 \cdot 10^{23} \, \mathrm{mol^{-1}}$
(inverse) fine structure constant	$\alpha^{-1} = 137.035\,999$
Bohr radius	$a_0 = \dfrac{\hbar}{\alpha m_e c} = 5.291\,772\,10 \cdot 10^{-11} \, \mathrm{m}$
electron Volt	$1\,\mathrm{eV} = 1.602\,176\,63 \cdot 10^{-19} \, \mathrm{J}$
	$= 11\,604.518\,1 \, k_B \mathrm{K}$
	$1\,\mathrm{GeV} = 1.782\,661\,92 \cdot 10^{-27} \mathrm{kg}$

Planck units

$$\text{Planck length} \quad 1\, l_P = \sqrt{\frac{\hbar G}{c^3}} \approx 1.616\,2 \cdot 10^{-35}\,\text{m}$$

$$\text{Planck time} \quad 1\, t_P = \sqrt{\frac{\hbar G}{c^5}} \approx 5.391\,2 \cdot 10^{-44}\,\text{s}$$

$$\text{Planck mass} \quad 1\, m_P = \sqrt{\frac{\hbar c}{G}} \approx 2.176\,4 \cdot 10^{-8}\,\text{kg}$$

Astronomical and cosmological quantities

Earth mass	$1\, M_\oplus \approx 5.972\,2 \cdot 10^{24}\,\text{kg}$
solar mass	$1\, M_\odot \approx 1.988\,416 \cdot 10^{30}\,\text{kg}$
radius of solar mass black hole	$2GM_\odot/c^2 \approx 2\,953\,\text{m}$
year	$31\,557\,600\,\text{s}$
light-year	$9.460\,73 \cdot 10^{15}\,\text{m}$
parsec	$1\,\text{pc} \approx 3.261\,563\,77\,\text{light-years}$
Sun–Earth distance	$8.3\,\text{light-minutes}$
Sun–Proxima Centauri distance	$4.24\,\text{light-years}$
(visible) diameter of Milky Way	$87\,400 \pm 3\,600\,\text{light-years}$
Milky Way–Andromeda distance	$2.54 \cdot 10^6\,\text{light-years}$
cosmological constant	$\Lambda \approx 1.3 \cdot 10^{-52}\,\text{m}^{-2}$
corresponding vacuum mass density	$\dfrac{\Lambda c^2}{8\pi G} \approx 7 \cdot 10^{-27}\,\text{kg}\,\text{m}^{-3}$
Hubble rate	$100\, h\,\text{km}\,\text{s}^{-1}\,\text{Mpc}^{-1}$
Hubble radius/time	$\dfrac{1}{h} \cdot 9.778 \cdot 10^9\,\text{(light-)years}$
10 billion light-years (size)	$\approx 10^{60}\, l_P$
10 billion years (age)	$\approx 10^{60}\, t_P$

Appendix B
Picard–Lefschetz theory and saddle-point approximation

The path integral approach to quantum mechanics, and to quantum gravity, involves integrals that are of the general form

$$I = \int_\mathcal{C} dx\, e^{\frac{i}{\hbar}S(x)}, \tag{B.1}$$

along with their generalizations. Here $S(x)$ is a real function (usually the action), $\hbar$ a real parameter, and $\mathcal{C}$ the integration contour. If $\mathcal{C}$ is the real line or half-line, then this kind of integral is only conditionally convergent. Its modulus is 1 everywhere, though because of the oscillations cancellations occur and a finite answer can be obtained.

However, the problem is that the values of such conditionally convergent integrals can be ambiguous. Let us look at a specific example to illustrate this: Take the Fresnel integral in 2 dimensions, and evaluate it in two different ways, once by using Cartesian coordinates as a regulator, and once using polar coordinates to regulate,

$$\int_{-\infty}^{+\infty} e^{ix^2} dx \int_{-\infty}^{+\infty} e^{iy^2} dy = \left(\lim_{R \to \infty} \int_{-R}^{+R} e^{ix^2} dx \right)^2$$
$$= \left((1+i)\sqrt{\pi/2} \right)^2$$
$$= i\pi, \tag{B.2}$$

$$\int_{-\infty}^{+\infty} e^{ix^2} dx \int_{-\infty}^{+\infty} e^{iy^2} dy = \lim_{R \to \infty} \int_0^R e^{ir^2} r\, dr \int_0^{2\pi} d\theta$$
$$= \lim_{R \to \infty} i\pi(1 - e^{iR^2}), \tag{B.3}$$

where one notices that the second version does not converge. The discrepancy between (B.2) and (B.3) is troubling, as we would like our physical theories to lead to definite outcomes. It is due to the fact that, for oscillating integrals, the result can depend on the order of integration – in other words, Fubini's theorem does not hold. Picard–Lefschetz theory remedies this problem by *defining* the integrals as (sums of) manifestly convergent integrals, essentially by extending the integrand $S(x)$ into the complex domain and using Cauchy's theorem to deform the integration contour into a contour along which the modulus of the integrand (which is then no longer unity) falls off as fast as possible. This guarantees unambiguous answers. Let us see in more detail how this works.

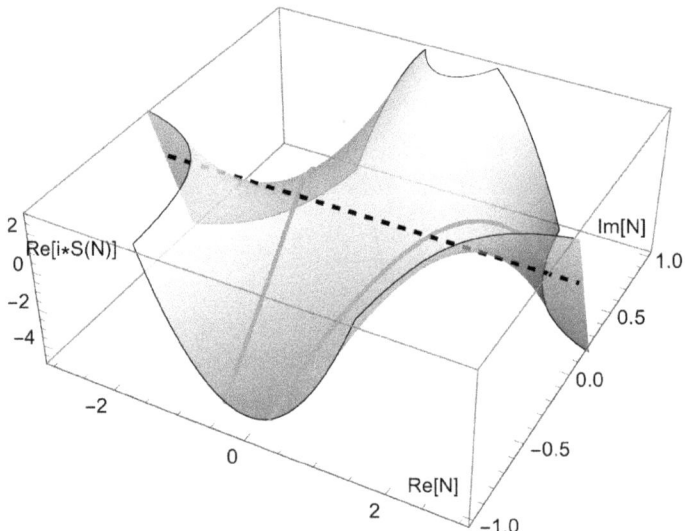

Figure B.1 The weighting $\mathrm{Re}[iS(N)]$ of the function $S(N) = N^3 - 3N$, shown as height above the complex N plane. For an integration contour along the real N line (black dashed line), the weighting/Morse function is zero, and the integral is oscillating (see also Fig. B.2). Meanwhile, along the two thimbles (steepest descent contours, in gray) the weighting falls off as fast as possible, and the integral does not contain oscillations. The real line contour can be deformed into a sum over the two thimbles, plus arcs at infinity along which the integrand vanishes. In this example, the saddle points reside at the intersection points of the thimbles with the real line, at $N_s = \pm 1$.

We start by extending the variable x to the complex plane, $x \in \mathbb{C}$, and by viewing $S(x)$ as a holomorphic function of x. We will also write $x = u^1 + iu^2$ for real u^1, u^2. Similarly, we decompose the integrand as $\mathcal{I} = iS/\hbar = W + iP$, where W is the *weighting* and P the *phase*. Mathematicians also call W the Morse function.

Downward flow (i.e., decreasing weighting) is defined by

$$\frac{\mathrm{d}u^i}{\mathrm{d}\lambda} = -g^{ij}\frac{\partial W}{\partial u^j}, \qquad (B.4)$$

where λ is a parameter along the flow and g_{ij} is a (Riemannian) metric on the complex plane. To check that the weighting falls off along such a flow, note that

$$\frac{\mathrm{d}W}{\mathrm{d}\lambda} = \sum_i \frac{\partial W}{\partial u^i}\frac{\mathrm{d}u^i}{\mathrm{d}\lambda} = -\sum_i \left(\frac{\partial W}{\partial u^i}\right)^2 < 0. \qquad (B.5)$$

Incidentally, this also implies that downward flows run from stationary points of the weighting ($\partial W/\partial x = 0$) to singularities of the integrand, where the weighting diverges to minus infinity. Stationary points of the weighting are necessarily saddle points ($\partial S/\partial x = 0$) of the full integrand.[1] The downward flows associated with saddle points are called *Lefschetz thimbles*. We will shortly show that thimbles are not only steepest descent contours, but also loci of stationary phase. In that sense, they minimize oscillations of the integrand. Along thimbles, the integral has the best chance of being convergent. In fact it is convergent if

$$\left|\int_{\mathcal{J}_\sigma} \mathrm{d}x\, e^{iS[x]/\hbar}\right| \leq \int_{\mathcal{J}_\sigma} |\mathrm{d}x|\left|e^{iS[x]/\hbar}\right| = \int_{\mathcal{J}_\sigma} |\mathrm{d}x| e^{W(x)} < \infty. \qquad (B.6)$$

We can denote the length along the thimble as $l = \int |\mathrm{d}x|$. Then the integral converges if $W(x(l)) < -\ln(l) + A$, for some constant A, as $l \to \infty$, which is not a very strong assumption.

In complete analogy with downward flow one can define upward flow, via

$$\frac{\mathrm{d}u^i}{\mathrm{d}\lambda} = +g^{ij}\frac{\partial W}{\partial u^j}. \qquad (B.7)$$

Perhaps the simplest example of these concepts is provided by a quadratic function $S(x) = x^2$, which has a saddle point at $x = 0$. Then $\mathrm{Re}[iS(x)] = -2u^1u^2$. The weighting decreases most rapidly along the contour $u^1 = u^2$, and this is the Lefschetz thimble. Conversely, the weighting increases fastest along $u^1 = -u^2$, and this is the steepest ascent contour. We will denote thimbles by $\mathcal{J}$ and steepest ascent contours by $\mathcal{K}$. The next simplest example, a cubic function, is illustrated in Figs. B.1 and B.2.

---

[1] The Cauchy–Riemann equations imply that if the real part of a holomorphic function is at an extremum, then the full function is at an extremum. That this is a saddle point can be seen by expanding the action near a critical point x_c, obtaining
$S(x) \approx S(x_c) + \frac{1}{2}S''(x_c)[(u^1)^2 - (u^2)^2 + 2iu^1u^2]$, demonstrating that setting off from the critical point there is always one direction along which the action is increasing and one along which it is decreasing.

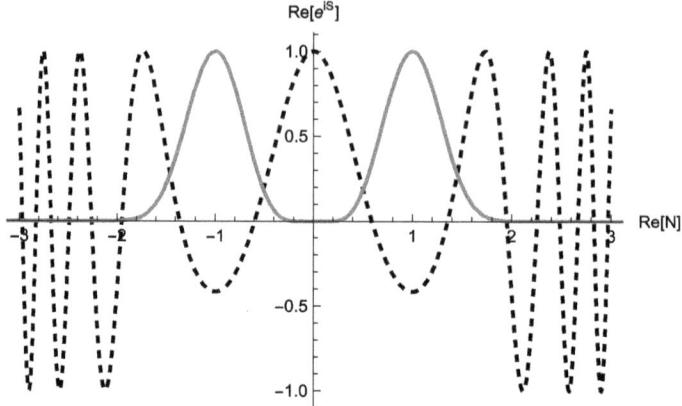

Figure B.2 This figure is connected with Fig. B.1 and uses the same conventions for the paths. It shows how the integrand oscillates along the original real line contour (black dashed line), but falls off rapidly on either side of the saddle points when the contour is shifted to lie along the thimbles (in gray).

The metric used in the definition of the flow is a matter of choice. In our case we will only need the simplest one, namely a flat Cartesian metric $ds^2 = |dx|^2$. In terms of complex coordinates, $(u, \bar{u}) = (u^1 + iu^2, u^1 - iu^2)$, this metric becomes $g_{uu} = g_{\bar{u}\bar{u}} = 0$, $g_{u\bar{u}} = g_{\bar{u}u} = 1/2$. Then $W = (\mathcal{I} + \bar{\mathcal{I}})/2$ and the flow equations (B.4) can conveniently be rewritten as

$$\frac{du}{d\lambda} = -\frac{\partial \bar{\mathcal{I}}}{\partial \bar{u}}, \quad \frac{d\bar{u}}{d\lambda} = -\frac{\partial \mathcal{I}}{\partial u}. \tag{B.8}$$

This reformulation allows for a quick proof of the statement made above, that the phase along steepest descent contours is constant:

$$\frac{dP}{d\lambda} = \frac{1}{2i}\frac{d(\mathcal{I} - \bar{\mathcal{I}})}{d\lambda} = \frac{1}{2i}\left(\frac{\partial \mathcal{I}}{\partial u}\frac{du}{d\lambda} - \frac{\partial \bar{\mathcal{I}}}{\partial \bar{u}}\frac{d\bar{u}}{d\lambda}\right) = 0. \tag{B.9}$$

This is an important property: On the one hand, it shows that there will be no oscillations along steepest descent contours, which is what we were aiming for! And on the other hand, it allows for an efficient numerical way of identifying a thimble, as it is simply given by the downwards curves that start at a saddle point and have the same phase as that of the saddle point.

There is a subtlety that one needs to be aware of: It can happen that two saddle points are connected by the same flow line, which is then an upward flow from the point of view of one saddle and a downward flow for the other saddle. This situation can arise when two saddles happen to have the same phase, which may occur when a symmetry is present. In such a case, one can add a small deformation (breaking the symmetry in question), then determine the new flows (which will now run to $\mathcal{W} \to \pm\infty$, as usual), and take the limit of vanishing deformation at the end of the calculation.

After all this preliminary setup, we are finally in a position to address the most crucial question: Which saddle points, along with their associated thimbles, contribute to the integral? And is there a systematic way of answering this question?

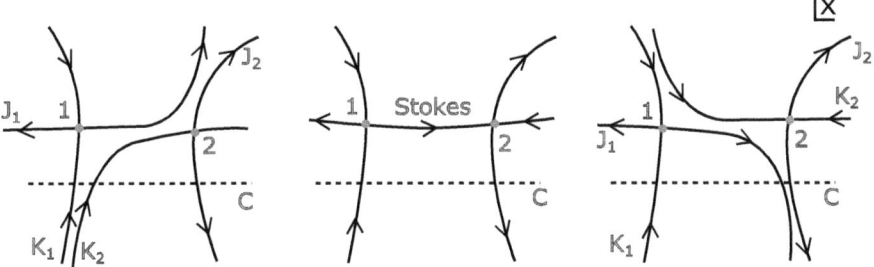

Figure B.3 On the left, both saddles contribute to the integral with integration contour $\mathcal{C}$, as the latter is crossed by both $\mathcal{K}_1$ and $\mathcal{K}_2$. This means that $\mathcal{C}$ can be deformed into a sum of $\mathcal{J}_1$ and $\mathcal{J}_2$. When a parameter of the integral is varied, there can be a topological change in the flow lines known as a Stokes phenomenon. At the Stokes transition, the steepest descent contour emanating from saddle 1 is simultaneously the steepest ascent contour of saddle 2 – it is known as a *Stokes ray*. On the right, we can see that after the transition only saddle 1 contributes to the integral. Arrows denote the direction of descent.

Start by noting that, if all degeneracies are removed as explained above, we may associate a unique saddle point σ to every thimble, and likewise to every steepest ascent contour. We may write this observation as an intersection

$$\text{Int}(\mathcal{J}_\sigma, \mathcal{K}_{\sigma'}) = \delta_{\sigma\sigma'}, \qquad (B.10)$$

where implicitly a direction is assigned to every thimble and ascent contour. What we would like to do now is rewrite the original integration contour $\mathcal{C}$ as an appropriate sum of thimbles,

$$\mathcal{C} = \sum_\sigma n_\sigma \mathcal{J}_\sigma, \qquad (B.11)$$

where the $n_\sigma \in \{0, \pm 1\}$ coefficients determine whether a saddle point contributes or not ($n_\sigma = -1$ when the orientations of $\mathcal{C}$ and the contributing thimble are opposite to each other). We can extract the n_σ coefficients by intersecting both sides of the equation with a steepest ascent contour,

$$n_\sigma = \text{Int}(\mathcal{C}, \mathcal{K}_\sigma). \qquad (B.12)$$

This gives us our desired result. A saddle point and its associated thimble contribute to the integral (B.1) if and only if the steepest ascent contour emanating from the saddle intersects the contour $\mathcal{C}$. For an illustration see the left panel in Fig. (B.3).

This is a very sensible result: Along $\mathcal{C}$, we have an integral with fast oscillations and thus numerous cancellations. We are replacing this with an integral over a nonoscillating function, which therefore must necessarily have smaller amplitude. Thus the thimbles must lie lower (in terms of the Morse function) than the original contour. Put differently, from the original contour we must flow downwards to reach the thimbles into which it is to be deformed. Note also that there will be no additional contribution to the integral from arcs at infinity (or near singularities) linking $\mathcal{C}$ to the thimbles, assuming that $\mathcal{C}$ itself runs between infinities and/or singularities. This is because asymptotically the weighting along the thimbles diverges to minus infinity, $\mathcal{W} \to -\infty$, and a saddle point only contributes to the integral if there is no intervening region of divergence between it and $\mathcal{C}$.

Using Eq. (B.12), we can now re-express the original integral as a sum over thimbles,

$$I = \int_{\mathcal{C}} \mathrm{d}x \, e^{\frac{i}{\hbar}S[x]} = \sum_{\sigma} n_{\sigma} \int_{\mathcal{J}_{\sigma}} \mathrm{d}x \, e^{\frac{i}{\hbar}S[x]} \,. \tag{B.13}$$

This provides the sought-after unambiguous definition of the integral. Each contributing integral along a thimble $\mathcal{J}_{\sigma}$ is manifestly convergent, and thus Fubini's theorem holds, meaning that in a higher-dimensional context all integrals can be evaluated in any order one chooses, always giving the same result. Note the important point that not all saddle points and their thimbles contribute, only those that give a nonvanishing n_{σ} in (B.12).

There is one interesting effect we should mention, called a *Stokes phenomenon*. When a parameter of the integral is varied, it can happen that the n_{σ} numbers suddenly change. This occurs when, as the parameter is varied across a critical value, the steepest ascent flow from one saddle merges with the descent flow from another, forming a Stokes ray – see Fig. B.3. After the transition, the flow lines have undergone a topological change, and a saddle point can start/cease to contribute to the integral. This phenomenon can completely change the asymptotic structure of the integral, and can lead to (otherwise unexpected) sudden changes in its value.

Picard–Lefschetz theory is not just an attractive formal rewriting of oscillating integrals – it also allows for a highly practical approximation scheme, called the saddle-point approximation. This more or less falls out of the rewriting by itself, in the limit that $\hbar$ is small. Indeed, note that we can extend the right-hand side of (B.13) as

$$I = \int_{\mathcal{C}} \mathrm{d}x \, e^{iS[x]/\hbar} = \sum_{\sigma} n_{\sigma} \int_{\mathcal{J}_{\sigma}} \mathrm{d}x \, e^{\frac{i}{\hbar}S[x]}$$

$$= \sum_{\sigma} n_{\sigma} \, e^{i\,P(x_{\sigma})} \int_{\mathcal{J}_{\sigma}} e^{W} \mathrm{d}x \qquad (B.14)$$

$$\approx \sum_{\sigma} n_{\sigma} \, e^{iS(x_{\sigma})/\hbar} \left[A_{\sigma} + \mathcal{O}(\hbar) \right] . \qquad (B.15)$$

The first line is the Picard–Lefschetz definition (B.13). Since the phase is constant along thimbles, it may be pulled out – this is the second line. Up to here the rewriting is exact. Now, as illustrated in Fig. B.2, the weighting is rather sharply peaked at the saddle point, and the more so the smaller $\hbar$ is. Therefore, in the limit of small $\hbar$, the integral along a thimble may be approximated by the value at the saddle point itself. This is the third line, which is the saddle-point approximation (also called the JWKB approximation, as this is essentially the semiclassical approximation). The factor A_{σ} encodes the leading corrections, and can be obtained by looking at small fluctuations around the saddle point. More precisely, it is obtained by evaluating a Gaussian integral over the action expanded to second order (an explicit example of such a calculation is provided in Section 5.2.1). And in principle, further subleading corrections can be calculated perturbatively.

The full integral is approximated by a sum of such saddle-point contributions, one for each contributing saddle. However, due to the smallness of $\hbar$, in realistic applications only the leading saddle-point contribution should be retained (also since corrections to the dominant saddle point can contribute more strongly than sub-dominant saddle points). Thus, in practice, the most important question that needs to be answered for a given integral is which of its saddle points is both relevant and dominant. In this way, Picard–Lefschetz theory not only helps in defining oscillating integrals properly, but also provides a means of extracting the leading behavior with relative ease.

Appendix C
Useful functions

Fourier transforms

Our conventions for Fourier transforms in 1 dimension are

$$f(x) = \int_{-\infty}^{\infty} \frac{dk}{2\pi} F(k)\, e^{ikx}, \qquad F(k) = \int_{-\infty}^{\infty} dx\, f(x)\, e^{-ikx}, \qquad \text{(C.1)}$$

while in 3 dimensions we use

$$f(\mathbf{x}) = \int \frac{d^3k}{(2\pi)^3} F(\mathbf{k})\, e^{i\mathbf{k}\mathbf{x}}, \qquad F(\mathbf{k}) = \int d^3x\, f(\mathbf{x})\, e^{-i\mathbf{k}\mathbf{x}}. \qquad \text{(C.2)}$$

A result that is used several times is that the Dirac delta function may be seen as the Fourier transform of unity,

$$\delta(x) = \frac{1}{2\pi} \int_{-\infty}^{\infty} e^{ikx}\, dk. \qquad \text{(C.3)}$$

Spherical harmonics in 3 dimensions

Distances on the unit 3-sphere are determined by the metric

$$ds^2 = d\chi^2 + \sin^2\chi \left(d\theta^2 + \sin^2\theta\, d\phi^2\right). \qquad \text{(C.4)}$$

Scalar harmonics then satisfy the equation

$$\Box^2 Q^{(n)} = -(n^2 - 1) Q^{(n)}, \quad n = 1, 2, 3 \ldots \qquad \text{(C.5)}$$

A general solution is a linear combination of n^2 terms,[1]

$$Q^{(n)} = \sum_{l=0}^{n-1} \sum_{m=-l}^{l} a_{lm}^n \Pi_l^n(\chi) Y_{lm}(\theta, \phi), \qquad \text{(C.6)}$$

---

[1] The number of terms is counted as follows: $\sum_{l=0}^{n-1} \sum_{m=-l}^{l} 1 = \sum_{l=0}^{n-1} (2l+1) = n^2$.

where a_{lm}^n are coefficients and $Y_{lm}(\theta, \phi)$ are the usual spherical harmonics on a 2-sphere. In a real representation, the first few are given by

$$Y_{00} = \frac{1}{2\sqrt{\pi}}, \tag{C.7}$$

$$Y_{1-1} = \sqrt{\frac{3}{4\pi}} \sin\theta \sin\phi, \tag{C.8}$$

$$Y_{10} = \sqrt{\frac{3}{4\pi}} \cos\theta, \tag{C.9}$$

$$Y_{11} = \sqrt{\frac{3}{4\pi}} \sin\theta \cos\phi, \tag{C.10}$$

$$Y_{2-2} = \frac{1}{4}\sqrt{\frac{15}{\pi}} \sin^2\theta \sin(2\phi), \tag{C.11}$$

$$Y_{2-1} = \frac{1}{4}\sqrt{\frac{15}{\pi}} \sin(2\theta) \sin\phi, \tag{C.12}$$

$$Y_{20} = \frac{1}{4}\sqrt{\frac{5}{\pi}} (3\cos^2\theta - 1), \tag{C.13}$$

$$Y_{21} = \frac{1}{4}\sqrt{\frac{15}{\pi}} \sin(2\theta) \cos\phi, \tag{C.14}$$

$$Y_{22} = \frac{1}{4}\sqrt{\frac{15}{\pi}} \sin^2\theta \cos(2\phi). \tag{C.15}$$

Meanwhile, the functions Π_l^n are known as the Fock polynomials,

$$\Pi_l^n(\chi) = \sin^l \chi \frac{d^{l+1}\cos(n\chi)}{d(\cos\chi)^{l+1}}. \tag{C.16}$$

For $l = 0$, these are simply $\Pi_0^n(\chi) = \sin(n\chi)/\sin\chi$.

Tensor harmonics are defined in a similar manner via

$$\Box G_{ij}^{(n)} = -(n^2 - 3) G_{ij}^{(n)}, \quad n = 3, 4, 5, \ldots \tag{C.17}$$

These are transverse and traceless,

$$\nabla^i G_{ij}^{(n)} = 0, \qquad G_i^{(n)i} = 0. \tag{C.18}$$

A general solution to (C.17) is given by a linear combination of $n^2 - 4$ terms,

$$G_{ij}^{(n)} = \sum_{l=2}^{n-1} \sum_{m=-l}^{l} c_{lm}^n (G_{ij})_{lm}^n, \tag{C.19}$$

where the c_{lm}^n are coefficients and explicit expressions for the $(G_{ij})_{lm}^n$ can be found in Gerlach and Sengupta (1978).

Hankel functions

The Hankel functions are linear combinations of the perhaps better-known Bessel J and Y functions,

$$H_\alpha^{(1,2)}(x) = J_\alpha(x) \pm i Y_\alpha(x). \tag{C.20}$$

All of the above functions solve the linear differential equation

$$x^2 f'' + x f' + (x^2 - \alpha^2) f = 0. \tag{C.21}$$

For us, the most important thing to know about the Hankel functions is their asymptotic limit at large and small arguments,

$$H_\alpha^{(1,2)}(x) \to \sqrt{\frac{2}{\pi x}} \exp\left[\pm i \left(x - \frac{\alpha \pi}{2} - \frac{\pi}{4}\right)\right] \quad \text{as} \quad x \to \infty, \tag{C.22}$$

$$H_\alpha^{(1,2)}(x) \to \mp \frac{i}{\pi} \Gamma(\alpha) \left(\frac{x}{2}\right)^{-\alpha} \quad \text{as} \quad x \to 0, \tag{C.23}$$

where $\Gamma(\alpha) = (\alpha - 1)!$ denotes the factorial function The leading imaginary part of the Hankel functions at small arguments comes from the asymptotic expression of the Bessel J function,

$$J_\alpha(x) \to \left(\frac{x}{2}\right)^\alpha \frac{1}{\Gamma(\alpha + 1)} \quad \text{as} \quad x \to 0. \tag{C.24}$$

Special values

We make use of the following special values for the Gamma function,

$$\Gamma\left(\frac{1}{2}\right) = \sqrt{\pi}, \qquad \Gamma\left(\frac{3}{2}\right) = \frac{\sqrt{\pi}}{2}, \tag{C.25}$$

and for the Riemann zeta function,

$$\zeta(0) = -\frac{1}{2}, \quad \zeta'(0) = -\frac{1}{2} \ln(2\pi). \tag{C.26}$$

Appendix D
Solutions to exercises

Exercise 1.1 – Solution
The Milky Way has a radius of about 10^5 light-years and a mass of about 10^{12} solar masses. The mass of the Sun is 10^{57} times that of a hydrogen atom. This translates into a Galactic density of about 10^{54} hydrogen atoms per cubic light-year, or 10^6 hydrogen atoms per cubic meter. The current critical density corresponds to 1 hydrogen atom per cubic meter. Thus the universe had a density equal to our Galactic density at redshift $z \approx 10^{6/3} \approx 100$. At the time of writing the most distant galaxies discovered by the James Webb Space Telescope are at redshifts of around $z = 14$, and thus comfortably later.

Exercise 1.2 – Solution
The Schwarzschild–de Sitter solution describes a black hole in an asymptotically de Sitter spacetime, so that it interpolates between a static black hole spacetime of mass M and an accelerating universe with constant vacuum energy Λ. It is given by

$$ds^2 = -\left(1 - \frac{2M}{r} - \frac{\Lambda}{3}r^2\right)dt^2 + \frac{dr^2}{1 - \frac{2M}{r} - \frac{\Lambda}{3}r^2} + r^2(d\theta^2 + \sin^2\theta\, d\phi^2). \quad \text{(D.1)}$$

In case you would like to verify that the Λ term induces accelerated expansion, set $M = 0$ and perform an appropriate change of coordinates; start by looking at $r^2 \gg \frac{3}{\Lambda}$ and set $r = e^{\sqrt{\Lambda/3}T}$. We may use the black hole part as a first approximation to the Local Group of galaxies, with mass $M \approx 4 \times 10^{42}$ kg. Then the $\Lambda r^2/3$ term, which induces the de Sitter acceleration, becomes dominant over the $2M/r$ term at a radius

$$r_{\text{cross-over}} \sim \left(\frac{6M}{\Lambda}\right)^{1/3} \sim 10^7 \text{ light-years}. \quad \text{(D.2)}$$

Exercise 1.3 – Solution
From (1.75), we may infer that the distance to an object at scale factor a_g is given by

$$r_g = \frac{1}{H_0} \int_{a_g}^{1} \frac{da}{\sqrt{\Omega_r + \Omega_m a + \Omega_\Lambda a^4}}, \tag{D.3}$$

which for scale factors $a_g = 1/1.1, 1/2, 1/11$ gives, respectively, 1.41, 11.1, 31.4 billion light-years. Analogously, the corresponding cosmic times are given by the integral

$$t_g = \int_{a=0}^{a=a_g} dt = \frac{1}{H_0} \int_0^{a_g} \frac{da}{\sqrt{\Omega_r a^{-2} + \Omega_m a^{-1} + \Omega_\Lambda a^2}}, \tag{D.4}$$

yielding $12.4, 5.85$, and 0.471 billion years, respectively.

Exercise 1.4 – Solution
During a short time interval Δt, the scale factor of the universe changes by (from (D.4))

$$\Delta a = \sqrt{\Omega_r a^{-2} + \Omega_m a^{-1} + \Omega_\Lambda a^2}\, H_0 \Delta t. \tag{D.5}$$

In Planck units, one year corresponds to $\Delta t \approx 10^{50}$, and we have $H_0 \approx 10^{-60}$ and $a \approx 10^{60}$, giving a receding volume per year of

$$\Delta \text{Volume} = 4\pi a^2 \Delta a \approx 10 \times 10^{120} \times 10^{50} = \left(10^{57}\right)^3. \tag{D.6}$$

This is on the order of a million light-years cubed, and hence corresponds to the loss of $\mathcal{O}(1)$ galaxies per year.

Exercise 1.5 – Solution
During radiation domination, the scale factor evolves as $a \propto t^{1/2}$. In Planck units, assuming a current age and size of both 10^{60}, we obtain the following evolution for the scale factor,

$$a(t) = 10^{60} \left(\frac{t}{10^{60}}\right)^{1/2}. \tag{D.7}$$

Using (1.55), we find $t_{\text{TeV}} \approx 10^{-12}\,\text{s} \approx 10^{31}\, t_P$. This gives a scale factor $a_{\text{TeV}} \approx 10^{45}\, l_P$, and thus a size 14 orders of magnitude larger than the time since the big bang. At 1 Planck time, the discrepancy is even more drastic, as it gives a size of $a_P \approx 10^{30}\, l_P$. This is another manifestation of the horizon puzzle.

Exercise 2.1 – Solution

In Planck units, 10^{13} GeV $\approx 10^{-6}$. The energy density has units of energy per volume, i.e., $[E]^4$. Thus, in Planck units, the potential is given by $\rho \approx V \approx (10^{-6})^4$. The Friedmann equation $3H^2 = \rho$ then implies a Hubble rate of $H \approx 10^{-12}$, or a time scale for 1 e-fold of expansion of $1/H \approx 10^{12} t_P \approx 10^{-31}$ s. The inflationary period thus lasts on the order of 10^{-29} s. During this short time, the universe grows by the huge factor of $e^{50} \approx 10^{21}$ in linear size, or 10^{63} in volume. (This is reminiscent of the "wheat and chessboard" problem.)

One may convert 1 eV to 11 600 $\approx 10^4$ K. Assuming a reheating temperature of 10^{13} GeV $\approx 10^{26}$ K, and a current temperature of about 3 K, we can see that from the start of inflation the universe grew by a factor of about $10^{21+26} = 10^{47}$. If the current size is on the order of 10^{60} Planck lengths, then at the beginning of inflation the observable part of the universe was just $10^{60-47} = 10^{13}$ Planck lengths in linear size, or about 10^{-22} m, significantly smaller than an atomic nucleus.

Exercise 2.2 – Solution

The slow-roll parameters are easily evaluated via direct computation,

$$\epsilon \approx \frac{V_{,\phi}^2}{2V^2} = \frac{4}{3\left(e^{\sqrt{2/3}\phi} - 1\right)^2}, \tag{D.8}$$

$$\eta \approx \frac{2V_{,\phi\phi}}{V} - \frac{2V_{,\phi}^2}{V^2} = -\frac{2}{3\sinh^2(\phi/\sqrt{6})}. \tag{D.9}$$

The important point to note is that the slow-roll parameters do not depend on the overall scale of the potential. Inflation ends when either one of the slow-roll parameters reaches unity. For ϵ this occurs when $\phi \approx 0.94$, while for $|\eta|$ this happens already a little earlier, at $\phi_{end} \approx 1.83$. The number of e-folds is given by the integral (2.41), which can be evaluated as

$$\mathcal{N} = \int_{\phi_{end}}^{\phi_{beg}} \frac{V}{V_{,\phi}} d\phi = \sqrt{\frac{8}{3}}(\phi_{end} - \phi_{beg}) + \frac{3}{4}\left(e^{\sqrt{2/3}\phi_{beg}} - e^{\sqrt{2/3}\phi_{end}}\right)$$
$$\approx \frac{3}{4} e^{\sqrt{2/3}\phi_{beg}}, \tag{D.10}$$

where the last approximation assumes $e^{\sqrt{2/3}\phi_{beg}} \gg 1$. In order to obtain $\mathcal{N} = 50$, inflation thus needs to start at $\phi_{beg} \approx 5.14$, implying a field range

$\Delta\phi \approx 3.31$. In the same limit, we may rewrite the slow-roll parameters as

$$\epsilon \approx \frac{3}{4\mathcal{N}^2}, \quad \eta \approx -\frac{2}{\mathcal{N}}. \tag{D.11}$$

These relations are useful in calculating the predictions for inflationary fluctuations, as we will see in Chapter 3.

Exercise 2.3 – Solution

By integration $\phi = ct + d$. We assume that ϕ is positive at early times and rolls to smaller values, so $c < 0$. Then the scalar equation of motion (2.19) implies $H = -\frac{M^2}{3}t - \frac{M^2 d}{3c}$. Plugging this into the Friedmann equation (2.17), one can see that an exact solution indeed exists for

$$\phi = -\sqrt{\frac{2}{3}}Mt + 1, \quad \Lambda = -\frac{M^2}{3}. \tag{D.12}$$

Integrating the Hubble rate, one finds that the scale factor follows a Gaussian $a = a_0 e^{\frac{M}{\sqrt{6}}t - \frac{M^2}{6}t^2}$, with integration constant a_0. Thus the universe expands until the scalar reaches the bottom of the potential, and then the universe recollapses.

It is useful to look at small perturbations $\delta H, \delta\phi$. Then, to linear order, the Friedmann equation yields $6H\delta H = \dot\phi\delta\dot\phi + M^2\phi\delta\phi$. Plugging this relation into the scalar field equation, we find

$$\delta\ddot\phi + (3H + \frac{\dot\phi^2}{2H})\delta\dot\phi = 0, \tag{D.13}$$

where the term proportional to $\delta\phi$ is found to have vanishing coefficient. The solution to this equation is

$$\delta\dot\phi \propto \phi e^{\frac{3}{4}\phi^2}. \tag{D.14}$$

This means that the solution is stable when $|\phi|$ is decreasing, i.e., during expansion, and (highly) unstable during the contracting phase when the scalar rolls up the potential.

Exercise 2.4 – Solution

To calculate the evolution of the Hubble rate, we have to consider the dark energy, ekpyrotic, and radiation phases in succession. During dark energy domination, the Hubble rate stays roughly constant. Then, during ekpyrosis, we have $H = 1/(\epsilon t)$ and $V = -1/(\epsilon t^2)$, so that $H \propto V^{1/2}$. During radiation domination, $a \propto t^{1/2}$ and hence $H \propto a^{-2} \propto \mathcal{T}^2$. Moreover, the energy density, the potential, and the temperature are related via $\rho \sim V \sim \mathcal{T}^4$ – then putting everything together we find

$$\frac{H_0}{H_{\text{last cycle}}} = \frac{H_{\text{ek-beg}}}{H_{\text{dark energy}}} \frac{H_{\text{ek-end}}}{H_{\text{ek-beg}}} \frac{H_{\text{rad-end}}}{H_{\text{rad-beg}}}$$

$$= 1 \cdot \frac{|V|_{\text{ek-end}}^{1/2}}{V_{\text{ek-beg}}^{1/2}} \cdot \frac{\mathcal{T}_0^2}{\mathcal{T}_r^2} = 1. \quad \text{(D.15)}$$

Meanwhile the scale factor grows by

$$\frac{a_0}{a_{\text{last cycle}}} = e^{\mathcal{N}_{\text{dark energy}}} \cdot \frac{V_{\text{ek-beg}}^{1/(2\epsilon)}}{|V|_{\text{ek-end}}^{1/(2\epsilon)}} \cdot \frac{\mathcal{T}_r}{\mathcal{T}_0}$$

$$= e^{\mathcal{N}_{\text{dark energy}}} \cdot 10^{-6} \cdot 10^{17}, \quad \text{(D.16)}$$

where $\epsilon \sim \mathcal{O}(10)$. This implies that one cycle ago, the currently observable part of the universe had a size of

$$a_{\text{last cycle}} \approx 10^{60-13} e^{-\mathcal{N}_{\text{dark energy}}} \approx e^{-\mathcal{N}_{\text{dark energy}}} \text{ light-hours}. \quad \text{(D.17)}$$

Thus the size of the currently observable part would have been at most 1 light-hour (about the distance to Saturn), but could be significantly smaller if there is/was a prolonged dark energy phase.

Exercise 3.1 – Solution

To solve the mode equations, we need to calculate the quantity z''/z, where $z^2 = 2a^2\epsilon$. When ϵ is constant, then we simply get $z''/z = a''/a$. The solution for the scale factor was given in (2.26), and leads to

$$a(\tau) = \bar{a}_0(-\tau)^{1/(\epsilon-1)}, \quad \rightarrow \quad \frac{a''}{a} = \frac{2-\epsilon}{(1-\epsilon)^2} \frac{1}{\tau^2}, \quad \text{(D.18)}$$

where $\bar{a}_0$ is an integration constant. The equation of motion we want to solve is thus

$$v_k'' + \left(k^2 - \frac{2-\epsilon}{(1-\epsilon)^2\tau^2}\right) v_k = 0, \quad \text{(D.19)}$$

and with Bunch–Davies initial conditions the solution is given by

$$v_k = \sqrt{-\tau} \sqrt{\frac{\pi}{4}} e^{i(2\alpha+1)\pi/4} H_\alpha^{(1)}(-k\tau), \quad \text{where} \quad \alpha = \frac{3-\epsilon}{2(1-\epsilon)}. \quad \text{(D.20)}$$

In terms of the comoving curvature perturbation $\mathcal{R} = v/z$, the late-time variance is thus given by

$$\Delta_\mathcal{R}^2 = \frac{k^3}{2\pi^2} |\mathcal{R}_k|^2 \quad \text{(D.21)}$$

$$= \frac{2^{2\alpha-4}\Gamma(\alpha)^2(1-\epsilon)^{2/(1-\epsilon)}}{\pi^3} \frac{H_\star^2}{\epsilon_\star}\left(\frac{k}{k_\star}\right)^{3-2\alpha}, \qquad (\text{D.22})$$

evaluated at $k_\star = a_\star H_\star$. This expression implies a spectral index

$$n_s = 4 - 2\alpha = \frac{1-3\epsilon}{1-\epsilon}, \qquad (\text{D.23})$$

while the tensor-to-scalar ratio is simply $r = 16\epsilon$. The observational bound on r implies the bound (3.160) on ϵ, which in turn implies that $n_s \gtrsim 0.993$. This is too large to agree with observations. Hence only models with a time-dependent equation of state ϵ can fit observations.

Exercise 3.2 – Solution

The spectral index and tensor-to-scalar ratio follow directly from the results of Exercise 2.2, and are given, to leading order in the number of e-folds $\mathcal{N}$, by

$$n_s \approx 1 - \frac{2}{\mathcal{N}}, \quad r \approx \frac{12}{\mathcal{N}^2}. \qquad (\text{D.24})$$

For an "average" value of $\mathcal{N} = 55$, one gets $n_s \approx 0.964$ and $r \approx 0.004$. These values are in good agreement with observations and indicate that, if one of these models is realized in nature, then the associated gravitational waves should be observed reasonably soon.

Matching the amplitude of temperature perturbations gives us the required scale of the potential, which comes out at

$$M \approx 3 \cdot 10^{13}\,\text{GeV}, \qquad (\text{D.25})$$

about 5 orders of magnitude below the Planck scale.

Exercise 3.3 – Solution

The spatial metric $h_{ij} = a(t)^2 \delta_{ij}$ can be kept fixed, since we can choose ξ^0 to eliminate ψ and ξ to eliminate E – see (3.46) and (3.47). At linear order, the constraints (3.67) and (3.68) then become

$$A = \frac{\dot\phi}{2H}\delta\phi = \sqrt{\frac{\epsilon}{2}}\delta\phi, \qquad (\text{D.26})$$

$$\partial^i \partial_i B = -\frac{1}{2H}(V_{,\phi} + \frac{\dot\phi}{H}V)\delta\phi - \frac{\dot\phi}{2H}\dot{\delta\phi} = -\epsilon\frac{\mathrm{d}}{\mathrm{d}t}\left(\frac{\delta\phi}{\sqrt{2\epsilon}}\right), \qquad (\text{D.27})$$

where in the constraint for B we have already used (D.26) to replace A. The constraints show that when the slow-roll parameter is very small, $\epsilon \ll 1$, the metric perturbations are negligible compared to the scalar field fluctuations

$\delta\phi$ since they are suppressed by factors of $\sqrt{\epsilon}$. This is the basis for the standard intuition that in slow-roll inflation one may think of the background spacetime as being constant, with only the scalar field fluctuating, but note that this is a gauge-dependent statement as, for example, in comoving gauge the scalar field fluctuation is eliminated.

In flat gauge the comoving curvature perturbation is given by $\mathcal{R} = \psi - \frac{H}{\dot\phi}\delta\phi = -\frac{H}{\dot\phi}\delta\phi \approx -\frac{1}{\sqrt{2\epsilon}}\delta\phi$, with variance $\Delta_\mathcal{R}^2 = \frac{H^2}{8\pi^2\epsilon}$. The relation between the curvature perturbation and the scalar field perturbation then implies that the variance of the scalar field is given by

$$\Delta\phi_{\mathrm{qu}} \equiv \langle(\delta\phi)^2\rangle^{1/2} = \frac{H}{2\pi}. \tag{D.28}$$

The condition for eternal inflation then translates into

$$\Delta\phi_{\mathrm{qu}} > \Delta\phi_{\mathrm{cl}} \quad \leftrightarrow \quad \frac{H^2}{2\pi|\dot\phi|} \approx \frac{H}{\sqrt{8\pi^2\epsilon}} > 1 \quad \leftrightarrow \quad \Delta_\mathcal{R}^2 > 1, \tag{D.29}$$

and thus the condition for eternal inflation is the same as requiring the variance of the comoving curvature perturbation to reach order unity.

Exercise 3.4 – Solution

In a matter-dominated phase, we have the equation of state $w = 0$, and consequently $\epsilon = 3(1 + w)/2 = 3/2$. Also, the scale factor evolves as $a \propto t^{2/3} \propto \tau^2$, where τ denotes conformal time. The comoving curvature perturbation is defined in (3.56), and we will work in comoving gauge where $\delta q = 0$. Each Fourier mode of the perturbations then obeys the equation (3.83) $v'' + \left(k^2 - \frac{z''}{z}\right)v = 0$, where $v = z\mathcal{R}$ is the Mukhanov–Sasaki variable. Here $z = \sqrt{2\epsilon}a = \sqrt{3}a$ and consequently $z''/z = 2/\tau^2$. Perturbations are amplified as $|\tau| \to 0$, which corresponds to contraction. The solution to the perturbation equation is the same as in (3.98), with $\alpha = 3/2$. This leads to an exactly scale invariant spectrum, since

$$\Delta_\mathcal{R}^2 \approx \frac{H^2}{8\pi^2\epsilon}\left(\frac{k}{k_\star}\right)^{3-2\alpha} = \frac{H^2}{12\pi^2}. \tag{D.30}$$

As shown in Section 2.2.1, anisotropies grow fast in a contracting universe (their energy density grows as a^{-6}) and come to dominate the evolution unless $\epsilon > 3$, which would correspond to an ekpyrotic phase. Here, however, $\epsilon = 3/2$ and consequently anisotropies take over during contraction. This means that even tiny perturbations will quickly spoil the isotropy of a matter contraction phase, rendering such a phase unrealistic.

Exercise 4.1 – Solution

According to (4.23) the momentum conjugate to the scale factor a is $p_a = -12\pi^2 a \dot{a} = -12\pi^2 a^2 H$. If one imposes standard commutation relations $[\hat{a}, \hat{p}_a] = i$, then one can deduce a corresponding uncertainty relation

$$\langle \delta a \rangle \langle \delta p_a \rangle \geq \frac{1}{2}. \tag{D.31}$$

From the definition of the momentum, we can infer that the scale factor and momentum, as well as the corresponding variations, are roughly equal when

$$a \sim 1/H. \tag{D.32}$$

This heuristic observation indicates that one might not be able to expect a classical description of spacetime below the de Sitter radius $a_{\min} = \frac{1}{H} = \sqrt{\frac{3}{\Lambda}}$. The interesting aspect is that this scale is set by the vacuum energy, and not the Planck scale. Thus quantum gravity effects may already operate at significantly lower energies than the naïvely expected Planck scale. We will return to this theme with more rigorous calculations.

Exercise 4.2 – Solution

From the explicit action Eq. (4.21) we can identify $K = -6\pi^2 a \dot{a}^2$ and $P = 2\pi^2 (a^3 \Lambda - 3a)$. The Hamiltonian constraint, obtained by varying the action with respect to the lapse, reads $\frac{K}{N^2} + P = 0$, and is thus solved by

$$N^2 = -\frac{K}{P} = \frac{3\dot{a}^2}{a^3 \Lambda - 3}. \tag{D.33}$$

The metric becomes

$$ds^2 = -\frac{da^2}{a^2 \frac{\Lambda}{3} - 1} + a^2 d\Omega_3^2, \tag{D.34}$$

and transforming to physical time T via $da/(a^2 \frac{\Lambda}{3} - 1)^{1/2} \equiv dT$ gives the standard expression for de Sitter spacetime in the closed slicing, namely $a = \sqrt{\frac{3}{\Lambda}} \cosh \sqrt{\frac{\Lambda}{3}} T$.

Exercise 4.3 – Solution

Using the scaling solution (2.26), one may immediately rewrite a_0 as $a_0 = a \left(\frac{\epsilon^2 V}{3-\epsilon} \right)^{1/(2\epsilon)}$. To calculate the on-shell action, we may make use of the Friedmann equation to obtain

$$S = 2\pi^2 \int N dt \left[-\frac{3}{N^2} a \dot{a}^2 + \frac{1}{2N^2} a^2 \dot{\phi}^2 + 3a - a^3 V(\phi) \right]$$

$$= 4\pi^2 \int dt \left[3a - a^3 V(\phi) \right]$$

$$\approx -4\pi^2 \int dt\, a^3 V(\phi) \approx -\frac{4\pi^2}{\epsilon} a_0^3 t_f^{-1+3/\epsilon}$$

$$= -4\pi^2 b^3 \left(\frac{V(\chi)}{3-\epsilon} \right)^{1/2}. \tag{D.35}$$

The weighting must satisfy

$$\nabla \mathcal{S} \cdot \nabla \mathcal{W} = G^{bb} \mathcal{S}_{,b} \mathcal{W}_{,b} + G^{\chi\chi} \mathcal{S}_{,\chi} \mathcal{W}_{,\chi} = 3b^{n_1} V^{n_2 + \frac{1}{2}}(-n_1 + 2\epsilon n_2) = 0, \tag{D.36}$$

which in fact implies that the weighting is asymptotically constant. Together with the rapid growth of the phase, this already shows that JWKB classicality is reached at large b values. We can be more precise, using $\mathcal{W} = b^{2\epsilon/(\epsilon-1)} V^{1/(\epsilon-1)}$. Then we obtain

$$|(\nabla \mathcal{W})^2 / (\nabla \mathcal{S})^2|^{1/2} \sim \left(\frac{b^{-3+4\epsilon/(\epsilon-1)} V^{2/(\epsilon-1)}}{b^3 V} \right)^{1/2} \sim b^{\epsilon-3}. \tag{D.37}$$

Thus classicality is reached roughly in inverse proportion to the volume of inflated space. In fact, the scaling of (D.37) is exactly the same as the fall-off of a nontrivial perturbation around the asymptotic solution, cf. Eq. (2.29).

Exercise 5.1 – Solution
With the FLRW metric $ds^2 = -N^2 dt^2 + a^2(t) ds_3^2$, where ds_3^2 stands for the spatial 3-metric with constant curvature $K = -1, 0, 1$, the total action reads

$$\frac{S + S_R}{V_3} = \int_0^1 dt\, N \left(3a^2 \frac{\ddot{a}}{N^2} + 3a \frac{\dot{a}^2}{N^2} + 3Ka - a^3 \Lambda \right) + -\xi a^3 |_{t=1}, \tag{D.38}$$

where V_3 denotes the volume of the spatial sections (we assume that for the case where $K = -1, 0$ the volume is made finite by having a nontrivial topology of the 3-space). Varying this action with respect to the scale factor leads to

$$\frac{\delta(S + S_R)}{V_3} = \int_0^1 dt\, 3\delta a \left(2\frac{a\ddot{a}}{N} + \frac{\dot{a}^2}{N} + KN - a^2 \Lambda N \right)$$

$$- \frac{3a^2 \delta \dot{a}}{N}\bigg|_{t=0} + 3a^2 \delta \left(\frac{\dot{a}}{N} - \xi a \right)\bigg|_{t=1}. \tag{D.39}$$

On the initial boundary we can thus fix $\dot{a}$, while on the final boundary we can indeed fix the Hubble rate $H = \frac{\dot{a}}{Na} = \xi$. The Robin boundary condition may thus be seen as a linear combination of Dirichlet and Neumann conditions.

Exercise 5.2 – Solution

In a flat universe, the Friedmann equation reads $\frac{3\dot{q}^2}{4N^2} = q\Lambda$. With the momentum $p = -\frac{3V_3}{2N}\dot{q}$, this can be recast as $\frac{p^2}{3V_3^2} = q\Lambda$ and implies that for an expanding universe we should take $p_i = -V_3\sqrt{3\Lambda q_i}$.

The flat-space version of the propagator equations (5.33) and (5.34) is given by

$$G[q_1; q_0] = \sqrt{\frac{3V_3 i}{4\pi\hbar}} \int_{0^+}^{\infty} \frac{d\tilde{N}}{\tilde{N}^{1/2}} e^{iS_0}, \qquad (D.40)$$

with

$$S_0 = V_3 \left[\tilde{N}^3 \frac{\Lambda^2}{36} + \tilde{N}\left(-\frac{\Lambda}{2}(q_0 + q_1)\right) + \frac{1}{\tilde{N}}\left(-\frac{3}{4}(q_1 - q_0)^2\right) \right]. \qquad (D.41)$$

The integral (5.93) is therefore Gaussian and can be performed explicitly, leading to the result

$$G[q_1, \bar{q}_0; \tilde{N}] = \frac{1}{\sqrt[4]{2\pi}} \sqrt{\frac{i3V_3\tilde{N}\Delta}{\hbar\tilde{N} + 3iV_3\Delta^2}} e^{\frac{i}{\hbar}\bar{S}^{(0)}[q_1;\bar{q}_0;\tilde{N}] + \frac{i}{\hbar}p_i(q_i)\bar{q}_0 - \frac{(\bar{q}_0 - q_i)^2}{4\Delta^2}}, \qquad (D.42)$$

with effective initial size

$$\bar{q}_0 = \frac{\tilde{N}q_i - i(\tilde{N}^2\Lambda V_3 - 2\tilde{N}p_i(q_i) - 3q_1 V_3)\Delta^2/\hbar}{\tilde{N} + 3i\Delta^2 V_3/\hbar}. \qquad (D.43)$$

In the limit of vanishing uncertainty in position $\Delta \to 0$, the initial size localizes to a position eigenstate, with $\bar{q}_0 \to q_i$.

Exercise 5.3 – Solution

The saddle points are the stationary points (when varying with respect to the lapse) of the exponent in (D.42). After some algebra, one finds that those with positive real part are given by

$$\tilde{N}_- = \sqrt{\frac{3}{\Lambda}}\left(\sqrt{q_1} - \sqrt{q_i}\right), \quad \tilde{N}_+ = \sqrt{\frac{3}{\Lambda}}\left(\sqrt{q_1} + \sqrt{q_i}\right) - \frac{i}{\hbar}6V_3\Delta^2. \qquad (D.44)$$

When $\Delta = 0$, the saddle point at $\tilde{N}_-$ corresponds to an expanding geometry, while the one at $\tilde{N}_+$ corresponds to a bouncing geometry, contracting to zero size before expanding to the final value q_1. The explicit solutions for the scale factor squared, when $\Delta = 0$, are given by

$$\bar{q}(\tilde{N}_\pm) = \left(\sqrt{\frac{\Lambda}{3}}\tilde{N}_\pm t \mp \sqrt{q_i}\right)^2. \tag{D.45}$$

When Δ is increased from zero, the expanding saddle point stays put, while the bouncing saddle point moves into the lower half $\tilde{N}$ plane. Correspondingly, the flow lines undergo a change – see Fig. D.1. For small Δ, both

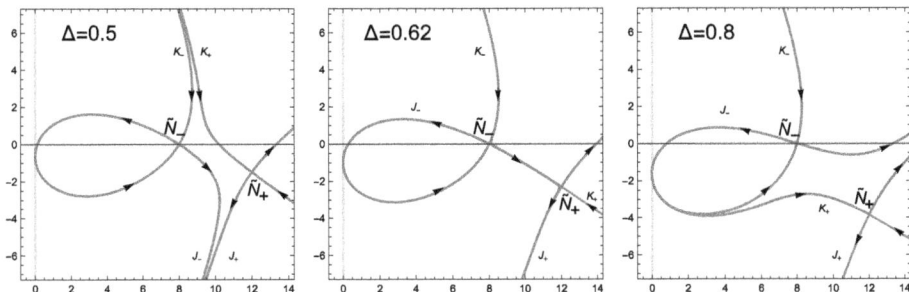

Figure D.1 Flow lines in the complexified $\tilde{N}$ plane, when an initial coherent state is assumed at the beginning of an inflationary phase, with $\Lambda = 3, q_i = 4, q_1 = 100$. The initial state has momentum corresponding to expansion, and an uncertainty Δ in the scale factor value. As the initial size uncertainty is increased, a Stokes phenomenon happens beyond which the bouncing saddle point $\tilde{N}_+$ no longer contributes to the evolution. *Left panel:* For small Δ, both saddles contribute as both $\mathcal{K}_-$ and $\mathcal{K}_+$ cross the real $\tilde{N}$ line. *Middle panel:* A Stokes ray connects the two saddle points. *Right panel:* For sufficiently large Δ, only $\mathcal{K}_-$ intersects the real line integration contour and $\tilde{N}_+$ has become irrelevant. The evolution then corresponds to pure expansion, and a semiclassical description in terms of quantum field theory on a single background spacetime is obtained.

saddle points contribute to the integral, and there is thus a superposition of an expanding and a bouncing saddle point. But as the uncertainty in the momentum becomes smaller, a Stokes phenomenon happens beyond which the bouncing saddle point no longer contributes. This is because the universe then "knows" with sufficient certainty that it is in an expanding state. At this point standard quantum field theory in curved spacetime is recovered.

Exercise 5.4 – Solution
Flow lines are characterized by having the same phase as that of the saddle point they are associated with. In order to find the Stokes line, we must

therefore equate the phases at the two saddle points. They are given by

$$\left[\bar{S}^{(0)} + p_i(q_i)\bar{q}_0 + i\frac{\hbar(\bar{q}_0 - q_i)^2}{4\Delta^2}\right]_{\tilde{N}_-} = -V_3\sqrt{\frac{\Lambda}{3}}\left(2q_1^{3/2} + q_i^{3/2}\right), \quad \text{(D.46)}$$

while for the bouncing saddle we have

$$\left[\bar{S}^{(0)} + p_i(q_i)\bar{q}_0 + i\frac{\hbar(\bar{q}_0 - q_i)^2}{4\Delta^2}\right]_{\tilde{N}_+}$$
$$= -V_3\sqrt{\frac{\Lambda}{3}}\left[2q_1^{3/2} + \sqrt{q_i}\left(\frac{5q_i\hbar^2 - 36\Lambda\Delta^4 V_3^2}{\hbar^2}\right)\right]$$
$$- i\left(\frac{\Lambda\Delta^4 V_3^2}{\hbar^2} - q_i\right)12\Lambda\Delta^2\frac{V_3}{\hbar}. \quad \text{(D.47)}$$

Therefore, we find that the two saddles are linked by a Stokes line if

$$4\frac{V_3}{\hbar^2}\sqrt{\frac{q_i\Lambda}{3}}(\hbar^2 q_i - 9V_3^2\Lambda\Delta^4) = 0, \quad \text{(D.48)}$$

implying that the critical localization is given by

$$\Delta_c = \left(\frac{q_i\hbar^2}{9\Lambda V_3^2}\right)^{1/4}. \quad \text{(D.49)}$$

Thus, appropriate initial conditions for a purely expanding universe require

$$q_i > \Delta_c, \quad \text{implying} \quad a_i^3 V_3 > \frac{\hbar}{3\sqrt{\Lambda}}. \quad \text{(D.50)}$$

This bound is independent of the final size q_1 reached. The surprising feature is that the volume of the universe must already be sufficiently large compared to the vacuum energy, rather than the naïvely expected Planck scale. This calculation therefore not only shows the need for a theory of the beginning of inflation, but also illustrates that quantum gravity may have significant effects far below Planckian curvatures.

Exercise 6.2 – Solution
The entropy is given by a quarter of the horizon area A, hence $\mathcal{S} = \frac{A}{4} = \pi R^2 = 4\pi M^2$. Thus we have that $d\mathcal{S} = 8\pi M dM$. If we identify the mass M with the energy E of the black hole, then using the Hawking temperature formula (6.80) we can rewrite this as $dE = T_H d\mathcal{S}$, which is the first law of thermodynamics.

Exercise 6.3 – Solution

1. Putting in the relevant constants, e.g., via dimensional analysis, the Unruh temperature reads $\mathcal{T}_U = \frac{\hbar a}{2\pi c k_B} \approx \frac{a}{[m/s^2]} \cdot 4.06 \cdot 10^{-21}$ K. To reach a room temperature of 300 K one would therefore need a staggering acceleration of $7.39 \cdot 10^{22}$ m s^{-2}.

2. The Hawking temperature is $\mathcal{T}_H = \frac{\hbar c^3}{8\pi G M k_B} = \frac{\hbar c}{4\pi k_B R} \approx \frac{1.82 \cdot 10^{-4}}{R}$ K · m, where M is the mass and R the Schwarzschild radius. A temperature of 450 K would thus require a black hole of radius $R \approx 4 \cdot 10^{-7}$ m, or a bit less than a micron in diameter, easily fitting in the oven.

3. The entropy is given by a quarter of the horizon area A in Planck units, as we derived in Eq. (6.133). With the physical constants explicitly inserted, this becomes $\mathcal{S} = \frac{k_B c^3 A}{4\hbar G} \approx 1.20 \cdot 10^{70} R^2 \text{m}^{-2}$ k_B. For a solar mass black hole, we get $\mathcal{S}_\odot \approx 1.05 \cdot 10^{77}$ k_B, while for a billion solar mass black hole this becomes $\mathcal{S}_{10^9 \odot} \approx 1.05 \cdot 10^{95}$ k_B. This suggests that black holes can be made up out of a huge number of different microstates.

4. The de Sitter temperature is given in terms of the Hubble rate H as $\mathcal{T}_{\text{dS}} = \frac{\hbar H}{2\pi k_B} = \frac{\hbar c}{2\pi k_B}\sqrt{\frac{\Lambda}{3}} \approx 3.64 \cdot 10^{-4}\sqrt{\frac{\Lambda}{3}}$ m · K. For the current cosmological constant $\Lambda \approx 1.09 \cdot 10^{-52}$ m^{-2}, this becomes $\mathcal{T}_{\text{dS}} \approx 2.20 \cdot 10^{-30}$ K. This is much lower than the current CMB temperature of 2.7 K.

These horizon-induced temperatures are extremely tiny, and thus completely unobservable at present. To observe Hawking radiation, one would have to discover very small, primordial, black holes. Meanwhile, the de Sitter temperature might actually have been observed already, in the sense that an early inflationary phase is a good approximation to a de Sitter phase. So if inflation can be observationally confirmed, it would mean that the (approximate) de Sitter temperature and its associated fluctuations are imprinted on the CMB sky.

Exercise 6.4 – Solution

The energy loss of a blackbody per unit area per unit time is given by $\sigma \gamma T^4$, which is the Stefan–Boltzmann law with σ the eponymous constant. Here γ denotes the number of different light degrees of freedom that can be radiated away. Thus the total rate of energy/mass loss is $\frac{dM}{dt} = -\sigma \gamma A T^4 = -\frac{\sigma \gamma}{256\pi^3 M^2}$. Integrating this relation, and assuming the final mass vanishes, we get a lifetime $t_{\text{bh}} = \frac{256 G^2}{3\sigma \gamma \hbar c^4} M^3$. Using the age of the universe $4.35 \cdot 10^{17}$ s for t_{bh} and guessing $\gamma \approx 10^3$, we get an initial mass of $M \approx 1.2 \cdot 10^{10}$ kg. This corresponds to the mass of an asteroid of size a little bigger than a football/soccer field. A black hole of this mass would only have had about the size of an atomic nucleus at the time of formation.

Exercise 6.5 – Solution

A massless scalar field in a fixed de Sitter background has the mode function given in Eq. (6.146), $\phi = \frac{H}{\sqrt{2k^3}} e^{-ik\tau} (1 + ik\tau)$, with the scale factor evolving as $a = -1/(H\tau)$ in the flat slicing. The power spectrum for each frequency is simply given by $|\phi|^2$ – see (3.6). And the variance (3.8) is thus given by

$$\Delta_\phi^2 = \frac{k^3}{2\pi^2} \mathcal{P}_\phi = \frac{H^2}{4\pi^2}(1 + k^2\tau^2). \tag{D.51}$$

At early times ($|k\tau| \gg 1$), we find $\Delta_\phi^2 = \frac{k^2}{4\pi^2 a^2}$, which agrees with the corresponding expression for vacuum fluctuations in Minkowski spacetime. Meanwhile, at late times ($|k\tau| \ll 1$) we obtain $\Delta_\phi^2 = \left(\frac{H}{2\pi}\right)^2 = \mathcal{T}_{\text{dS}}^2$. Thus a late-time, super-Hubble fluctuation has a typical size given by the de Sitter temperature.

The comoving curvature perturbation was defined in (3.55), $\mathcal{R} \equiv \psi - H\frac{\delta\phi}{\dot\phi}$. Here we do not want to use comoving gauge, in which the scalar fluctuation $\delta\phi$ is set to zero, as we would in fact like to use the scalar fluctuation and relate it to $\mathcal{R}$. To this end, we will rather use spatially flat gauge, in which the spatial metric fluctuation ψ vanishes. Using the results above, we can immediately infer that the variance of the curvature perturbation at late times is given by

$$\Delta_\mathcal{R}^2 = \frac{H^2}{\dot\phi^2} \Delta_\phi^2 = \frac{H^4}{4\pi^2 \dot\phi^2} = \frac{H^2}{8\pi^2 \epsilon}, \tag{D.52}$$

where $\epsilon = \dot\phi^2/(2H^2)$ is the first slow-roll parameter. Thus we recover the result (3.106), and can see that the generation of perturbations during an inflationary phase may also be seen as a direct consequence of the existence of an inflationary quasi–de Sitter horizon. Note that in the present calculation, the spectrum is exactly scale invariant as we assumed an exact de Sitter background.

Exercise 7.1 – Solution

We will follow the same strategy as that used in deriving approximate solutions for a massive scalar field in Section 7.1.2, but this time allowing for a general slow-roll potential. The Taylor expansion (7.39) implies that near the South Pole the metric is approximately given by $a = \sqrt{\frac{3}{V}} \sin(\sqrt{\frac{V}{3}}\sigma)$, where $V = V(\phi_{\rm SP}^R)$ and where we assume that $|\phi_{\rm SP}^I| \ll |\phi_{\rm SP}^R|$. The equator of the approximate hemisphere is reached at $\sigma_{\max} \approx \frac{\pi}{2}\sqrt{\frac{3}{V}}$. Meanwhile, at late Lorentzian times, the slow-roll solution (2.32) can be written as $\phi \approx \phi_{\rm SP} + i\sqrt{\frac{2\epsilon V}{3}}\sigma$. Thus, at late times we can obtain an approximately real-valued Lorentzian evolution if we choose the initial imaginary part of ϕ to be given by

$$\phi_{\rm SP}^I \approx -\sqrt{\frac{2\epsilon V}{3}}\sigma_{\max} \approx -\frac{\pi}{2}\sqrt{2\epsilon} = -\frac{\pi}{2}\frac{V_{,\phi}}{V}, \qquad (D.53)$$

where all quantities on the right-hand side are evaluated at $\phi = \phi_{\rm SP}^R$.

Exercise 7.3 – Solution
As we saw in Exercise 5.1, fixing the Hubble rate corresponds to a Robin boundary condition, which can be implemented by adding a surface term of the form

$$S_R = -H_1 \int_{t=1} d^3y \sqrt{h}, \qquad (D.54)$$

where H_1 is the Hubble rate. The solution to the equation of motion $\ddot{q} = 2\tilde{N}^2 \mathsf{H}^2$, where $\Lambda \equiv 3\mathsf{H}^2$, is

$$\bar{q}(t) = \mathsf{H}^2\tilde{N}^2(t^2-1) + 2\tilde{N}i(t-1) + \frac{(\mathsf{H}^2\tilde{N}+i)^2}{H_1^2}, \qquad (D.55)$$

leading to the lapse action

$$\frac{S_0}{2\pi^2} = -\mathsf{H}^2\left(\frac{\mathsf{H}^2}{H_1^2}-1\right)\left(\tilde{N}+\frac{i}{\mathsf{H}^2}\right)^3 + 3\left(\tilde{N}+\frac{i}{\mathsf{H}^2}\right) - i\frac{2}{\mathsf{H}^2}. \qquad (D.56)$$

Note that in de Sitter spacetime $H_1 < \mathsf{H}$. The saddle points are located at

$$\tilde{N}_\pm = \frac{1}{\mathsf{H}^2}\left(\pm\frac{H_1}{\sqrt{\mathsf{H}^2-H_1^2}}-i\right). \qquad (D.57)$$

The action contains no singularity, and the asymptotic regions of convergence are the wedges between angles $(\frac{\pi}{3}, \frac{2\pi}{3})$, $(\pi, \frac{4\pi}{3})$, and $(\frac{5\pi}{3}, 2\pi)$. The thimbles interpolate between two such wedges. For $\tilde{N}_-$ the interpolation is between $(\pi, \frac{4\pi}{3})$ and $(\frac{\pi}{3}, \frac{2\pi}{3})$, while for $\tilde{N}_+$ it is between $(\frac{5\pi}{3}, 2\pi)$ and $(\frac{\pi}{3}, \frac{2\pi}{3})$. If we integrate over both, the wave function is approximated by

$$\Psi_{NR}(H_1) \approx e^{\frac{4\pi^2}{\hbar \mathsf{H}^2}} \cos\left[\frac{4\pi^2 H_1}{\hbar \mathsf{H}^2\sqrt{\mathsf{H}^2-H_1^2}}\right]. \qquad (D.58)$$

It corresponds to a sum of two complex conjugate contributions, each one having a constant weighting and a phase changing ever more rapidly as $H_1 \to H$. As they grow, the two universes will decohere, as we will discuss in Chapter 8.

Exercise 9.1 – Solution

We can follow the steps taken in Section 9.1.2, but now with gravity included. Our goal is to calculate the action of the bounce solution. The action (9.32) can be simplified using integration by parts (we can drop surface terms as we are calculating the difference between the bounce and the false vacuum, but asymptotically these agree), as well as the constraint (9.31), to yield

$$S_E^{\text{on-shell}} = 4\pi^2 \int d\sigma \left(\rho^3 V - \frac{3}{\kappa^2} \rho \right). \tag{D.59}$$

The integral can again be divided into three parts. Outside of the wall, the difference in actions remains zero, $B_{\text{outside}} = 0$. If we again denote the location of the wall by $\bar{\rho}$, then the wall action is also unchanged, $B_{\text{wall}} = 2\pi^2 \bar{\rho}^3 S_1$, with S_1 being the wall tension. But inside the wall, the expression changes. The scalar $\phi \approx -a$ is approximately constant there. The constraint (9.31) implies that $d\sigma = d\rho(1 - \frac{\kappa^2}{3}\rho^2 V)^{-1/2}$, so that

$$\begin{aligned} B_{\text{inside}} &\approx -\frac{12\pi^2}{\kappa^2} \int_0^{\bar{\rho}} \rho \, d\rho (1 - \frac{\kappa^2}{3}\rho^2 V(-a))^{1/2} \\ &+ \frac{12\pi^2}{\kappa^2} \int_0^{\bar{\rho}} \rho \, d\rho (1 - \frac{\kappa^2}{3}\rho^2 V(+a))^{1/2} \\ &\approx \frac{12\pi^2}{\kappa^4 V(-a)} (1 - \frac{\kappa^2}{3}\rho^2 V(-a))^{3/2} \Big|_0^{\bar{\rho}} \\ &- \frac{12\pi^2}{\kappa^4 V(+a)} (1 - \frac{\kappa^2}{3}\rho^2 V(+a))^{3/2} \Big|_0^{\bar{\rho}}. \end{aligned} \tag{D.60}$$

Using $V(-a) \approx 0, V(a) \approx \Delta V$, one can then see that the total action is extremized at

$$\bar{\rho} = \frac{\bar{\rho}_0}{1 + \kappa^2 \Delta V \bar{\rho}_0^2 / 12}, \tag{D.61}$$

with $\bar{\rho}_0 = 3 S_1 / \Delta V$, and with the extremal action being

$$B = \frac{B_0}{(1 + \kappa^2 \Delta V \bar{\rho}_0^2 / 12)^2}, \tag{D.62}$$

with $B_0 = 27\pi^2 S_1^3 / (2(\Delta V)^3)$. Thus we see that gravity causes the bubble to be a little smaller, but also a little more likely to materialize.

Exercise 9.2 – Solution
The equations of motion (9.30) and (9.31) can be Taylor expanded near $\rho = 0$ to yield

$$\phi(\sigma) = \phi_0 + \frac{V'(\phi_0)}{8}\sigma^2 + \frac{V'(\phi_0)}{192}[V''(\phi_0) + \frac{2\kappa^2 V(\phi_0)}{3}]\sigma^4 + O(\sigma^6), \quad (D.63)$$

$$\rho(\sigma) = \sigma - \frac{\kappa^2}{18}V(\phi_0)\sigma^3 - \frac{\kappa^2}{120}[\frac{3}{8}V'(\phi_0)^2 - \frac{\kappa^2}{9}V(\phi_0)^2]\sigma^5 + O(\sigma^7), \quad (D.64)$$

where $V'(\phi_0) \equiv \frac{\partial V}{\partial \phi}|_{\phi=\phi_0}$, etc. This series expansion can be used to start the numerical integration near the pole P of the bubble solution. Similar power-law behavior will also be valid near the other pole as $\xi \to 0$, where $\xi = \sigma_{\max} - \sigma$.

Exercise 9.4 – Solution
This potential has local minima that are gradually becoming smaller, and eventually reach negative values, regardless of V_0. The minima are located at intervals $\Delta\phi = 2\pi f$, and the difference in height between adjacent minima is ϵ. Thus, if we would like to be sure that at least one minimum is in the range of the observed value of dark energy, then ϵ needs to be of that order or smaller. The value of f remains unconstrained by these considerations.

When the energy is high, then the local de Sitter vacuum has a high temperature, and it is easy for the scalar field to thermally jump out of a local vacuum and into a lower one. Thus, at first the field drops down the potential rather fast. However, when the temperature $\mathsf{T} = \mathsf{H}/2\pi \sim \sqrt{V}$ falls below the height of the barriers, i.e., when $V_{\min} < M^2$, where $V_{\min}$ denotes the potential height at a local minimum, then the field can only tunnel out via CDL bubble nucleation. If we approximate the potential minima via $V_{\min,n} \approx \beta n M^4$ for a coefficient β and an integer n, then it follows from (D.60) that the nucleation rate goes roughly as

$$\frac{\Gamma}{V} \approx e^{-B} \quad \text{with} \quad B \sim n^{-3/2}. \quad (D.65)$$

This implies that the field stays longer and longer in each local vacuum before tunneling to the next lowest one. Thus it will spend the longest time in the last local minimum at positive values of the potential. (Once it tunnels to negative values, the universe will collapse.) In this model, it is therefore to be expected to live in a bubble with very small vacuum energy, just as we observe in our universe. The problem with this model is that, after nucleation, the bubble is essentially empty. Inflation, followed by reheating, cannot take place as the vacuum energy is already tiny. Thus the model suffers from the so-called *empty universe* problem.

If we now consider this model in the context of a cyclic universe, most of the discussion above remains unchanged. The only difference is that inside of each bubble the universe undergoes cycles of evolution before tunneling out. Thus, even though right after nucleation the universe is empty, it can subsequently collapse, bounce, and fill with matter and radiation at the bounce (if a convincing model of a bounce can be developed). Thus, in the context of a (quasi-)cyclic model, a washboard potential may provide a setting where it is natural to expect a small cosmological constant.

Exercise 9.5 – Solution

In terms of the w radial coordinate, the Schwarzschild solution becomes

$$\mathrm{d}s^2 = -\frac{w^2}{w^2 + 2M}\mathrm{d}t^2 + 4(w^2 + 2M)\mathrm{d}w^2 + (w^2 + 2M)^2 \mathrm{d}\Omega^2. \tag{D.66}$$

If we look at the $t = 0$ spatial slice, for specificity, we see that there are two asymptotic regions $w \to \pm\infty$ in which the 2-sphere becomes large. Meanwhile, as w interpolates between these extremes, the sphere shrinks and reaches its minimum area of $4\pi(2M)^2$ at $w = 0$. This is the throat of the wormhole. This wormhole connects the two asymptotic regions of the full Schwarzschild spacetime, which are outside of the horizon – see Fig. 6.3. The wormhole is not traversable, as any time-like or null ray starting in one of the outside regions can never reach the other outside region (this would require faster-than-light travel). One can trace this back to the presence of the horizons at $w = 0$. This wormhole does not require any matter (it is a vacuum solution of the Einstein equations), and thus also does not require a violation of the null energy condition.

If one were to look at other space-like slices, one would find that the wormhole is also not static, but changes in shape and size as one moves up or down the spacetime diagram.

Exercise 10.1 – Solution

In varying the Nambu–Goto action, one may use the chain rule and vary with respect to $\gamma_{\alpha\beta}$ in an intermediate step, recalling $\delta\sqrt{-\gamma} = -\frac{1}{2}\sqrt{-\gamma}\gamma_{\alpha\beta}\delta\gamma^{\alpha\beta} = \frac{1}{2}\sqrt{-\gamma}\gamma^{\alpha\beta}\delta\gamma_{\alpha\beta}$. Using

$$\delta\gamma_{\alpha\beta} = 2\eta_{\mu\nu}\partial_\alpha X^\mu \partial_\beta \delta X^\nu, \tag{D.67}$$

one finds the equation of motion

$$\partial_\alpha \left(\sqrt{-\gamma} \gamma^{\alpha\beta} \partial_\beta X^\mu \right) = 0 . \qquad (D.68)$$

This is the equation of motion of a relativistic string. Similarly, the Polyakov action (10.2) can be varied with respect to X^μ and $g_{\alpha\beta}$, leading to the equations

$$\partial_\alpha \left(\sqrt{-g} g^{\alpha\beta} \partial_\beta X^\mu \right) = 0 , \qquad (D.69)$$

$$g_{\alpha\beta} = 2 \frac{\partial_\alpha X^\mu \partial_\beta X_\mu}{\partial^\lambda X^\nu \partial_\lambda X_\nu} . \qquad (D.70)$$

When substituting (D.70) into (D.69) one recovers the equation (D.68) – the conformal factor from (D.70) drops out as we are in 2 dimensions. This is a reflection of Weyl invariance.

Exercise 10.2 – Solution

For its use in other places, let us write out the general formula for the transformation of the Ricci scalar under a conformal rescaling,

$$\tilde{G}_{MN} = e^{2\Omega} G_{MN} , \qquad (D.71)$$

$$\tilde{R} = e^{-2\Omega} \left[R - 2(D-1) \Box \Omega = (D-2)(D-1) G^{MN} \Phi_{,M} \Phi_{,N} \right] . \qquad (D.72)$$

Using (10.70), and keeping in mind that the 3-form kinetic term contains three inverse metrics, one finds that (10.7) becomes

$$S_{10} = \frac{1}{2\kappa_0^2} \int d^{10}X \sqrt{-G_E} \left(R_E - \frac{1}{12} e^{-\Phi} H^{\mu\nu\lambda} H_{\mu\nu\lambda} - \frac{1}{2} \nabla^\mu \Phi \nabla_\mu \Phi \right) , \qquad (D.73)$$

where we have dropped the total derivative term. This shows that the dilaton field is in fact canonically normalized, with the proper, nonghost-like sign in front of the kinetic term.

Exercise 10.3 – Solution

We consider a probe brane in the background (10.32). The brane action is given by (10.37). It is easiest to choose coordinates such that the brane worldvolume directions are described by the same coordinates as the corresponding background directions (e.g., $z^1 = x^1$), and to let the radial position of the brane be $r(t)$. Then the pullback of the metric is straightforward to evaluate. The most interesting component is

$$\gamma_{00} = G_{00} + \partial_0 r \partial_0 r G_{rr} = -H^{-\frac{(7-p)}{8}} + H^{\frac{(p+1)}{8}} \dot{r}^2 , \qquad (D.74)$$

while the others are simply $\gamma_{ij} = G_{ij}$. This leads to
$$\det \gamma = H^{-(p+1)(7-p)/8}(-1 + H\dot{r}^2). \tag{D.75}$$

Also, we have that
$$e^{\frac{n-5}{4}\phi} = e^{\frac{p-3}{4}\phi} = H^{-\frac{(3-p)^2}{16}}. \tag{D.76}$$

Putting it all together, we obtain
$$S_{Dp} = -T_p \int d^{p+1}z \, H^{-1}\sqrt{1 - H\dot{r}^2} + C_p \int d^{p+1}z \, H^{-1}, \tag{D.77}$$

which can be expanded to yield
$$S_{Dp} \approx \int d^{p+1}z \left[T_p \frac{1}{2}\dot{r}^2 - (T_p - C_p) H^{-1} \right]. \tag{D.78}$$

This shows that Dp-branes do not feel any force between them, since $C_p = T_p$, while antibranes (with $C_p = -T_p$) moving in a brane background do feel a force, approximated by the potential $V(r) = 2T_p H^{-1}$.

Exercise 10.4 – Solution
If one writes the complex scalar in terms of two real scalars, $A = \frac{1}{\sqrt{2}}(\phi + i\chi)$, then the Kähler potential $K = -\frac{1}{2}(A - A^\star)^2$ leads to the kinetic term
$$-K_{,AA^\star}\partial^\mu A \partial_\mu A^\star = -\partial^\mu A \partial_\mu A^\star = -\frac{1}{2}(\partial\phi)^2 - \frac{1}{2}(\partial\chi)^2, \tag{D.79}$$

which is canonically normalized. The potential then becomes
$$V(\phi, \chi) = w^2(c^2 + 2\chi^2 - 3)e^{\sqrt{2}c\phi + \chi^2}. \tag{D.80}$$

The χ field will be expected to roll to its minimum at $\chi = 0$. Then the potential for ϕ simplifies to
$$V(\phi) = w^2(c^2 - 3)e^{\sqrt{2}c\phi}. \tag{D.81}$$

A positive potential is obtained for $c^2 > 3$, but this implies a "slow-roll" parameter $\epsilon > 3$, which is much too steep for inflation. By contrast, a negative potential is obtained for $c^2 < 3$, implying $\epsilon < 3$, and this is now too shallow for an ekpyrotic phase. Therefore, one typically obtains exactly the opposite of what one would like for these cosmological models. The general formula (10.71) for the potential illustrates the problem: The first term in the brackets is positive, but it requires larger derivatives to be dominant, while the second term in brackets is negative, but requires small derivatives to be dominant. This is in no way saying that it is impossible to construct inflationary or ekpyrotic potentials in supergravity (and examples of such potentials exist), but it illustrates the challenge of keeping such potentials under control when additional fields are included.

Exercise 10.5 – Solution
The rewriting in Einstein frame can be obtained by plugging in the redefinition given above. The procedure can also be slightly simplified by first rewriting the theory using an auxiliary field as

$$S_8 = \frac{1}{2}\int d^8x \sqrt{-\hat{g}}\left(f_{,\hat{R}}\hat{R} - U - \frac{1}{2\cdot 4!}q^2 F_{(4)}^2\right), \tag{D.82}$$

with $f = \hat{R} + \alpha \hat{R}^4$ and $U = (f_{,\hat{R}}\hat{R} = \hat{R})$, and then making use of the field redefinition. Either way, the result is

$$S_8 = \int d^8x \sqrt{-g}\left(\frac{R}{2} - \frac{1}{2}(\partial\phi)^2 - V(\phi) - \frac{1}{4\cdot 4!}q^2 F_{(4)}^2\right), \tag{D.83}$$

where the scalar was rescaled as $\sqrt{42}\varphi \equiv \phi$ and its potential is given by

$$V(\phi) = \tilde{\alpha}\left(1 - e^{-\sqrt{\frac{6}{7}}\phi}\right)^{4/3}, \quad \tilde{\alpha} \equiv \frac{3}{2^{11/3}\alpha^{1/3}}. \tag{D.84}$$

This theory can be reduced to 4 dimensions using the metric ansatz

$$ds_8^2 = e^{-\frac{2}{\sqrt{3}}\chi}\left(-N^2 dt^2 + a^2 d\Omega_3^2\right) + e^{\frac{1}{\sqrt{3}}\chi}d\Omega_4^2, \tag{D.85}$$

and we end up with a theory containing the scale factor a and two scalar fields ϕ, χ,

$$S_4 = \frac{16\pi^4}{3}\int dt \left(-3\frac{a\dot{a}^2}{N} + \frac{a^3}{2N}\left(\dot{\phi}^2 + \dot{\chi}^2\right) + 3Na - Na^3 V(\phi,\chi)\right), \tag{D.86}$$

where we neglected a surface term that vanishes when $a = 0$. The scalar potential is given by

$$V(\phi,\chi) = \tilde{\alpha}\left(1 - e^{-\sqrt{\frac{6}{7}}\phi}\right)^{\frac{4}{3}} e^{-\frac{2}{\sqrt{3}}\chi} + n_4^2 e^{-2\sqrt{3}\chi} - 6e^{-\sqrt{3}\chi}. \tag{D.87}$$

The ϕ direction provides a suitably flat potential which can be used for an inflationary phase, as long as χ is stabilized. To this end, we can look at the potential for χ (at fixed $\phi = 5$, say) for varying values of the flux n_4. This is shown in Fig. D.2 with $\tilde{\alpha} = 1$. One can see that the amount of flux greatly influences the potential. For small n_4, there is a local minimum, but it resides at negative values of the potential, where inflation cannot take place. Then, as n_4 is raised, the minimum moves to positive values and no-boundary instantons can exist, nucleating a universe with topology $S^4 \times S^4$. At higher values of the flux $n_4 \gtrsim 17$, there is no local minimum anymore, and all solutions would decompactify (as χ rolls to infinity, the 4-sphere radius grows without bound).

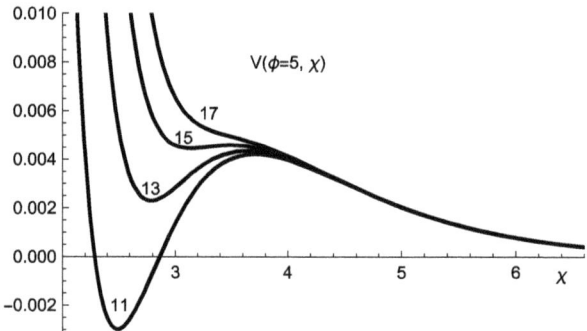

Figure D.2 A slice through the scalar potential (D.87) at fixed $\phi = 5$, with $\tilde{\alpha} = 1$ and for the flux values $n_4 = 11, 13, 15, 17$.

Thus, in this model, the no-boundary wave function provides a probability distribution over fluxes, and thus over compactifications. Moreover, if we insist that there should be only 3 large spatial dimensions on the final hypersurface, then the values of the flux are cut off at around $n_4 = 17$. The most likely universe is one that nucleates at the smallest possible flux value (in string theory, the values of the flux are expected to be quantized). Additional factors, such as the allowability of the metrics, might influence this result. Note that this model provides a concrete illustration of the general properties discussed around Fig. 10.2.

Appendix E
Guide to the bibliography

Papers that I would recommend reading as background material are

- S. Coleman's review of quantum mechanics, Coleman (1993)
- J. Wheeler's insights into quantum gravity, in Woolf (1981)
- S. Hawking's original description of the no-boundary idea, Hawking (1982)

For each chapter, I provide a few references that either constitute the original literature on which the chapter is based, or provide further reading opportunities for a more in-depth study of a particular subject. These are just pointers to papers and books that I have personally found valuable. I do not make an attempt at completeness here (which, in any case, for a subject spanning half a century, would be an overwhelming enterprise), given that the main aim of the book is to be pedagogical.

Chapter 1

A great paper describing the CMB and our motion through the universe is the semi-popular account by Muller (1978). The Planck satellite experiment and its most important observational data are described in Planck Collaboration, Aghanim et al. (2020a). For further details about observational cosmology, see also the book by Dodelson (2003).

The conformal map of the universe is constructed in detail and beautifully explained in the paper by Gott et al. (2005).

The first comprehensive description of the puzzles of the big bang model can be found in the still highly readable paper "Enigmas and nostrums" by R. Dicke and J. Peebles in the collection by Hawking and Israel (1979).

Chapter 2

The seminal paper on chaotic inflation is the one by Linde (1983). Higgs inflation was introduced in Bezrukov and Shaposhnikov (2008), while the original Starobinsky model is described in Starobinsky (1980).

Puzzles and open issues of inflation are described very clearly in Ijjas et al. (2013).

The smoothing properties of an ekpyrosic phase are elucidated in Erickson et al. (2004), while the original paper on the cyclic model is Steinhardt and Turok (2002). A general review of ekpyrotic and cyclic models can be found in Lehners (2008).

The model of inflation with a negative potential minimum, explored in the exercises, was devised by Lin (2023).

Chapter 3

Details about the observations of the cosmic microwave background are explained very clearly in the book by Dodelson (2003). It may also be worthwhile reading the Planck paper on inflationary constraints in Planck Collaboration, Akrami et al. (2020b).

The ADM formalism was introduced in Arnowitt et al. (1962/2008), and their presentation is still up to date. Gauge-invariant cosmological variables were discussed in Bardeen et al. (1983). A modern perspective, including a calculation of non-Gaussian corrections in simple inflationary models, can be found in Maldacena (2003).

How to generate realistic perturbations in a contracting phase is shown in Lehners et al. (2007).

Chapter 4

For an old, but still valuable, introduction to canonical quantization and general properties thereof, see Misner et al. (1973).

If you are looking for an approach that tries to avoid quantizing gravity (and an example of what lengths one has to go to in order to possibly achieve this), see Oppenheim (2023).

A good exposition of variational principles for gravity, including surface terms, is provided in Padmanabhan (2014).

The interpretation of the wave function of the universe is developed to a large extent in Vilenkin (1989). The recovery of the Schrödinger equation is discussed in Kiefer (1987). Meanwhile, the approach to classicality during inflation and ekpyrosis is derived in Lehners (2015).

For the exercise on the lapse function see the paper Woodard and Yesilyurt (2025).

Chapter 5

The gravitational path integral, and gauge fixing issues, are discussed in Teitelboim (1982). How to derive the WDW equation from a concrete symmetry-reduced path integral is shown in Halliwell (1989). A general reference that is useful for the types of integrals we are dealing with is Grosche and Steiner (1998).

Many of the now standard minisuperspace techniques are developed in Halliwell and Louko (1989). The Lorentzian approach to gravitational path integrals is described in Feldbrugge et al. (2017).

For the exercise on initial conditions for inflation, see Di Tucci et al. (2019b).

Chapter 6

The link between gravitational path integrals and thermodynamics is laid out very generally in Gibbons and Hawking (1977a). And a good reference on quantum fields on curved backgrounds is Birrell and Davies (1982). The Unruh and Hawking effects are carefully described in Mukhanov and Winitzki (2007).

For more recent work on Hawking radiation and the associated information puzzle, see Penington (2020). A highly readable popular account of the new ideas involving so-called islands, and entanglement between the black hole exterior and interior regions, can be found in Cox and Forshaw (2022).

The section on the Hawking–Page transition, as derived from a gravitational path integral, is based on Di Tucci et al. (2020).

The original reference on de Sitter thermodynamics is Gibbons and Hawking (1977b), and it still offers interesting insights to date.

Chapter 7

The idea of a quantum origin of the universe was formulated (almost oracularly) already by Lemaître (1931) in a one-page paper. This idea resurfaced periodically – see for instance Tryon (1973). A more precise formulation then arose with the no-boundary proposal, first detailed in Hartle and Hawking (1983). For the motivation behind the closely related tunneling proposal, see Vilenkin (1982).

For minisuperspace implementations, see Halliwell and Louko (1989). For the inclusion of scalar fields, see Lyons (1992) and Hartle et al. (2008). Boundary conditions are discussed in Di Tucci et al. (2019a).

Allowability conditions on complex metrics are proposed in Kontsevich and Segal (2021), extended to gravity in Witten (2021) and analyzed in quantum cosmology in Lehners (2022).

Chapter 8

A general overview of decoherence and the development of associated ideas over the last half century can be found in Zurek (2025).

In the context of quantum cosmology, decoherence was discussed by Kiefer (1987) and extended to multiple saddle points by Halliwell (1989). The effect of gravitational couplings on perturbations was analyzed in Nelson (2016k).

For a viewpoint on quantum theory that works well in the cosmological context see Rovelli (1996).

Chapter 9

The approach to tunneling using complexified time (or complexified energy) is described from different points of view in Bender et al. (2010), Turok (2014) and Bramberger et al. (2016).

The classic paper discussing the decay of a false vacuum is Coleman (1977), and its extension to include gravity can be found in Coleman and De Luccia (1980). For a criterion on the existence of CDL instantons, see Jensen and Steinhardt (1984), and for a discussion of the relation to thermal fluctuations, see Brown and Weinberg (2007).

After several false starts, the treatment of negative modes was clarified in Khvedelidze et al. (2000). For numerical examples of CDL solutions, as well as an analysis of negative modes, see Battarra et al. (2013).

Bubble collisions and their potential cosmological implications are reviewed in Aguirre and Johnson (2011).

A related mechanism to reduce the value of the cosmological constant was originally devised in Abbott (1985), and implemented in the cyclic model in Steinhardt and Turok (2006).

The bubble of nothing was discovered, and beautifully described, in Witten (1982).

A useful pedagogical introduction to Lorentzian wormholes can be found in Morris and Thorne (1988). For a discussion of the Raychaudhuri equation, see chapter 5 of Padmanabhan (2014). Traversable wormholes were explicitly constructed in Maldacena et al. (2023). For a conjectured link between nontraversable wormholes and entanglement, see Maldacena and Susskind (2013).

Euclidean wormholes supported by axions are constructed in Giddings and Strominger (1988). For a discussion of the nucleation of baby universes, see Lavrelashvili et al. (1987), and for physical implications of wormholes and alpha vacua see Coleman (1988). These papers constitute required background reading when trying to follow the fast-developing current literature on wormholes.

Chapter 10

For a general introduction to string theory I recommend the lecture notes by D. Tong, and for Kaluza–Klein theory the notes by C. Pope. These can be found on their respective websites (at the time of writing these are at www.damtp.cam.ac.uk/user/tong/string.html and at https://people.tamu.edu/~c-pope/).

The no-go result for de Sitter from smooth compactifications was discussed in old talks by G. Gibbons. The version presented in this book is based on Maldacena and Nunez (2000).

Standard clocks and their uses are described in Chen et al. (2015).

For a general review of branes, see Stelle (1998). Models of brane inflation are developed in Kachru et al. (2003) and Conlon and Quevedo (2006). The conformal coupling of the brane to curvature is explained in Seiberg and Witten (1999).

The link between 11-dimensional supergravity and heterotic $E_8 \times E_8$ theory is presented in Horava and Witten (1996). For the resulting 5-dimensional braneworld model see Lukas et al. (1999). This setting inspired the ekpyrotic model – see Khoury et al. (2001). A description of how to generate cosmological perturbations in such a model is found in Lehners et al. (2007). And an attempt at treating the brane collision can be found in Turok et al. (2004).

The exercise on 8-dimensional no-boundary solutions is based on Lehners et al. (2023).

Bibliography

Abbott, Laurence F. 1985. A Mechanism for Reducing the Value of the Cosmological Constant. *Phys. Lett. B* **150**, 427–430.

Aguirre, Anthony, and Johnson, Matthew C. 2011. A Status Report on the Observability of Cosmic Bubble Collisions. *Rept. Prog. Phys.* **74**, 074901.

Arnowitt, Richard L., Deser, Stanley, and Misner, Charles W. 1962/2008. The Dynamics of General Relativity. *Gen. Rel. Grav.* **40**, 1997–2027.

Bardeen, James M., Steinhardt, Paul J., and Turner, Michael S. 1983. Spontaneous Creation of Almost Scale-Free Density Perturbations in an Inflationary Universe. *Phys. Rev. D* **28**, 679.

Battarra, Lorenzo, Lavrelashvili, George, and Lehners, Jean-Luc. 2013. Zoology of Instanton Solutions in Flat Potential Barriers. *Phys. Rev. D* **88**, 104012.

Bender, Carl M., Hook, Daniel W., Meisinger, Peter N., and Wang, Qing-hai. 2010. Complex Correspondence Principle. *Phys. Rev. Lett.* **104**, 061601.

Bezrukov, Fedor L., and Shaposhnikov, Mikhail. 2008. The Standard Model Higgs Boson as the Inflaton. *Phys. Lett. B* **659**, 703–706.

Birrell, Nicholas D., and Davies, Paul C. W. 1982. *Quantum Fields in Curved Space*. Cambridge Monographs on Mathematical Physics. Cambridge: Cambridge University Press.

BOSS Collaboration; Alam, Shadab, et al. 2017. The clustering of galaxies in the completed SDSS-III Baryon Oscillation Spectroscopic Survey: cosmological analysis of the DR12 galaxy sample. *Mon. Not. Roy. Astron. Soc.* **470**, 2617.

Bramberger, Sebastian F., Lavrelashvili, George, and Lehners, Jean-Luc. 2016. Quantum Tunneling from Paths in Complex Time. *Phys. Rev. D* **94**(6), 064032.

Brown, Adam R., and Weinberg, Erick J. 2007. Thermal Derivation of the Coleman–De Luccia Tunneling Prescription. *Phys. Rev. D* **76**, 064003.

Chen, Xingang, Namjoo, Mohammad H., and Wang, Yi. 2016. Quantum Primordial Standard Clocks. *JCAP* **02**, 013.

Coleman, Sidney R. 1977. The Fate of the False Vacuum. 1. Semiclassical Theory. *Phys. Rev. D* **15**, 2929–2936. [Erratum: *Phys. Rev. D* **16** 1248 (1977)].

Coleman, Sidney R. 1988. Black Holes as Red Herrings: Topological Fluctuations and the Loss of Quantum Coherence. *Nucl. Phys. B* **307**, 867–882.

Coleman, Sidney R. 1993. Quantum Mechanics with the Gloves Off. Dirac Memorial Lecture, Cambridge University, Cambridge, UK. www.damtp.cam.ac.uk/user/ho/Coleman.pdf.

Coleman, Sidney R., and De Luccia, Frank. 1980. Gravitational Effects on and of Vacuum Decay. *Phys. Rev. D* **21**, 3305.

Conlon, Joseph P., and Quevedo, Fernando. 2006. Kahler Moduli Inflation. *JHEP* **01**, 146.

Cox, Brian, and Forshaw, Jeff. 2022. *Black Holes*. London: William Collins.

DESI Collaboration; Abdul Karim, M., et al. 2025. DESI DR2 Results. II. Measurements of Baryon Acoustic Oscillations and Cosmological Constraints. *Phys. Rev. D* **112**, 083515.

Di Tucci, Alice, Feldbrugge, Job, Lehners, Jean-Luc, and Turok, Neil. 2019b. Quantum Incompleteness of Inflation. *Phys. Rev. D* **100**(6), 063517.

Di Tucci, Alice, Heller, Michal P., and Lehners, Jean-Luc. 2020. Lessons for Quantum Cosmology from Anti–de Sitter Black Holes. *Phys. Rev. D* **102**(8), 086011.

Di Tucci, Alice, Lehners, Jean-Luc, and Sberna, Laura. 2019a. No-Boundary Prescriptions in Lorentzian Quantum Cosmology. *Phys. Rev. D* **100**(12), 123543.

Dodelson, Scott. 2003. *Modern Cosmology*. Amsterdam: Academic Press.

Erickson, Joel K., Wesley, Daniel H., Steinhardt, Paul J., and Turok, Neil. 2004. Kasner and Mixmaster Behavior in Universes with Equation of State w $>=$ 1. *Phys. Rev. D* **69**, 063514.

Feldbrugge, Job, Lehners, Jean-Luc, and Turok, Neil. 2017. Lorentzian Quantum Cosmology. *Phys. Rev. D* **95**(10), 103508.

Gerlach, Ulrich H., and Sengupta, Uday K. 1978. Homogeneous Collapsing Star: Tensor and Vector Harmonics for Matter and Field Asymmetries. *Phys. Rev. D* **18**, 1773–1784.

Gibbons, Gary W., and Hawking, Stephen W. 1977a. Action Integrals and Partition Functions in Quantum Gravity. *Phys. Rev. D* **15**, 2752–2756.

Gibbons, Gary W., and Hawking, Stephen W. 1977b. Cosmological Event Horizons, Thermodynamics, and Particle Creation. *Phys. Rev. D* **15**, 2738–2751.

Giddings, Steven B., and Strominger, Andrew. 1988. Axion Induced Topology Change in Quantum Gravity and String Theory. *Nucl. Phys. B* **306**, 890–907.

Gott, III, J. Richard, Juric, Mario, Schlegel, David, Hoyle, Fiona, Vogeley, Michael, Tegmark, Max, Bahcall, Neta A., and Brinkmann, Jon. 2005. A Map of the Universe. *Astrophys. J.* **624**, 463.

Grosche, Christian, and Steiner, Frank. 1998. *Handbook of Feynman Path Integrals*. Springer Tracts in Modern Physics **145**. Berlin: Springer.

Halliwell, Jonathan J. 1989. Decoherence in Quantum Cosmology. *Phys. Rev. D* **39**, 2912.

Halliwell, Jonathan J., and Louko, Jorma. 1989. Steepest Descent Contours in the Path Integral Approach to Quantum Cosmology. 1. The De Sitter Minisuperspace Model. *Phys. Rev. D* **39**, 2206.

Hartle, James B., and Hawking, Stephen W. 1983. Wave Function of the Universe. *Phys. Rev. D* **28**, 2960–2975.

Hartle, James B., Hawking, Stephen W., and Hertog, Thomas. 2008. The Classical Universes of the No-Boundary Quantum State. *Phys. Rev. D* **77**, 123537.

Hawking, Stephen W. 1982. The Boundary Conditions of the Universe. *Pontif. Acad. Sci. Scr. Varia* **48**, 563–574.

Hawking, Stephen W., and Israel, Werner (eds). 1979. *General Relativity: An Einstein Centenary Survey*. Cambridge: Cambridge University Press.

Horava, Petr, and Witten, Edward. 1996. Eleven-Dimensional Supergravity on a Manifold with Boundary. *Nucl. Phys. B,* **475**, 94–114.

Huterer, Dragan, and Shafer, Daniel L. 2018. Dark Energy Two Decades After: Observables, Probes, Consistency Tests. *Rept. Prog. Phys.* **81**(1), 016901.

Ijjas, Anna, Steinhardt, Paul J., and Loeb, Abraham. 2013. Inflationary Paradigm in Trouble After Planck2013. *Phys. Lett. B* **723**, 261–266.

Jensen, Lars G., and Steinhardt, Paul J. 1984. Bubble Nucleation and the Coleman–Weinberg Model. *Nucl. Phys. B* **237**, 176–188.

Jonas, Caroline, Lehners, Jean-Luc, and Quintin, Jerome. 2022. Uses of Complex Metrics in Cosmology. *JHEP* **08**, 284.

Kachru, Shamit, Kallosh, Renata, Linde, Andrei D., Maldacena, Juan M., McAllister, Liam P., and Trivedi, Sandip P. 2003. Towards Inflation in String Theory. *JCAP* **10**, 013.

Khoury, Justin, Ovrut, Burt A., Steinhardt, Paul J., and Turok, Neil. 2001. The Ekpyrotic Universe: Colliding Branes and the Origin of the Hot Big Bang. *Phys. Rev. D* **64**, 123522.

Khvedelidze, Arsen, Lavrelashvili, George V., and Tanaka, Takahiro. 2000. On Cosmological Perturbations in Closed FRW Model with Scalar Field and False Vacuum Decay. *Phys. Rev. D* **62**, 083501.

Kiefer, Claus. 1987. Continuous Measurement of Minisuperspace Variables by Higher Multipoles. *Class. Quant. Grav.* **4**, 1369.

Kontsevich, Maxim, and Segal, Graeme. 2021. Wick Rotation and the Positivity of Energy in Quantum Field Theory. *Quart. J. Math. Oxford Ser.* **72**(1–2), 673–699.

Lavrelashvili, George V., Rubakov, Valery A., and Tinyakov, Peter G. 1987. Disruption of Quantum Coherence upon a Change in Spatial Topology in Quantum Gravity. *JETP Lett.* **46**, 167–169.

Lehners, Jean-Luc. 2008. Ekpyrotic and Cyclic Cosmology. *Phys. Rept.* **465**, 223–263.

Lehners, Jean-Luc. 2015. Classical Inflationary and Ekpyrotic Universes in the No-Boundary Wavefunction. *Phys. Rev. D* **91**(8), 083525.

Lehners, Jean-Luc. 2022. Allowable Complex Metrics in Minisuperspace Quantum Cosmology. *Phys. Rev. D* **105**(2), 026022.

Lehners, Jean-Luc, Leung, Rahim, and Stelle, Kellogg S. 2023. How to Create Universes with Internal Flux. *Phys. Rev. D* **107**(4), 046006.

Lehners, Jean-Luc, McFadden, Paul, Turok, Neil, and Steinhardt, Paul J. 2007. Generating Ekpyrotic Curvature Perturbations before the Big Bang. *Phys. Rev. D* **76**, 103501.

Lehners, Jean-Luc, and Quintin, Jerome. 2024. A Small Universe. *Phys. Lett. B* **850**, 138488.

Lemaitre, Georges. 1931. The Beginning of the World from the Point of View of Quantum Theory. *Nature* **127**, 706.

Lin, Chia-Min. 2023. Uniform Rate Inflation. *JCAP* **04**, 037.

Linde, Andrei D. 1983. Chaotic Inflation. *Phys. Lett. B* **129**, 177–181.

Lukas, Andre, Ovrut, Burt A., Stelle, Kellogg S., and Waldram, Daniel. 1999. The Universe as a Domain Wall. *Phys. Rev. D* **59**, 086001.

Lyons, Glenn W. 1992. Complex Solutions for the Scalar Field Model of the Universe. *Phys. Rev. D* **46**, 1546–1550.

Maldacena, Juan M. 2003. Non-Gaussian Features of Primordial Fluctuations in Single Field Inflationary Models. *JHEP* **05**, 013.

Maldacena, Juan, Milekhin, Alexey, and Popov, Fedor. 2023. Traversable Wormholes in Four Dimensions. *Class. Quant. Grav.* **40**(15), 155016.

Maldacena, Juan, and Nunez, Carlos. 2001. Supergravity Description of Field Theories on Curved Manifolds and a No Go Theorem. *Int. J. Mod. Phys. A* **16**, 822–855.

Maldacena, Juan, and Susskind, Leonard. 2013. Cool Horizons for Entangled Black Holes. *Fortsch. Phys.* **61**, 781–811.

Misner, Charles W., Thorne, Kip S., and Wheeler, John A. 1973. *Gravitation*. San Francisco: W. H. Freeman.

Morris, Michael S., and Thorne, Kip S. 1988. Wormholes in Space-Time and Their Use for Interstellar Travel: A Tool for Teaching General Relativity. *Am. J. Phys.* **56**, 395–412.

Mukhanov, Viatcheslav, and Winitzki, Sergei. 2007. *Introduction to Quantum Effects in Gravity*. Cambridge: Cambridge University Press.

Muller, Richard A. 1978. The Cosmic Background Radiation and the New Aether Drift. *Scientific American Magazine* **238**, 64.

Nelson, Elliot. 2016. Quantum Decoherence during Inflation from Gravitational Nonlinearities. *JCAP* **03**, 022.

Oppenheim, Jonathan. 2023. A Postquantum Theory of Classical Gravity? *Phys. Rev. X* **13**(4), 041040.

Padmanabhan, Thanu. 2014. *Gravitation: Foundations and Frontiers*. Cambridge: Cambridge University Press.

Penington, Geoffrey. 2020. Entanglement Wedge Reconstruction and the Information Paradox. *JHEP* **09**, 002.

Planck Collaboration; Aghanim, N., et al. 2020a. Planck 2018 Results. I. Overview and the Cosmological Legacy of Planck. *Astron. Astrophys.* **641**, A1.

Planck Collaboration; Akrami, Y., et al. 2020b. Planck 2018 Results. X. Constraints on Inflation. *Astron. Astrophys.* **641**, A10.

Rovelli, Carlo. 1996. Relational Quantum Mechanics. *Int. J. Theor. Phys.* **35**, 1637–1678.

Scolnic, Daniel M., et al. 2015. Supercal: Cross-calibration of Multiple Photometric Systems to Improve Cosmological Measurements with Type Ia Supernovae. *Astrophys. J.* **815**, 117.

Seiberg, Nathan, and Witten, Edward. 1999. The D1 / D5 System and Singular CFT. *JHEP* **04**, 017.

Starobinsky, Alexei A. 1980. A New Type of Isotropic Cosmological Models without Singularity. *Phys. Lett. B* **91**, 99–102.

Steinhardt, Paul J., and Turok, Neil. 2002. A Cyclic Model of the Universe. *Science* **296**, 1436–1439.

Steinhardt, Paul J., and Turok, Neil. 2006. Why the Cosmological Constant is Small and Positive. *Science* **312**, 1180–1182.

Stelle, Kellogg S. 1998 (3). BPS Branes in Supergravity. In *1997 Summer School in High Energy Physics and Cosmology, ICTP, Trieste, Italy, 2 June–4 July 1997*. The ICTP Series in Theoretical Physics **14**, eds. Gava, E., Masiero, A., Narain, K. S., Randjbar-Daemi, S., Senjanovic, G., Smirnov, A., and Shafi, Q. Singapore: World Scientific.

Teitelboim, Claudio. 1982. Quantum Mechanics of the Gravitational Field. *Phys. Rev. D* **25**, 3159.

Tryon, Edward P. 1973. Is the Universe a Vacuum Fluctuation? *Nature* **246**, 396.

Turok, Neil. 2014. On Quantum Tunneling in Real Time. *New J. Phys.* **16**, 063006.

Turok, Neil, Perry, Malcolm, and Steinhardt, Paul J. 2004. M Theory Model of a Big Crunch / Big Bang Transition. *Phys. Rev. D* **70**, 106004. [Erratum: *Phys. Rev. D* **71**, 029901 (2005)].

Vilenkin, Alexander. 1982. Creation of Universes from Nothing. *Phys. Lett. B* **117**, 25.

Vilenkin, Alexander. 1989. The Interpretation of the Wave Function of the Universe. *Phys. Rev. D* **39**, 1116.

Witten, Edward. 1982. Instability of the Kaluza–Klein Vacuum. *Nucl. Phys. B* **195**, 481–492.

Witten, Edward. 2021. A Note on Complex Spacetime Metrics. Preprint, submitted November 12, 2021. https://arxiv.org/abs/2111.06514.

Woodard, Richard P., and Yesilyurt, Boran. 2025. The Other ADM. *J. Phys. A* **58**(19), 195401.

Woolf, Harry (ed.). 1981. *Some Strangeness in the Proportion*. Reading, MA: Addison-Wesley.

Zurek, Wojciech H. 2025. *Decoherence and Quantum Darwinism*. Cambridge: Cambridge University Press.

Index

ΛCDM model, 13
ADM decomposition of metric, 75, 105
allowability criterion, 240, 245
   maximal curve, 243
   minisuperspace, 247
big bang model, 13
   puzzles, 23
Bogolyubov transformations, 160, 168
boundary condition, 104, 232
   Dirichlet, 126, 222
   Neumann, 141, 224
   Robin, 152, 253, 353, 359
bubble of nothing, 289
   evolution of, 290
Bunch–Davies vacuum, 83, 192, 203
   preparation in Euclidean time, 193
Coleman–DeLuccia instantons, 276
   negative modes, 287, 288
   oscillating instantons, 279
comoving observer, 4
complex metrics, 136, 182, 222, 224, 235, 239
continuity, equation of, 9, 212
cosmic microwave background radiation (CMB), 20, 58
   polarization, 65
   power spectrum, 63
cosmological constant, 10, 11, 14, 124, 175, 187, 199, 212
cosmological principle, 1
curvature perturbation
   action, 78
   bispectrum, 90
   comoving, 74, 78
   conservation on large scales, 79
   gauge transformations, 72
   in classical physics, 67
   non-Gaussian corrections, 94
   quantization, 81
   scale invariance, 64
   slow-roll solutions, 84
   spectral index, 64, 85
   variance, 64, 85
   wave function, 86
cyclic universe
   see ekpyrosis,
dark energy, 14, 19, 29, 55
   and extra dimensions, 309
   and tunneling, 302, 361
dark matter
   from branes, 318
   from Kaluza–Klein modes, 314
de Sitter horizon, 35
de Sitter mode functions
   closed slicing, 190, 202
   flat slicing, 189, 194, 358
   wave function, 192
de Sitter spacetime, 34
decoherence
   see also double-slit experiment, 254
   and quantum Darwinism, 267
   between different saddles, 260
   for a single saddle, 261
   functionals, 259
   of cosmological fluctuations, 262
   pointer states, 256
Dirac
   comments on initial conditions, 198
double-slit experiment, 119, 254
ekpyrosis
   and WDW equation, 116
   basic idea, 51
   bounce, 56
   cyclic model, 54
extra dimensions, 289, 304
   and coupling constants, 309
   and dark energy, 309
   and gauge symmetries, 307
flatness puzzle, 24
fluctuation integral, 127, 142
fluctuations

see curvature perturbation,
see negative modes,
see tensor perturbation,
Friedmann equation, 8, 10
gauge
   comoving, 73
   transformations, 72
general relativity
   variational principle, 104
Gibbons–Hawking temperature, 188
Hawking radiation, 170
Hawking–Page phase transition, 175
higher-derivative curvature terms, 236
horizon, 155, 165, 170, 188, 260
   thermodynamics of, 159
horizon puzzle, 25
hot big bang model, 15
Hubble law, 7
Hubble rate, 6
inflation
   and WDW equation, 116
   basic idea, 31
   brane inflation, 319
   eta problem, 321
   eternal, 99
   Higgs model, 44
   power-law, 38
   puzzles of, 98
   quadratic potential, 43
   reheating, 49
   slow-roll parameters, 41
   Starobinsky model, 46
JWKB classicality, 87, 113, 117, 134, 193, 216, 257, 340, 352
Kaluza–Klein vacuum, 289, 307
Kontsevich–Segal criterion, 240
lapse contour, 131, 151, 184, 223, 226, 249
   nonclassical transition with Dirichlet conditions, 136
lapse function, 75
lapse integral
   complex, 136, 139, 145, 182, 223, 226
   Euclidean, 135
   Lorentzian, 133, 138, 145, 226
metric
   complex, 182, 239
   conformal, 5
   convenient minisuperspace form, 126
   de Sitter, 34, 187
   Euclidean, 156, 171
   Euclidean Anti-de Sitter, 177
   Kantowski–Sachs, 176
   Rindler, 166
   Robertson–Walker, 3
   Schwarzschild, 7, 170
   Schwarzschild–de Sitter, 29, 345
minisuperspace, 109, 124, 220, 247

anisotropic model, 230
moduli, 308, 311, 320, 326
moduli space approximation, 326
Mukhanov–Sasaki variable, 81, 190
negative modes, 283, 292
   of wormholes, 298
   with gravity, 285
no-boundary proposal, 195
   approach to JWKB classicality, 217
   biaxial Bianchi IX model, 230
   de Sitter example, 199
   difficulties with initial Dirichlet condition, 223
   higher-derivative curvature corrections, 236
   in string theory, 332, 365
   initial Neumann condition, 224
   massive scalar field, 205
   merits and puzzles, 249
   minisuperspace, 220
   numerical examples, 210
   probabilities for inflationary histories, 219, 246
   seen from WDW equation, 228, 230
   selection of Bunch–Davies state, 203
null energy condition
   implications for bounces, 56
   implications for wormholes, 294
Olbers's paradox, 23
path integral
   definition in minisuperspace, 121
   effects of wormholes, 299
   heuristic motivation, 119
   relation to WDW, 145
   thermodynamic aspects, 175
perturbations
   see curvature perturbation,
   see negative modes,
   see tensor perturbation,
Picard–Lefschetz theory, 335
propagator
   as solution of inhomogeneous WDW equation, 145
   composition law, 148
quantization
   canonical, 106
   curvature perturbations, 81
   de Sitter mode functions, 189
   Hawking radiation, 170
   minisuperspace models, 220
   path integral, 119, 121, 124
   scalar field, 160
   strings, 305
   Unruh effect, 164
Raychaudhuri equation, 294
redshift, 5
scalar field

dynamics, 37
no-boundary solutions, 206
quantization, 160
shift function, 75
singularity puzzle, 23
standard clocks, 313
Stokes phenomenon, 153, 235, 338, 355
Stokes ray, 138, 339
string theory, 303
   and 11-dimensional supergravity, 322
   and the no-boundary proposal, 332, 365
   brane collision as big bang, 328
   brane dynamics, 326
   brane inflation, 319
   dark matter, 314
   difficulties with de Sitter space, 310
   Dine–Seiberg problem, 311
   Dp-branes, 315, 316, 319, 331, 364
   eta problem, 321
   Hořava–Witten theory, 322
   Kaluza–Klein tower, 313
   orbifolds, 322
   review, 305
tensor perturbation, 88
   power spectrum, 66
   scale invariance, 66
   spectral index, 66, 89
   tensor-to-scalar ratio, 90
   variance, 66, 89
tensor-to-scalar ratio, 66, 90
thermal history, 13
time
   conformal, 5
   imaginary, 154, 269
   proper, 4
   recovery of effective time, 115
tunneling
   and dark energy, 302, 361
   bounce solution, 273

bubbles of nothing, 289
evolution of bubble, 275, 280
in complex time, 269
in field theory, 271
negative modes, 283
thin wall approximation, 273, 301, 360
via Coleman–DeLuccia instantons, 276
tunneling proposal, 198
   instability, 204

universe
   age of, 16
   composition of, 15
   conformal map of, 26
Unruh effect, 164

wave function
   approach to JWKB classicality, 116, 117, 217, 352
   interpretation, 267
   JWKB limit, 112
   no-boundary, 197, 203, 210, 224
   of no-boundary perturbations, 203
   of perturbations, 86
   tunneling, 204
Wheeler–DeWitt (WDW) equation
   and path integral, 145
   basic properties, 106
   in momentum space, 228
   minisuperspace, 109
   no-boundary condition, 230
Wigner function, 257
wormhole
   alpha vacua, 300
   effect on coupling constants, 299
   Einstein–Rosen bridge, 302, 362
   Euclidean, 295
   Giddings–Strominger, 297
   Lorentzian, 292
   Morris–Thorne, 293
   traversibility, 295

For EU product safety concerns, contact us at Calle de José Abascal, 56–1°, 28003 Madrid, Spain or eugpsr@cambridge.org.

www.ingramcontent.com/pod-product-compliance
Ingram Content Group UK Ltd.
Pitfield, Milton Keynes, MK11 3LW, UK
UKHW050833130426
469865UK00009B/98